Stacks Project Expository Collection (SPEC)

The Stacks Project Expository Collection (SPEC) compiles expository articles in advanced algebraic geometry that are intended to bring graduate students and researchers up to speed on recent developments in the geometry of algebraic spaces and algebraic stacks. The articles in the text make explicit in modern language many results, proofs, and examples that were previously only implicit, incomplete, or expressed in classical terms in the literature. Where applicable this is done by explicitly referring to the Stacks project for preliminary results. Topics include the construction and properties of important moduli problems in algebraic geometry (such as the Deligne–Mumford compactification of the moduli of curves, the Picard functor, or moduli of semistable vector bundles and sheaves) and arithmetic questions for fields and algebraic spaces.

PIETER BELMANS is Assistant Professor of Mathematics at the University of Luxembourg. He studies algebraic geometry and noncommutative algebra from the point of view of derived categories. He developed the infrastructure that runs the Stacks project.

WEI HO is Associate Professor of Mathematics at the University of Michigan. Her research interests are primarily in arithmetic geometry, number theory, and algebraic geometry. She first became involved with the Stacks project during her postdoctoral fellowship at Columbia University.

AISE JOHAN DE JONG is Professor at Columbia University. He has worked at Harvard University, Princeton University, and MIT. Currently he spends most of his research time advising his graduate students and working on the Stacks project. He received the 2022 AMS Leroy P. Steele Prize for Mathematical Exposition for his work on the Stacks project.

LONDON MATHEMATICAL SOCIETY LECTURE NOTE SERIES

Managing Editor: Professor Endre Süli, Mathematical Institute, University of Oxford, Woodstock Road, Oxford OX2 6GG, United Kingdom

The titles below are available from booksellers, or from Cambridge University Press at www.cambridge.org/mathematics

367 Random matrices: High dimensional phenomena, G. BLOWER
368 Geometry of Riemann surfaces, F.P. GARDINER, G. GONZÁLEZ-DIEZ & C. KOUROUNIOTIS (eds)
369 Epidemics and rumours in complex networks, M. DRAIEF & L. MASSOULIÉ
370 Theory of p-adic distributions, S. ALBEVERIO, A.YU. KHRENNIKOV & V.M. SHELKOVICH
371 Conformal fractals, F. PRZYTYCKI & M. URBAŃSKI
372 Moonshine: The first quarter century and beyond, J. LEPOWSKY, J. MCKAY & M.P. TUITE (eds)
373 Smoothness, regularity and complete intersection, J. MAJADAS & A. G. RODICIO
374 Geometric analysis of hyperbolic differential equations: An introduction, S. ALINHAC
375 Triangulated categories, T. HOLM, P. JØRGENSEN & R. ROUQUIER (eds)
376 Permutation patterns, S. LINTON, N. RUŠKUC & V. VATTER (eds)
377 An introduction to Galois cohomology and its applications, G. BERHUY
378 Probability and mathematical genetics, N. H. BINGHAM & C. M. GOLDIE (eds)
379 Finite and algorithmic model theory, J. ESPARZA, C. MICHAUX & C. STEINHORN (eds)
380 Real and complex singularities, M. MANOEL, M.C. ROMERO FUSTER & C.T.C WALL (eds)
381 Symmetries and integrability of difference equations, D. LEVI, P. OLVER, Z. THOMOVA & P. WINTERNITZ (eds)
382 Forcing with random variables and proof complexity, J. KRAJÍČEK
383 Motivic integration and its interactions with model theory and non-Archimedean geometry I, R. CLUCKERS, J. NICAISE & J. SEBAG (eds)
384 Motivic integration and its interactions with model theory and non-Archimedean geometry II, R. CLUCKERS, J. NICAISE & J. SEBAG (eds)
385 Entropy of hidden Markov processes and connections to dynamical systems, B. MARCUS, K. PETERSEN & T. WEISSMAN (eds)
386 Independence-friendly logic, A.L. MANN, G. SANDU & M. SEVENSTER
387 Groups St Andrews 2009 in Bath I, C.M. CAMPBELL *et al* (eds)
388 Groups St Andrews 2009 in Bath II, C.M. CAMPBELL *et al* (eds)
389 Random fields on the sphere, D. MARINUCCI & G. PECCATI
390 Localization in periodic potentials, D.E. PELINOVSKY
391 Fusion systems in algebra and topology, M. ASCHBACHER, R. KESSAR & B. OLIVER
392 Surveys in combinatorics 2011, R. CHAPMAN (ed)
393 Non-abelian fundamental groups and Iwasawa theory, J. COATES *et al* (eds)
394 Variational problems in differential geometry, R. BIELAWSKI, K. HOUSTON & M. SPEIGHT (eds)
395 How groups grow, A. MANN
396 Arithmetic differential operators over the p-adic integers, C.C. RALPH & S.R. SIMANCA
397 Hyperbolic geometry and applications in quantum chaos and cosmology, J. BOLTE & F. STEINER (eds)
398 Mathematical models in contact mechanics, M. SOFONEA & A. MATEI
399 Circuit double cover of graphs, C.-Q. ZHANG
400 Dense sphere packings: a blueprint for formal proofs, T. HALES
401 A double Hall algebra approach to affine quantum Schur–Weyl theory, B. DENG, J. DU & Q. FU
402 Mathematical aspects of fluid mechanics, J.C. ROBINSON, J.L. RODRIGO & W. SADOWSKI (eds)
403 Foundations of computational mathematics, Budapest 2011, F. CUCKER, T. KRICK, A. PINKUS & A. SZANTO (eds)
404 Operator methods for boundary value problems, S. HASSI, H.S.V. DE SNOO & F.H. SZAFRANIEC (eds)
405 Torsors, étale homotopy and applications to rational points, A.N. SKOROBOGATOV (ed)
406 Appalachian set theory, J. CUMMINGS & E. SCHIMMERLING (eds)
407 The maximal subgroups of the low-dimensional finite classical groups, J.N. BRAY, D.F. HOLT & C.M. RONEY-DOUGAL
408 Complexity science: the Warwick master's course, R. BALL, V. KOLOKOLTSOV & R.S. MACKAY (eds)
409 Surveys in combinatorics 2013, S.R. BLACKBURN, S. GERKE & M. WILDON (eds)
410 Representation theory and harmonic analysis of wreath products of finite groups, T. CECCHERINI-SILBERSTEIN, F. SCARABOTTI & F. TOLLI
411 Moduli spaces, L. BRAMBILA-PAZ, O. GARCÍA-PRADA, P. NEWSTEAD & R.P. THOMAS (eds)
412 Automorphisms and equivalence relations in topological dynamics, D.B. ELLIS & R. ELLIS
413 Optimal transportation, Y. OLLIVIER, H. PAJOT & C. VILLANI (eds)
414 Automorphic forms and Galois representations I, F. DIAMOND, P.L. KASSAEI & M. KIM (eds)
415 Automorphic forms and Galois representations II, F. DIAMOND, P.L. KASSAEI & M. KIM (eds)
416 Reversibility in dynamics and group theory, A.G. O'FARRELL & I. SHORT
417 Recent advances in algebraic geometry, C.D. HACON, M. MUSTAŢĂ & M. POPA (eds)
418 The Bloch–Kato conjecture for the Riemann zeta function, J. COATES, A. RAGHURAM, A. SAIKIA & R. SUJATHA (eds)
419 The Cauchy problem for non-Lipschitz semi-linear parabolic partial differential equations, J.C. MEYER & D.J. NEEDHAM
420 Arithmetic and geometry, L. DIEULEFAIT *et al* (eds)

421 O-minimality and Diophantine geometry, G.O. JONES & A.J. WILKIE (eds)
422 Groups St Andrews 2013, C.M. CAMPBELL *et al* (eds)
423 Inequalities for graph eigenvalues, Z. STANIĆ
424 Surveys in combinatorics 2015, A. CZUMAJ *et al* (eds)
425 Geometry, topology and dynamics in negative curvature, C.S. ARAVINDA, F.T. FARRELL & J.-F. LAFONT (eds)
426 Lectures on the theory of water waves, T. BRIDGES, M. GROVES & D. NICHOLLS (eds)
427 Recent advances in Hodge theory, M. KERR & G. PEARLSTEIN (eds)
428 Geometry in a Fréchet context, C.T.J. DODSON, G. GALANIS & E. VASSILIOU
429 Sheaves and functions modulo p, L. TAELMAN
430 Recent progress in the theory of the Euler and Navier–Stokes equations, J.C. ROBINSON, J.L. RODRIGO, W. SADOWSKI & A. VIDAL-LÓPEZ (eds)
431 Harmonic and subharmonic function theory on the real hyperbolic ball, M. STOLL
432 Topics in graph automorphisms and reconstruction (2nd Edition), J. LAURI & R. SCAPELLATO
433 Regular and irregular holonomic D-modules, M. KASHIWARA & P. SCHAPIRA
434 Analytic semigroups and semilinear initial boundary value problems (2nd Edition), K. TAIRA
435 Graded rings and graded Grothendieck groups, R. HAZRAT
436 Groups, graphs and random walks, T. CECCHERINI-SILBERSTEIN, M. SALVATORI & E. SAVA-HUSS (eds)
437 Dynamics and analytic number theory, D. BADZIAHIN, A. GORODNIK & N. PEYERIMHOFF (eds)
438 Random walks and heat kernels on graphs, M.T. BARLOW
439 Evolution equations, K. AMMARI & S. GERBI (eds)
440 Surveys in combinatorics 2017, A. CLAESSON *et al* (eds)
441 Polynomials and the mod 2 Steenrod algebra I, G. WALKER & R.M.W. WOOD
442 Polynomials and the mod 2 Steenrod algebra II, G. WALKER & R.M.W. WOOD
443 Asymptotic analysis in general relativity, T. DAUDÉ, D. HÄFNER & J.-P. NICOLAS (eds)
444 Geometric and cohomological group theory, P.H. KROPHOLLER, I.J. LEARY, C. MARTÍNEZ-PÉREZ & B.E.A. NUCINKIS (eds)
445 Introduction to hidden semi-Markov models, J. VAN DER HOEK & R.J. ELLIOTT
446 Advances in two-dimensional homotopy and combinatorial group theory, W. METZLER & S. ROSEBROCK (eds)
447 New directions in locally compact groups, P.-E. CAPRACE & N. MONOD (eds)
448 Synthetic differential topology, M.C. BUNGE, F. GAGO & A.M. SAN LUIS
449 Permutation groups and cartesian decompositions, C.E. PRAEGER & C. SCHNEIDER
450 Partial differential equations arising from physics and geometry, M. BEN AYED *et al* (eds)
451 Topological methods in group theory, N. BROADDUS, M. DAVIS, J.-F. LAFONT & I. ORTIZ (eds)
452 Partial differential equations in fluid mechanics, C.L. FEFFERMAN, J.C. ROBINSON & J.L. RODRIGO (eds)
453 Stochastic stability of differential equations in abstract spaces, K. LIU
454 Beyond hyperbolicity, M. HAGEN, R. WEBB & H. WILTON (eds)
455 Groups St Andrews 2017 in Birmingham, C.M. CAMPBELL *et al* (eds)
456 Surveys in combinatorics 2019, A. LO, R. MYCROFT, G. PERARNAU & A. TREGLOWN (eds)
457 Shimura varieties, T. HAINES & M. HARRIS (eds)
458 Integrable systems and algebraic geometry I, R. DONAGI & T. SHASKA (eds)
459 Integrable systems and algebraic geometry II, R. DONAGI & T. SHASKA (eds)
460 Wigner-type theorems for Hilbert Grassmannians, M. PANKOV
461 Analysis and geometry on graphs and manifolds, M. KELLER, D. LENZ & R.K. WOJCIECHOWSKI
462 Zeta and L-functions of varieties and motives, B. KAHN
463 Differential geometry in the large, O. DEARRICOTT *et al* (eds)
464 Lectures on orthogonal polynomials and special functions, H.S. COHL & M.E.H. ISMAIL (eds)
465 Constrained Willmore surfaces, Á.C. QUINTINO
466 Invariance of modules under automorphisms of their envelopes and covers, A.K. SRIVASTAVA, A. TUGANBAEV & P.A. GUIL ASENSIO
467 The genesis of the Langlands program, J. MUELLER & F. SHAHIDI
468 (Co)end calculus, F. LOREGIAN
469 Computational cryptography, J.W. BOS & M. STAM (eds)
470 Surveys in combinatorics 2021, K.K. DABROWSKI *et al* (eds)
471 Matrix analysis and entrywise positivity preservers, A. KHARE
472 Facets of algebraic geometry I, P. ALUFFI *et al* (eds)
473 Facets of algebraic geometry II, P. ALUFFI *et al* (eds)
474 Equivariant topology and derived algebra, S. BALCHIN, D. BARNES, M. KĘDZIOREK & M. SZYMIK (eds)
475 Effective results and methods for Diophantine equations over finitely generated domains, J.-H. EVERTSE & K. GYŐRY
476 An indefinite excursion in operator theory, A. GHEONDEA
477 Elliptic regularity theory by approximation methods, E.A. PIMENTEL
478 Recent developments in algebraic geometry, H. ABBAN, G. BROWN, A. KASPRZYK & S. MORI (eds)
479 Bounded cohomology and simplicial volume, C. CAMPAGNOLO, F. FOURNIER- FACIO, N. HEUER & M. MORASCHINI (eds)
480 Stacks Project Expository Collection (SPEC), P. BELMANS, W. HO & A.J. DE JONG (eds)

Stacks Project Expository Collection (SPEC)

EDITED BY

PIETER BELMANS
University of Luxembourg

WEI HO
University of Michigan, Ann Arbor

AISE JOHAN DE JONG
Columbia University, New York

Shaftesbury Road, Cambridge CB2 8EA, United Kingdom

One Liberty Plaza, 20th Floor, New York, NY 10006, USA

477 Williamstown Road, Port Melbourne, VIC 3207, Australia

314–321, 3rd Floor, Plot 3, Splendor Forum, Jasola District Centre,
New Delhi – 110025, India

103 Penang Road, #05–06/07, Visioncrest Commercial, Singapore 238467

Cambridge University Press is part of Cambridge University Press & Assessment,
a department of the University of Cambridge.

We share the University's mission to contribute to society through the pursuit of
education, learning and research at the highest international levels of excellence.

www.cambridge.org
Information on this title: www.cambridge.org/9781009054850

DOI: 10.1017/9781009051897

First published 2022

A catalogue record for this publication is available from the British Library.

ISBN 978-1-009-05485-0 Paperback

Contents

Contributors

Jarod Alper, *Department of Mathematics, University of Washington, Box 354350, C-138 Padelford, Seattle, WA 98195-4350, United States*

Pieter Belmans, *Department of Mathematics, University of Luxembourg, 6, Avenue de la Fonte, L-4364 Esch-sur-Alzette, Luxembourg*

Daniel Bragg, *Department of Mathematics, University of California, Berkeley, 970 Evans Hall, Berkeley, CA 94720-3840, United States*

Raymond Cheng, *Department of Mathematics, Columbia University, 2990 Broadway, New York, NY 10027, United States*

Elsa Corniani, *Dipartimento di Matematica e Informatica, Università di Ferrara, Via Machiavelli 30, 44121 Ferrara, Italy*

Neeraj Deshmukh, *Institut für Mathematik, Universität Zürich, Winterthurerstrasse 190, CH-8057 Zürich, Switzerland*

Haoyang Guo, *Max-Planck-Institut für Mathematik, Vivatsgasse 7, 53111 Bonn, Germany*

Lena Ji, *Department of Mathematics, University of Michigan, 530 Church Street, Ann Arbor, MI 48109, United States*

Nikolas Kuhn, *Max-Planck-Institut für Mathematik, Vivatsgasse 7, 53111 Bonn, Germany*

Matt Larson, *Department of Mathematics, Stanford University, 450 Jane Stanford Way, Stanford, CA 94305, United States*

Shizhang Li, *Department of Mathematics, University of Michigan, 530 Church Street, Ann Arbor, MI 48109, United States*

Carl Lian, *Institut für Mathematik, Humboldt-Universität zu Berlin, Rudower Chaussee 25, Raum 1.412, 12489 Berlin, Germany*

Jason Liang, *Department of Mathematics, University of Michigan, 530 Church Street, Ann Arbor, MI 48109, United States*

Devlin Mallory, *Department of Mathematics, University of Utah, 155 South 1400 East, Salt Lake City, Utah 84112, United States*

Patrick McFaddin, *Department of Mathematics, Fordham University, 113 West 60th Street, New York, NY 10023, United States*

Stephen McKean, *Department of Mathematics, Harvard University, Cambridge, MA 02138, United States*

Drew Moore, *Department of Mathematics, University of Chicago, Chicago, IL, United States*

Takumi Murayama, *Department of Mathematics, Princeton University, Princeton, NJ 08544-1000, USA*

Brett Nasserden, *Department of Mathematics, Western University, London, ON N6A 5B7, Canada*

Noah Olander, *Department of Mathematics, Columbia University, 2990 Broadway, New York, NY 10027, United States*

Emanuel Reinecke, *Max-Planck-Institut für Mathematik, Vivatsgasse 7, 53111 Bonn, Germany*

Soumya Sankar, *Department of Mathematics, The Ohio State University, 231 West 18th Avenue, Columbus, OH 43210, United States*

Sanal Shivaprasad, *Department of Mathematics, University of Michigan, 530 Church Street, Ann Arbor, MI 48109, United States*

Dylan Spence, *Department of Mathematics, University of Wisconsin-Whitewater, Laurentide Hall, 800 West Main Street, Whitewater, WI 53190-1790, United States*

Matthew Stevenson, *Google, Sunnyvale, CA, United States*

Nawaz Sultani, *Department of Mathematics, University of Michigan, 530 Church Street, Ann Arbor, MI 48109, United States*

Tuomas Tajakka, *Matematiska Institutionen, Stockholm Universitet, Kräftriket 5A, 114 19 Stockholm, Sweden*

Vaidehee Thatte, *Department of Mathematics, King's College London, Strand, London WC2R 2LS, United Kingdom*

Rachel Webb, *Department of Mathematics, University of California Berkeley, Berkeley, CA 94720, United States*

Kirsten Wickelgren, *Department of Mathematics, Duke University, Box 90320, Durham, NC 27708, United States*

Yueqiao Wu, *Department of Mathematics, University of Michigan, 530 Church Street, Ann Arbor, MI 48109, United States*

Yifei Zhao, *Mathematics Münster, University of Münster, Einsteinstrasse 62, 48149 Münster, Germany*

Preface

The *Stacks project* is "an open source textbook and reference work on algebraic geometry" that has been consistently growing since 2008. At the time of writing, the material encompasses more than 7500 pages, and it is regularly expanded. It contains background material in commutative algebra and algebraic geometry, advanced material, and even previously unpublished work. Widely used as a reference in algebraic geometry, the Stacks project is most easily accessed via its website

```
https://stacks.math.columbia.edu
```

The project is collaborative, with more than 500 contributors, listed on the website. Every submission and comment is reviewed carefully, for correctness and coherence with the rest of the text, by the second editor of this volume.

Although the Stacks project covers most basic and some advanced algebraic geometry, many interesting and useful advanced topics are not yet part of the Stacks project. To add a new topic often requires a large amount of work, partially due to the requirement that all prerequisites are built from scratch. Moreover, for some subjects, the existing literature may not use the newest foundations or machinery available to mathematicians now. Before blindly adding a new result or concept, it thus makes sense to explore it and find efficient proof strategies using, as much as possible, already existing material in the Stacks project. The chapters in this volume are first and foremost expository presentations in their own right. But they can also be considered as explorations on subjects which deserve inclusion in the Stacks project; we hope they can someday be used as the initial versions of new chapters of the Stacks project.

Each chapter grew out of group work at one of two workshops affiliated with the Stacks project that we organized. The first workshop took place in person at the University of Michigan in summer 2017, and the second – although originally intended to take place in Ann Arbor again – became a virtual workshop during summer 2020. At each of these workshops, senior algebraic geometers led groups of graduate students and postdoctoral scholars in learning and exploring advanced topics in algebraic geometry, with the goal of writing careful expositions at a level appropriate for advanced graduate students. We hope that this book will serve as a useful reference for students and researchers interested in learning about these topics. While most of the chapters are primarily expository, many contain new examples and proofs not found elsewhere in the literature.

We now give a brief description of the chapters in this volume. The first several are related to moduli problems in algebraic geometry. In Chapter 1, "Projectivity of the moduli of curves", following a method of Kollár that avoids geometric invariant theory, the authors give an exposition of the projectivity (over $\operatorname{Spec}\mathbb{Z}$) of the Deligne–Mumford moduli space $\overline{M}_g$ of stable curves of genus $g \geq 2$. Chapter 2, "The stack of admissible covers is algebraic", gives a proof of the algebraicity of the stack of classifying stable genus-g curves equipped with "admissible" G-covers, following a result of Abramovich, Corti, and Vistoli (and going back to ideas of Harris and Mumford); this chapter also has an exposition of group actions on algebraic spaces. Chapter 3, "Projectivity of the moduli space of vector bundles on a curve", explains a proof that the moduli space of semistable vector bundles (of fixed rank and degree) on a curve of genus $g \geq 2$ is projective; unlike the classical proof via geometric invariant theory, the proof here relies on the modern notion of good moduli spaces and uses a method due to Esteves and Popa to prove the projectivity.

Going from bundles on curves to sheaves on varieties, in Chapter 4, "Boundedness of semistable sheaves", the authors explain, following work of Langer, why the moduli space of semistable torsion-free sheaves with fixed Hilbert polynomial is bounded for any projective variety in any characteristic; the key result is an upper bound for the maximal slope of the restriction of the sheaf under consideration to a general hypersurface. Chapter 5, "Theorem of the Base", contains a modern proof of a fundamental theorem in algebraic geometry: the Néron–Severi group of a proper variety is finitely generated. The proof relies on a Weil cohomology (such as ℓ-adic cohomology) to prove finite generation up to torsion, and a reduction to the smooth projective case via de Jong's alterations to handle torsion. In Chapter 6, "Weil restriction for schemes and beyond", the authors discuss, with several examples, the notion of Weil restriction for both schemes and algebraic spaces; amongst other things, one finds a proof of the preservation of quasi-projectivity under this operation. Chapter 7, "Heights over finitely generated fields", contains an exposition of several notions of heights in modern arithmetic geometry: naive heights over number fields, geometric heights over function fields, and finally Moriwaki heights over finitely generated fields.

Chapter 8, "An explicit self-duality", gives an explicit construction of a consequence of Grothendieck duality: if $A \longrightarrow B$ is a finite flat complete intersection map with A noetherian and B local, then B is self-dual as an A-module (and canonically so after choosing a presentation). In Chapter 9, "Tannakian reconstruction of coalgebroids", the author gives an exposition of a recent result of Schäppi generalizing classical Tannakian reconstruction theorems: there is an explicit criterion allowing one to reconstruct a group scheme G over a commutative ring R from its category of representations of finite projective R-modules.

We thank the authors for their contributions to this volume. We also thank the mentors (Jarod Alper, Bhargav Bhatt, Brian Conrad, Matthew Emerton, Max Lieblich, Davesh Maulik, Martin Olsson, Alex Perry, Ravi Vakil, Kirsten Wickelgren) and the participants of the Stacks project workshops for making these workshops such pleasant and productive events, and we thank the National Science Foundation, Compositio, and the University of Michigan for funding these workshops. Finally, we thank the contributors to and users of the Stacks project.

Pieter Belmans
Wei Ho
Aise Johan de Jong

1

Projectivity of the moduli of curves

Raymond Cheng
Columbia University

Carl Lian
Humboldt-Universität zu Berlin

Takumi Murayama
Princeton University

In collaboration with Yordanka Kovacheva *and* Monica Marinescu

Abstract In this expository paper, we show that the Deligne–Mumford moduli space of stable curves is projective over Spec$(\mathbf{Z})$. The proof we present is due to Kollár. Ampleness of a line bundle is deduced from nefness of a related vector bundle via the ampleness lemma, a classifying map construction. The main positivity result concerns the pushforward of relative dualizing sheaves on families of stable curves over a smooth projective curve.

Introduction

Let $\overline{\mathcal{M}}_g$ be the moduli stack of stable curves of genus $g \geq 2$ and write $\overline{M}_g$ for its corresponding moduli space. We prove that the moduli of stable curves is projective in the following sense, see Theorem 1.7.2:

Theorem *The Deligne–Mumford moduli space $\overline{M}_g$ of stable curves of genus $g \geq 2$ is a projective scheme over* Spec$(\mathbf{Z})$.

In particular, this means that $\overline{M}_g$, which is *a priori* just an algebraic space, is actually a projective scheme over $\mathbf{Z}$. Together with the work of Deligne and Mumford [9] (see also [30, Theorem 0E9C]) this means that $\overline{M}_g$ is actually an irreducible projective scheme over $\mathbf{Z}$.

We explain a proof due to Kollár in [21]. Specifically, the task of showing that a certain line bundle on $\overline{M}_g$ is ample is transferred, via

Kollár's ampleness lemma, to the problem of showing that a related vector bundle is nef on $\overline{M}_g$. Since nefness is a condition that only depends on the behaviour of the vector bundle upon restriction to curves, projectivity is thus reduced to a problem regarding positivity of one-parameter families of stable curves.

Kollár's method differs from other existing proofs of projectivity of $\overline{M}_g$ in at least two main ways: First, the technique is independent of the methods of geometric invariant theory (GIT), on which the proofs of [29, 11, 7] rely. In a similar spirit, in [1], which is Chapter 3 of this volume, the projectivity of the moduli space of semistable vector bundles on a curve is established without using GIT.

Second, Kollár's criterion does not require one to directly check that a line bundle on the moduli space is ample, in contrast to the approach of Knudsen and Mumford [19, 17, 18]; rather, one only needs to show that some vector bundle on the moduli space is nef. As such, this method has since been used in other settings, such as in the moduli of weighted stable curves [14], of stable varieties [22], and, recently, of K-polystable Fano varieties [6, 33].

An outline of this article is as follows. We set up notation in regards to the moduli of curves in Section 1.1, after which we begin in Sections 1.2–1.4 with some material on positivity of sheaves. In Section 1.5, we explain Kollár's ampleness lemma; see Proposition 1.5.4. In Section 1.6, we prove the main positivity statement: the pushforward of the relative dualizing sheaf of a 1-parameter family of stable curves of genus at least 2 is nef; see Theorem 1.6.10. Finally, we put everything together in Section 1.7 to show that $\overline{M}_g$ is projective over $\mathbf{Z}$ when $g \geq 2$.

Conventions Throughout, k will denote a field. Following the conventions of the Stacks project, a *variety* is a separated integral scheme of finite type over a field k and a *curve* is a variety of dimension 1, see [30, Definitions 020D and 0A23]. Given a scheme X over k and a sheaf $\mathcal{F}$ of $\mathcal{O}_X$-modules, we write

$$h^i(X, \mathcal{F}) := \dim_k(H^i(X, \mathcal{F})) \quad \text{for all } i \in \mathbf{Z}.$$

1.1 Stable curves

In this section, we record the definition of the moduli problem in which we are primarily interested, namely that of the moduli space of stable curves. The main references are [9] and [30, Chapter 0DMG].

First we define what we mean by a family of curves. Compare the following with [30, Situation 0D4Z], and with [30, Definitions 0C47, 0C5A, and 0E75]. We diverge slightly from the Stacks project in that we require our families of nodal curves to have geometrically connected fibres. Caution: the closed fibres of a family of nodal curves are *not* curves in the sense of our conventions, as they may be reducible. See [30, Section 0C58] for a discussion on such terminology.

Definition 1.1.1 Let S be a scheme.

(i) A *family of nodal curves over* S is a flat, proper, finitely presented morphism of schemes $f: X \to S$ of relative dimension 1 such that all geometric fibres are connected and smooth except at possibly finitely many nodes.
(ii) A *family of stable curves over* S is a family of nodal curves such that the geometric fibres have arithmetic genus ≥ 2 and do not contain rational tails or bridges.
(iii) A family of stable curves over S is said to *have genus* g if all geometric fibres have genus g.

Condition (ii) is equivalent to ampleness of the dualizing sheaf, and also finiteness of automorphism groups. See [30, Section 0E73] for details. For the following, see [30, Definition 0E77].

Definition 1.1.2 For $g \geq 2$, the *moduli stack of stable curves of genus* g is the category $\overline{\mathcal{M}}_g$ fibred in groupoids whose category of sections over a scheme S has objects given by families of stable curves of genus g over S, and morphisms given by isomorphisms of families over S.

The stack $\overline{\mathcal{M}}_g$ is a smooth, proper Deligne–Mumford stack over $\mathrm{Spec}(\mathbf{Z})$; see [30, Theorem 0E9C]. Classically, and in many geometric applications such as [13], it is convenient to work with a space rather than the stack. As such, it is useful to extract an algebraic space which is, in some sense, the closest approximation of the stack, obtained by "forgetting" the automorphism groups: this is the notion of a *uniform categorical moduli space* or simply a *moduli space* of a stack; see [30, Definition 0DUG].

Lemma 1.1.3 *The stack* $\overline{\mathcal{M}}_g$ *admits a uniform categorical moduli space* $f_g: \overline{\mathcal{M}}_g \to \overline{M}_g$ *such that* f_g *is separated, quasi-compact, and a universal homeomorphism.*

Proof The stack $\overline{\mathcal{M}}_g$ has finite inertia by [30, Lemmas 0E7A and

0DSW], so the existence of f_g follows from the Keel–Mori theorem [30, Theorem 0DUT]. □

Definition 1.1.4 The space $\overline{M}_g$ is the *moduli space of curves of genus* g.

Our primary goal is to show that $\overline{M}_g$ is projective over $\mathbf{Z}$; see Theorem 1.7.2. Thus we must exhibit an ample invertible sheaf on $\overline{M}_g$. We obtain invertible sheaves on the moduli space by taking powers of invertible sheaves on the stack $\overline{\mathcal{M}}_g$, via the following general fact:

Lemma 1.1.5 *Let $\mathcal{X}$ be an algebraic stack. Assume the inertia $\mathcal{I}_{\mathcal{X}} \to \mathcal{X}$ is finite and let $f\colon \mathcal{X} \to M$ be its moduli space, as in [30, Theorem 0DUT]. Then*

$$f^*\colon \mathrm{Pic}(M) \to \mathrm{Pic}(\mathcal{X})$$

is injective. If $\mathcal{X}$ is furthermore quasi-compact, then the cokernel of f^* is annihilated by a positive integer.

Proof For the injectivity, note that $f_*\mathcal{O}_{\mathcal{X}} \cong \mathcal{O}_M$ as M is initial for morphisms from $\mathcal{X}$ to algebraic spaces and the structure sheaf represents the functor $\mathrm{Hom}(-, \mathbf{A}^1)$. Thus if $\mathcal{N} \in \mathrm{Pic}(M)$ is such that $f^*\mathcal{N} \cong \mathcal{O}_{\mathcal{X}}$, the canonical map $\mathcal{N} \to f_*f^*\mathcal{N} \to \mathcal{O}_M$ is an isomorphism as $\mathcal{N}$ is locally trivial. This further shows that if $\mathcal{N}_1, \mathcal{N}_2 \in \mathrm{Pic}(M)$ are such that there exists an isomorphism $\varphi\colon f^*\mathcal{N}_1 \to f^*\mathcal{N}_2$, then there is a unique isomorphism $\psi\colon \mathcal{N}_1 \to \mathcal{N}_2$ such that $f^*\psi = \varphi$.

We now show that, if $\mathcal{X}$ is furthermore quasi-compact, then there is a positive integer n such that, for every $\mathcal{L} \in \mathrm{Pic}(\mathcal{X})$, $\mathcal{L}^{\otimes n} \cong f^*\mathcal{N}$ for some $\mathcal{N} \in \mathrm{Pic}(M)$. For this, we may replace $\mathcal{X}$ by any $\mathcal{X}'$ with a surjective separated étale morphism $h\colon \mathcal{X}' \to \mathcal{X}$ of algebraic stacks inducing isomorphisms on automorphism groups. Indeed, [30, Lemma 0DUV] gives the cartesian square

$$\begin{array}{ccc} \mathcal{X}' & \xrightarrow[h]{} & \mathcal{X} \\ {\scriptstyle f'}\downarrow & & \downarrow{\scriptstyle f} \\ M' & \longrightarrow & M \end{array}$$

where M' is the moduli space of $\mathcal{X}'$. If there were $\mathcal{N}' \in \mathrm{Pic}(M')$ such that $h^*\mathcal{L}^{\otimes n} \cong f'^*\mathcal{N}'$, then the injectivity of $f'^*\colon \mathrm{Pic}(M') \to \mathrm{Pic}(\mathcal{X}')$ shows that the étale descent datum for $h^*\mathcal{L}^{\otimes n}$ over $\mathcal{X}$ induces an étale descent datum for $\mathcal{N}'$ over M, yielding $\mathcal{N} \in \mathrm{Pic}(M)$ as above.

Choose such a cover $h\colon \mathcal{X}' \to \mathcal{X}$ as in [30, Lemma 0DUE]: $\mathcal{X}' = \coprod_{i \in I} \mathcal{X}_i$ where each $\mathcal{X}_i$ is a quotient stack $[U_i/R_i]$, $(U_i, R_i, s_i, t_i, c_i)$ is a groupoid scheme with U_i and R_i affine, and $s_i, t_i\colon R_i \to U_i$ are finite locally free of some constant rank; see [30, Lemmas 0DUM and 03BI]. Since $\mathcal{X}$ is quasi-compact, we are reduced to the case where $\mathcal{X}$ is a finite disjoint union of such stacks $\mathcal{X}_i$. Let $f_i\colon \mathcal{X}_i \to M_i$ be the moduli space. If there exists a positive integer n_i annihilating the cokernel of f_i^*, then the least common multiple n of the n_i annihilates the cokernel of f^*.

Thus it suffices to consider the case where $\mathcal{X} = [U/R]$ is as above. By [30, Proposition 06WT], an invertible $\mathcal{O}_\mathcal{X}$-module may be represented as a pair $(\mathcal{L}, \alpha)$ consisting of an invertible $\mathcal{O}_U$-module $\mathcal{L}$ together with an isomorphism $\alpha\colon t^*\mathcal{L} \to s^*\mathcal{L}$ of $\mathcal{O}_R$-modules as in [30, Definition 03LI]. We claim that if n is the rank of the morphisms $s, t\colon R \to U$, then $(\mathcal{L}^{\otimes n}, \alpha^n)$ is in the image of f^*. Namely, writing $\pi\colon U \to M$, there exists an invertible $\mathcal{O}_M$-module $\mathcal{N}$ and an isomorphism of invertible modules $(\pi^*\mathcal{N}, \alpha_{\mathrm{can}}) \cong (\mathcal{L}^{\otimes n}, \alpha^n)$ on the groupoid (U, R, s, t, c), where α_{can} is the identity map; this makes sense since $\pi \circ t = \pi \circ t$ as maps $R \to M$.

Construct $\mathcal{N}$ as follows. First, if $U = \bigcup U_i$ is any affine open cover, then the $V_i := \pi(U_i)$ together form an affine open cover of M. That the V_i form an open cover follows from the fact that π is the composition of the faithfully flat and finitely presented morphism $U \to \mathcal{X}$ and the universal homeomorphism $\mathcal{X} \to M$; see [30, Lemmas 01UA and 0DUP]. That the V_i are affine is because π is integral; see [30, Lemmas 03BJ and 05YU]. Next, since $t\colon R \to U$ is finite locally free, [30, Lemma 0BCY] constructs an invertible $\mathcal{O}_U$-module $\mathcal{L}' := \mathrm{Norm}_t(s^*\mathcal{L})$ as follows. Let $(\{U_i\}, \{u_{ij}\})$ be a system of cocycles locally defining $\mathcal{L}$, so that $U = \bigcup U_i$ is an affine open cover and $u_{ij} \in \mathcal{O}_U^*(U_i \cap U_j)$ are units. Then $\mathcal{L}'$ is defined by the cocycles $(\{U_i\}, \{u'_{ij}\})$ with $u'_{ij} := \mathrm{Norm}_{t^\sharp}(s^\sharp(u_{ij}))$. Finally, setting $V_i := \pi(U_i)$, [30, Lemma 03BH] implies that the u'_{ij} lie in the subgroup $\mathcal{O}_M^*(V_i \cap V_j) \subseteq \mathcal{O}_U^*(U_i \cap U_j)$ of R-invariant units, so $(\{V_i\}, \{u'_{ij}\})$ forms a system of cocycles on M defining an invertible module $\mathcal{N}$.

On the one hand, the construction implies $\mathcal{L}' \cong \pi^*\mathcal{N}$. On the other hand, [30, Lemma 0BCZ] yields an isomorphism

$$\mathrm{Norm}_t(\alpha)\colon \mathcal{L}^{\otimes n} \cong \mathrm{Norm}_t(t^*\mathcal{L}) \to \mathrm{Norm}_t(s^*\mathcal{L}) = \mathcal{L}' \cong \pi^*\mathcal{N}.$$

Thus it suffices to show that the diagram of isomorphisms

$$\begin{array}{ccc} t^*\mathcal{L}^{\otimes n} & \xrightarrow{\alpha^n} & s^*\mathcal{L}^{\otimes n} \\ {\scriptstyle t^*\mathrm{Norm}_t(\alpha)}\downarrow & & \downarrow{\scriptstyle s^*\mathrm{Norm}_t(\alpha)} \\ t^*\pi^*\mathcal{N} & \overset{\alpha_{\mathrm{can}}}{=\!=\!=} & s^*\pi^*\mathcal{N} \end{array}$$

is commutative. By properties of the norm, the compatibilities of α from [30, Definition 03LH(1)], and the diagram of [30, Lemma 03BH], we have

$$\begin{aligned}\alpha^n &= \mathrm{Norm}_c(c^*\alpha) = \mathrm{Norm}_c(\mathrm{pr}_1^*\alpha \circ \mathrm{pr}_0^*\alpha)\\ &= \mathrm{Norm}_c(\mathrm{pr}_1^*\alpha) \circ \mathrm{Norm}_c(\mathrm{pr}_0^*\alpha) = s^*\,\mathrm{Norm}_s(\alpha) \circ t^*\,\mathrm{Norm}_t(\alpha).\end{aligned}$$

Since $s = t \circ i$ where $i\colon R \to R$ is the inverse, $\mathrm{Norm}_s(\alpha) = \mathrm{Norm}_t(i^*\alpha)$. Therefore

$$s^*\,\mathrm{Norm}_t(\alpha) \circ \alpha^n \circ t^*\,\mathrm{Norm}_t(\alpha)^{-1} = s^*(\mathrm{Norm}_t(\alpha \circ i^*\alpha)).$$

This is the identity since, by [30, Lemma 077Q], $i^*\alpha$ is the inverse of α. □

We now specify some invertible sheaves on $\overline{\mathcal{M}}_g$. By [30, Definition 06TR and Lemma 06WI], the data of such a sheaf $\mathcal{L}$ are the following: for each family of stable curves $X \to S$, an invertible $\mathcal{O}_S$-module $\mathcal{L}(X \to S)$, and, for every cartesian square

$$\begin{array}{ccc} X' & \xrightarrow{g'} & X \\ {\scriptstyle f'}\downarrow & & \downarrow{\scriptstyle f} \\ S' & \xrightarrow{g} & S, \end{array}$$

an isomorphism of invertible $\mathcal{O}_{S'}$-modules

$$\varphi_g\colon g^*\mathcal{L}(X \to S) \cong \mathcal{L}(X' \to S')$$

such that for every composition of cartesian squares

$$\begin{array}{ccccc} X'' & \longrightarrow & X' & \longrightarrow & X \\ \downarrow & & \downarrow & & \downarrow \\ S'' & \xrightarrow{h} & S' & \xrightarrow{g} & S \end{array}$$

the isomorphisms are subject to the cocycle condition

$$\begin{array}{ccc} h^*(g^*\mathcal{L}(X \to S)) & \xrightarrow[h^*\varphi_g]{} & h^*\mathcal{L}(X' \to S') \\ \cong\downarrow & & \downarrow{\scriptstyle \varphi_h} \\ (gh)^*\mathcal{L}(X \to S) & \xrightarrow{\varphi_{gh}} & \mathcal{L}(X'' \to S''). \end{array}$$

Definition 1.1.6 For each integer $m \geq 1$, define an invertible sheaf λ_m on $\overline{\mathcal{M}}_g$ as follows. Given a family of stable curves $f\colon X \to S$, let $\omega_{X/S}^{\otimes m}$ be its relative dualizing sheaf; see [30, Definition 0E6Q]. This is an invertible $\mathcal{O}_X$-module. Note that the sheaves $f_*\omega_{X/S}^{\otimes m}$ are locally free on S. Set

$$\lambda_m(f\colon X \to S) := \det(f_*\omega_{X/S}^{\otimes m}).$$

Given a cartesian square as above, we have isomorphisms φ_g given by

$$g^* \det(f_*\omega_{X/S}^{\otimes m}) \cong \det(g^* f_*\omega_{X/S}^{\otimes m}) \to \det(f'_* g'^* \omega_{X/S}^{\otimes m}) \cong \det(f'_*\omega_{X'/S'}^{\otimes m}),$$

the functorial base change maps, and the fact that the formation of $\omega_{X/S}$ commutes with arbitrary base change; see [30, Lemma 0E6R]. Functoriality ensures that these satisfy the required cocycle condition.

Our goal will be to show that there is some m such that λ_m descends to an ample invertible sheaf on $\overline{M}_g$.

1.2 Nakai–Moishezon criterion for ampleness

In this section, we discuss the Nakai–Moishezon criterion for ampleness, relating the ampleness of an invertible sheaf with the positivity of intersection numbers. We directly prove the criterion for proper algebraic spaces over a field in Proposition 1.2.4 (compare with [21, Theorem 3.11]); the proof closely follows that of [16, Section III.1, Theorem 1], with suitable modifications. Using [30, Lemma 0D3A], one can also formulate a relative version; see, for example, [15, Proposition 2.10].

In the following, we work with proper algebraic spaces over a field. For generalities on algebraic spaces, see [30, Part 0ELT].

We will use numerical intersection theory on spaces as developed in [30, Section 0DN3]; see also [30, Section 0BEL] and [26, Section 1.1.C] for the situation of varieties. The main construction is the *intersection number* $(\mathcal{L}_1 \cdots \mathcal{L}_d \cdot Z)$ between a closed subspace $\iota\colon Z \to X$ of positive dimension d and invertible $\mathcal{O}_X$-modules $\mathcal{L}_1, \ldots, \mathcal{L}_d$: this is the coefficient of $n_1 \cdots n_d$ in the numerical polynomial

$$\chi(X, \iota_*\mathcal{O}_Z \otimes \mathcal{L}_1^{\otimes n_1} \otimes \cdots \otimes \mathcal{L}_d^{\otimes n_d}) = \chi(Z, \mathcal{L}_1^{\otimes n_1} \otimes \cdots \otimes \mathcal{L}_d^{\otimes n_d}|_Z).$$

See [30, Definition 0EDF].

The Nakai–Moishezon criterion relates ampleness to the positivity of intersection numbers. To formulate this succinctly, we make a definition. In the following, recall that a separated algebraic space Z is integral if and only if it is reduced and $|Z|$ is irreducible; see [30, Definition 0AD4] and [30, Section 03I7].

Definition 1.2.1 Let X be a proper algebraic space over k and let $\mathcal{L}$ be an invertible $\mathcal{O}_X$-module. We say that $\mathcal{L}$ *has positive degree* if, for every integral closed subspace Z of X of positive dimension d, $(\mathcal{L}^d \cdot Z) > 0$.

Note that the Stacks project only defines the degree of an invertible sheaf $\mathcal{L}$ either when $\mathcal{L}$ is ample or when $\dim(X) \leq 1$; see [30, Definitions 0BEW and 0AYR]. The content of the Nakai–Moishezon criterion is that if $\mathcal{L}$ has positive degree, then $\mathcal{L}$ is ample. Thus this is *a fortiori* compatible with the conventions of the Stacks project.

The main technical property we need is the permanence of positivity under finite morphisms.

Lemma 1.2.2 *Let X be a proper algebraic space over k. Let $f: Y \to X$ be a finite morphism of algebraic spaces. Let $\mathcal{L}$ be an invertible $\mathcal{O}_X$-module. If $\mathcal{L}$ has positive degree, then $f^*\mathcal{L}$ has positive degree.*

Proof This follows from the compatibility of numerical intersection numbers and pullbacks: if $Z \subset Y$ is a proper integral closed subspace of dimension d, then

$$(f^*\mathcal{L}^d \cdot Z) = \deg(Z \to f(Z))(\mathcal{L}^d \cdot f(Z))$$

where $\deg(Z \to f(Z))$ is positive as f is finite; see [30, Lemma 0EDJ]. □

The following is the core of the inductive proof of the criterion:

Lemma 1.2.3 *Let X be a proper algebraic space over k and let D be an effective Cartier divisor of X. If $\mathcal{O}_X(D)|_D$ is ample, then $\mathcal{O}_X(mD)$ is globally generated for all $m \gg 0$.*

Proof For each $m \geq 0$, there is a short exact sequence

$$0 \to \mathcal{O}_X((m-1)D) \to \mathcal{O}_X(mD) \to \mathcal{O}_X(mD)|_D \to 0.$$

Since $\mathcal{O}_X(D)|_D$ is ample, Serre vanishing [30, Lemma 0GFA] gives an integer m_1 such that $H^1(D, \mathcal{O}_X(mD)|_D) = 0$ for $m \geq m_1$. Hence the maps

$$\rho_m : H^1(X, \mathcal{O}_X((m-1)D)) \to H^1(X, \mathcal{O}_X(mD)),$$

arising from the long exact sequence on cohomology, are surjective for all $m \geq m_1$, yielding a nonincreasing sequence of nonnegative integers

$$h^1(X, \mathcal{O}_X(mD)) \geq h^1(X, \mathcal{O}_X((m+1)D)) \geq \cdots.$$

There is some $m_2 \geq m_1$ after which the sequence stabilizes, so that, for all $m \geq m_2$, the ρ_m are bijective and the restriction maps

$$H^0(X, \mathcal{O}_X(mD)) \to H^0(D, \mathcal{O}_X(mD)|_D)$$

are surjective. Finally, since $\mathcal{O}_X(D)|_D$ is ample, there exists some m_3 such that $\mathcal{O}_X(mD)|_D$ is generated by its global sections for all $m \geq m_3$.

Let $m_0 := \max(m_2, m_3)$. We show that the evaluation maps

$$H^0(X, \mathcal{O}_X(mD)) \otimes_k \mathcal{O}_X \to \mathcal{O}_X(mD)$$

are surjective for all $m \geq m_0$. We verify this on stalks. For $x \in |X \setminus D|$, a global section defining mD restricts to a unit in $\mathcal{O}_X(mD)_x$ and thus generates the stalk. So consider $x \in |D|$ and let $\kappa(x)$ be the residue field of D at x; see [30, Definition 0EMW]. Since $D \to X$ is a monomorphism, $\kappa(x)$ is also the residue field at x of X by [30, Lemma 0EMX]. Consider the diagram

$$\begin{array}{ccc} H^0(X, \mathcal{O}_X(mD)) \otimes_k \kappa(x) & \longrightarrow & \mathcal{O}_X(mD) \otimes_{\mathcal{O}_X} \kappa(x) \\ \downarrow & & \downarrow \simeq \\ H^0(D, \mathcal{O}_X(mD)|_D) \otimes_k \kappa(x) & \twoheadrightarrow & \mathcal{O}_X(mD)|_D \otimes_{\mathcal{O}_D} \kappa(x) \end{array}$$

obtained from the evaluation and restriction maps upon taking the fibre at x. By our choice of m_0, the restriction map on the left is surjective and $\mathcal{O}_X(mD)|_D$ is globally generated, so the bottom map is surjective. Since the right map is an isomorphism, commutativity of the diagram implies that the top map is surjective. Nakayama's lemma then implies that the evaluation map is surjective on the local ring $\mathcal{O}_X(mD)_x$. Hence the evaluation map is surjective, meaning $\mathcal{O}_X(mD)$ is globally generated. □

Proposition 1.2.4 (Nakai–Moishezon criterion) *Let X be a proper algebraic space over k. Let $\mathcal{L}$ be an invertible $\mathcal{O}_X$-module. Then $\mathcal{L}$ is ample on X if and only if $\mathcal{L}$ has positive degree.*

Proof If $\mathcal{L}$ is ample, then X is a scheme, $\mathcal{L}$ is ample in the schematic sense, and $\mathcal{L}$ has positive degree; see [30, Lemmas 0D32 and 0BEV].

Assuming $\mathcal{L}$ has positive degree, we show it is ample. We proceed by induction on $\dim(X)$. When $\dim(X) = 0$, since X is separated it is a scheme by [30, Theorem 086U], in which case the result is clear. When $\dim(X) = 1$, our assumption simplifies to $\deg(\mathcal{L}) > 0$. Now apply [30, Proposition 09YC] to obtain a finite surjective morphism $f\colon Y \to X$ from a scheme Y. Lemma 1.2.2 shows that $\deg(f^*\mathcal{L}) > 0$ and so [30, Lemma 0B5X] gives the ampleness of $f^*\mathcal{L}$. Since f is finite, [30, Lemma

0GFB] shows $\mathcal{L}$ is also ample. So we assume that $\dim(X) \geq 2$ and that the criterion holds for all proper spaces over k of lower dimension.

Step 1 Using [30, Lemmas 0GFB, 0GFA], and Lemma 1.2.2, we may replace X by the reduction of an irreducible component and $\mathcal{L}$ by its restriction to assume that X is integral.

Step 2 We show that some power of $\mathcal{L}$ is effective. As X is integral, the discussion of [30, Section 0ENV] shows that $\mathcal{L}$ has a regular meromorphic section s. Consider its sheaf of denominators $\mathcal{I}_1$, i.e., the ideal sheaf in $\mathcal{O}_X$ whose sections over $V \in X_{\text{étale}}$ are

$$\mathcal{I}_1(V) := \{f \in \mathcal{O}_X(V) \mid fs \in \mathcal{L}(V)\};$$

compare [30, Definition 02P1]. Set $\mathcal{I}_2 := \mathcal{I}_1 \otimes \mathcal{L}^\vee$. Since the formation of the $\mathcal{I}_j$, $j = 1, 2$, is étale local, their properties may be reduced to the schematic case. Thus [30, Lemma 02P0] shows that the $\mathcal{I}_j$ are quasi-coherent sheaves of ideals and the corresponding closed subspaces $Y_j = V(\mathcal{I}_j)$ satisfy $\dim(Y_j) < \dim(X)$. By Lemma 1.2.2, induction applies so the $\mathcal{L}|_{Y_j}$ are ample. By construction, for each $m \geq 0$, there are exact sequences

$$\begin{array}{ccccccccc} 0 & \longrightarrow & \mathcal{I}_1 \otimes \mathcal{L}^{\otimes m} & \longrightarrow & \mathcal{L}^{\otimes m} & \longrightarrow & \mathcal{L}^{\otimes m}|_{Y_1} & \longrightarrow & 0 \\ & & \| & & & & & & \\ 0 & \longrightarrow & \mathcal{I}_2 \otimes \mathcal{L}^{\otimes (m-1)} & \longrightarrow & \mathcal{L}^{\otimes (m-1)} & \longrightarrow & \mathcal{L}^{\otimes (m-1)}|_{Y_2} & \longrightarrow & 0. \end{array}$$

Serre vanishing, [30, Lemma 0B5U], gives some $m_0 \geq 0$ such that, for all $m \geq m_0$, $H^i(Y_j, \mathcal{L}^{\otimes m}|_{Y_j}) = 0$ for all $i > 0$ and $j = 1, 2$. Thus comparing the long exact sequences in cohomology for the sequences above yields

$$\begin{aligned} h^i(X, \mathcal{L}^{\otimes m}) &= h^i(X, \mathcal{I}_1 \otimes \mathcal{L}^{\otimes m}) \\ &= h^i(X, \mathcal{I}_2 \otimes \mathcal{L}^{\otimes (m-1)}) = h^i(X, \mathcal{L}^{\otimes (m-1)}) \end{aligned}$$

for all $i \geq 2$ and $m \geq m_0$. Hence, for all $m \geq m_0$,

$$N := \sum_{i=2}^{\dim(X)} (-1)^i h^i(X, \mathcal{L}^{\otimes m})$$

is a constant. By definition of the intersection numbers, the leading coefficient of the numerical polynomial $\chi(X, \mathcal{L}^{\otimes m})$ is $(\mathcal{L}^{\dim X} \cdot X)$ and this is positive by assumption. Thus

$$\chi(X, \mathcal{L}^{\otimes m}) = h^0(X, \mathcal{L}^{\otimes m}) - h^1(X, \mathcal{L}^{\otimes m}) + N \to \infty \quad \text{as } m \to \infty.$$

So $h^0(X, \mathcal{L}^{\otimes m}) \to \infty$ and $\mathcal{L}^{\otimes m}$ is effective for $m \gg 0$. Ampleness is insensitive to powers (see [30, Lemma 01PT]), so we may replace $\mathcal{L}$ by $\mathcal{L}^{\otimes m}$ to assume $\mathcal{L} = \mathcal{O}_X(D)$ for some effective Cartier divisor D.

Step 3 By induction, $\mathcal{L}|_D = \mathcal{O}_X(D)|_D$ is ample, so Lemma 1.2.3 implies that $\mathcal{L}^{\otimes m}$ is generated by its global sections for $m \gg 0$. We may replace $\mathcal{L}$ by $\mathcal{L}^{\otimes m}$ to assume that $\mathcal{L}$ is generated by its global sections.

Step 4 Via [30, Lemmas 01NE and 085D], a basis of global sections of $\mathcal{L}$ induces a proper morphism

$$f\colon X \to \mathbf{P}^n_k \quad \text{with } n := h^0(X, \mathcal{L}) - 1$$

such that $f^*\mathcal{O}_{\mathbf{P}^n_k}(1) = \mathcal{L}$. We now claim that f is finite, from which we may conclude: X is then a scheme as f is then representable, and the pullback of an ample by an affine morphism is ample; see [30, Lemmas 03ZQ and 0892]. By [30, Lemma 0A4X], it suffices to show that f has discrete fibres. But if there were $y \in \mathbf{P}^n_k$ such that the fibre X_y was positive dimensional, then we would obtain a commutative diagram

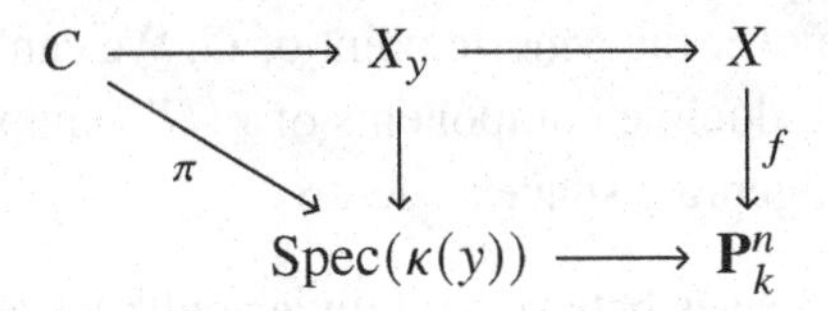

where the right square is cartesian and C is some complete curve in X_y. By commutativity of the diagram, we see that

$$\mathcal{L}|_C = (f^*\mathcal{O}_{\mathbf{P}^n}(1))|_C \simeq \pi^*\mathcal{O}_{\mathrm{Spec}(k(y))} = \mathcal{O}_C.$$

But now we reach a contradiction: on the one hand, $\mathcal{L}$ has positive intersection numbers with C; however, on the other hand, by [30, Lemma 0EDK],

$$0 < (\mathcal{L} \cdot C) = \deg_C(\mathcal{L}|_C) = \deg_C(\mathcal{O}_C) = 0,$$

the degree on the right being the usual degree on a curve; see [30, Definition 0AYR]. Thus f is a finite morphism, as claimed. □

1.3 Positivity of invertible sheaves

We next prove some preliminary results about nef invertible sheaves on proper algebraic spaces and about big invertible sheaves on proper schemes over arbitrary fields. See [26] for the theory for varieties over algebraically closed fields.

We start with the definition of nefness.

Definition 1.3.1 Let X be a proper algebraic space over k. An invertible $\mathcal{O}_X$-module is *nef* if $(\mathcal{L} \cdot C) \geq 0$ for every integral closed subspace $C \subset X$ of dimension 1.

To show that nef invertible sheaves behave well under pullbacks, we show that we may lift curves along surjective morphisms; compare with [16, Section I.4, Lemma 1]:

Lemma 1.3.2 *Let X be a proper algebraic space over k. Let $f: Y \to X$ be a surjective morphism of algebraic spaces and let $C \subset X$ be an integral closed subspace of dimension 1. Then there exists an integral closed subspace $C' \subset Y$ of dimension 1 such that $C = f(C')$.*

Proof By the weak version of Chow's lemma, [30, Lemma 089J], there exists a proper surjective morphism $g: Y' \to f^{-1}(C)$ from a scheme Y' projective over k. Taking $\dim(Y') - 1$ general hyperplane sections, we obtain a scheme $C'' \subset Y'$ of dimension 1 mapping onto C, since C'' intersects the fibre over the generic point of C. We can then take $C' \subset Y$ to be one of the irreducible components of $g(C'')$ mapping onto C with reduced induced algebraic space structure. □

Nef invertible sheaves behave well under pullbacks.

Lemma 1.3.3 *Let X be a proper algebraic space over k. Let $f: Y \to X$ be a proper morphism of algebraic spaces. Let $\mathcal{L}$ be an invertible $\mathcal{O}_X$-module.*

(i) *If $\mathcal{L}$ is nef, then $f^*\mathcal{L}$ is nef.*
(ii) *If f is surjective and $f^*\mathcal{L}$ is nef, then $\mathcal{L}$ is nef.*

Proof For (i), let $C \subset Y$ be an integral closed subspace of dimension 1. By the projection formula, [30, Lemma 0EDJ], we have

$$(f^*\mathcal{L} \cdot C) = \deg(C \to f(C))(\mathcal{L} \cdot f(C)) \geq 0,$$

where we set $\deg(C \to f(C)) = 0$ by convention if $\dim(f(C)) = 0$.

For (ii), let $C \subset X$ be an integral closed subspace of dimension 1. By Lemma 1.3.2, there exists an integral closed subspace $C' \subset Y$ such that $C = f(C')$. The projection formula again gives

$$(\mathcal{L} \cdot C) = (\mathcal{L} \cdot f(C')) = \deg(C' \to C)^{-1}(f^*\mathcal{L} \cdot C') \geq 0. \qquad \square$$

Nef invertible sheaves are also well behaved under field extensions.

Lemma 1.3.4 *Let X be a proper algebraic space over k. Let $\mathcal{L}$ be* *an invertible $\mathcal{O}_X$-module. Then $\mathcal{L}$ is nef if and only if, for every field extension $k \subseteq k'$, the pullback of $\mathcal{L}$ to $X \otimes_k k'$ is nef.*

Proof $\Leftarrow$ holds by setting $k = k'$, and hence it suffices to show $\Rightarrow$. By the weak version of Chow's lemma, [30, Lemma 089J], there exists a proper surjective morphism $g\colon Y \to X$ from a scheme Y proper over k. Since $\mathcal{L}$ is nef, $g^*\mathcal{L}$ is nef by Lemma 1.3.3, and hence the pullback of $g^*\mathcal{L}$ to $Y \otimes_k k'$ is nef by [15, Lemma 2.18(1)]. Finally, the pullback of $\mathcal{L}$ to $X \otimes_k k'$ is nef by applying Lemma 1.3.3 again. □

We will need the following result about nef invertible sheaves on curves that are not necessarily integral.

Lemma 1.3.5 *Let X be a proper scheme of dimension* 1 *over k. Let $\mathcal{L}$ be an invertible $\mathcal{O}_X$-module. If $\mathcal{L}$ is nef, then* $\deg_X(\mathcal{L}) \geq 0$.

Proof When X is integral, the conclusion follows from [30, Lemma 0BEY] and the definitions. In general, let $C_1, C_2, \ldots, C_t$ be the irreducible components of X viewed as subschemes of X with the reduced induced subscheme structure. By [30, Lemma 0AYW], we have

$$\deg_X(\mathcal{L}) = \sum_{i=1}^{t} m_i \deg_{C_i}(\mathcal{L}|_{C_i}) \quad \text{for some positive integers } m_i.$$

The integral case gives $\deg_X(\mathcal{L}|_{C_i}) \geq 0$ and thus $\deg_X(\mathcal{L}) \geq 0$. □

We adopt the following definition for big invertible sheaves on proper schemes, following Kollár [21, (i) on pp. 236–237].

Definition 1.3.6 Let X be a proper scheme over k. An invertible $\mathcal{O}_X$-module $\mathcal{L}$ is *big* if there exists a constant $C > 0$ such that

$$h^0(X, \mathcal{L}^{\otimes n}) > C \cdot n^{\dim(X)} \quad \text{for all sufficiently large } n.$$

By the asymptotic Riemann–Roch theorem, [30, Proposition 0BJ8], ample invertible sheaves are big. We show that unlike ampleness, the property of being big behaves well under birational morphisms.

Lemma 1.3.7 *Let $f\colon Y \to X$ be a birational morphism of proper schemes over k. Let $\mathcal{L}$ be an invertible $\mathcal{O}_X$-module on X. Then $\mathcal{L}$ is big if and only if $f^*\mathcal{L}$ is big.*

Proof Consider the short exact sequence

$$0 \to \mathcal{O}_X \to f_*\mathcal{O}_Y \to Q \to 0.$$

Then $\dim(Q) \leq \dim(X) - 1$ as f is birational, so, upon twisting by $\mathcal{L}^{\otimes n}$

and taking global sections, we see that, by [8, Proposition 1.31(a)], there exists a constant $C' > 0$ such that

$$h^0(Y, f^*\mathcal{L}^{\otimes n}) - h^0(X, \mathcal{L}^{\otimes n}) \leq h^0(X, Q \otimes_{\mathcal{O}_X} \mathcal{L}^{\otimes n}) \leq C' \cdot n^{\dim(X)-1}$$

for all sufficiently large n. Thus $\mathcal{L}$ is big if and only if $f^*\mathcal{L}$ is big. □

Our next goal is to give an alternative characterization of big invertible sheaves on projective varieties. We start with the following result, known as Kodaira's lemma; see [20, p. 42] and [26, Proposition 2.2.6].

Lemma 1.3.8 *Let X be a proper scheme over k. Let $\mathcal{L}$ be a big invertible $\mathcal{O}_X$-module. Then for every closed subscheme $Z \subset X$ of dimension less than* $\dim(X)$*, there exists an integer $m > 0$ for which*

$$H^0(X, \mathcal{I}_Z \otimes_{\mathcal{O}_X} \mathcal{L}^{\otimes m}) \neq 0.$$

Proof Consider the twisted ideal sheaf sequence

$$0 \to \mathcal{I}_Z \otimes_{\mathcal{O}_X} \mathcal{L}^{\otimes n} \to \mathcal{L}^{\otimes n} \to \mathcal{L}^{\otimes n}|_Z \to 0.$$

Since Z is a proper scheme of dimension $< \dim(X)$ over k, there exists a constant $C' > 0$ such that

$$h^0(Z, \mathcal{L}^{\otimes n}|_Z) \leq C' \cdot n^{\dim(Z)}$$

for all sufficiently large n by [8, Proposition 1.31(a)]. Since $\mathcal{L}$ is big,

$$h^0(X, \mathcal{L}^{\otimes m}) > h^0(Z, \mathcal{L}^{\otimes m}|_Z)$$

for some $m > 0$. Taking global sections in the twisted ideal sheaf sequence then gives $H^0(X, \mathcal{I}_Z \otimes_{\mathcal{O}_X} \mathcal{L}^{\otimes m}) \neq 0$. □

We now prove that a variant of the conclusion in Kodaira's lemma 1.3.8 characterizes big invertible sheaves on projective varieties.

Lemma 1.3.9 *Let X be a projective variety over k. Let $\mathcal{L}$ be an invertible $\mathcal{O}_X$-module. Then the following are equivalent:*

(i) *$\mathcal{L}$ is big.*
(ii) *For every ample invertible $\mathcal{O}_X$-module $\mathcal{A}$, there exists an integer $m > 0$ for which $H^0(X, \mathcal{A}^{-1} \otimes_{\mathcal{O}_X} \mathcal{L}^{\otimes m}) \neq 0$.*

Proof (i) ⇒ (ii). Let r be sufficiently large that there are effective Cartier divisors $H_r \in |\mathcal{A}^{\otimes r}|$ and $H_{r+1} \in |\mathcal{A}^{\otimes(r+1)}|$. By Lemma 1.3.8, there exists an integer $m > 0$ for which $H^0(X, \mathcal{O}_X(-H_{r+1}) \otimes_{\mathcal{O}_X} \mathcal{L}^{\otimes m}) \neq 0$. Since the composition

$$\mathcal{O}_X(-H_{r+1}) \cong \mathcal{A}^{\otimes -(r+1)} \cong \mathcal{O}_X(-H_r) \otimes_{\mathcal{O}_X} \mathcal{A}^{-1} \hookrightarrow \mathcal{A}^{-1}$$

is injective, we then have

$$0 \neq H^0(X, \mathcal{O}_X(-H_{r+1}) \otimes_{\mathcal{O}_X} \mathcal{L}^{\otimes m}) \hookrightarrow H^0(X, \mathcal{A}^{-1} \otimes_{\mathcal{O}_X} \mathcal{L}^{\otimes m}).$$

(ii) $\Rightarrow$ (i). Let $\mathcal{A}$ be a very ample invertible sheaf on X' and choose an effective Cartier divisor $H \in |\mathcal{A}|$. By (ii), there exists an integer $m > 0$ such that $H^0(X, \mathcal{O}_X(-H) \otimes_{\mathcal{O}_X} \mathcal{L}^{\otimes m}) \neq 0$. We can therefore find an effective Cartier divisor $E \in |\mathcal{O}_X(-H) \otimes_{\mathcal{O}_X} \mathcal{L}^{\otimes m}|$ which satisfies

$$\mathcal{O}_X(E) \cong \mathcal{O}_X(-H) \otimes_{\mathcal{O}_X} \mathcal{L}^{\otimes m} \cong \mathcal{A}^{-1} \otimes_{\mathcal{O}_X} \mathcal{L}^{\otimes m}.$$

By the asymptotic Riemann–Roch theorem, [8, Proposition 1.31(b)], there exists a constant $C' > 0$ such that for n sufficiently large,

$$h^0(X, \mathcal{L}^{-i} \otimes_{\mathcal{O}_X} \mathcal{A}^{\otimes n}) > C' \cdot n^{\dim(X)} \quad \text{for every } i \in \{0, 1, \ldots, m-1\}.$$

Writing $n = m \cdot \lceil n/m \rceil - i$ for $i \in \{0, 1, \ldots, m-1\}$, we then have

$$\begin{aligned} h^0(X, \mathcal{L}^{\otimes n}) &= h^0(X, \mathcal{L}^{-i} \otimes_{\mathcal{O}_X} \mathcal{A}^{\otimes \lceil n/m \rceil}(\lceil n/m \rceil E)) \\ &\geq h^0(X, \mathcal{L}^{-i} \otimes_{\mathcal{O}_X} \mathcal{A}^{\otimes \lceil n/m \rceil}) \\ &> C' \cdot \lceil n/m \rceil^{\dim(X)} > \frac{C'}{m^{\dim(X)}} \cdot n^{\dim(X)}; \end{aligned}$$

hence, choosing $C = C'/m^{\dim(X)}$, we see that $\mathcal{L}$ is big. □

1.4 Nef locally free sheaves

In this section, we define and study basic properties of nef locally free sheaves; note that these are referred to as *semipositive* in [21]. See [27, Part Two] for the theory for varieties over algebraically closed fields.

First, a definition. Compare with [21, Definition–Proposition 3.3].

Definition 1.4.1 Let X be a proper algebraic space over k. A finite locally free $\mathcal{O}_X$-module $\mathcal{E}$ is *ample* (resp. *nef*) if $\mathcal{O}_{\mathbf{P}(\mathcal{E})}(1)$ is ample (resp. nef) on $\mathbf{P}(\mathcal{E})$ in the sense of [30, Definition 0D31] (resp. Definition 1.3.1).

Here, $\mathbf{P}(\mathcal{E})$ denotes the projective bundle of one-dimensional quotients of $\mathcal{E}$. In other words, we set

$$\mathbf{P}(\mathcal{E}) := \underline{\mathrm{Proj}}_X(\mathrm{Sym}^\bullet(\mathcal{E})),$$

where $\underline{\mathrm{Proj}}_X$ is defined as in [30, Definition 084C]. By [30, Lemma 085D], $\mathbf{P}(\mathcal{E})$ satisfies the following universal property: for an algebraic space $g\colon Y \to X$, giving a morphism $r\colon Y \to \mathbf{P}(\mathcal{E})$ is the same as giving an invertible sheaf $\mathcal{L}$ on Y and a surjective morphism $g^*\mathcal{E} \to \mathcal{L}$. Here, $\mathcal{L} \cong r^*\mathcal{O}_{\mathbf{P}(\mathcal{E})}(1)$.

We show that locally free quotients of ample or nef locally free sheaves are ample or nef. See [21, Corollary 3.4(i)].

Lemma 1.4.2 *Let X be a proper algebraic space over k. Let $\mathcal{E} \to \mathcal{F}$ be a surjection of finite locally free $\mathcal{O}_X$-modules. If $\mathcal{E}$ is ample (resp. nef), then $\mathcal{F}$ is ample (resp. nef).*

Proof The surjection $\mathcal{E} \to \mathcal{F}$ induces a closed embedding $\mathbf{P}(\mathcal{F}) \hookrightarrow \mathbf{P}(\mathcal{E})$ such that $\mathcal{O}_{\mathbf{P}(\mathcal{E})}(1)$ restricts to $\mathcal{O}_{\mathbf{P}(\mathcal{F})}(1)$ by the functoriality of Proj; see [30, Lemma 085H]. The ample case follows from the fact that $\mathbf{P}(\mathcal{E})$ is a projective k-scheme by the assumption that $\mathcal{O}_{\mathbf{P}(\mathcal{E})}(1)$ is ample, and ampleness is preserved under restriction; see [30, Lemma 01PU]. The nef case follows from Lemma 1.3.3(i). □

We now focus our attention on nef locally free sheaves. First, nef locally free sheaves behave well under pullbacks, as was the case for invertible sheaves in Lemma 1.3.3.

Lemma 1.4.3 *Let X be a proper algebraic space over k. Let $f : Y \to X$ be a proper morphism of algebraic spaces. Let $\mathcal{E}$ be a finite locally free $\mathcal{O}_X$-module.*

(i) *If $\mathcal{E}$ is nef, then $f^*\mathcal{E}$ is nef.*
(ii) *If f is surjective and $f^*\mathcal{E}$ is nef, then $\mathcal{E}$ is nef.*

Proof By [30, Lemma 085C], we have a cartesian diagram

$$\begin{array}{ccc} \mathbf{P}(f^*\mathcal{E}) & \xrightarrow[f']{} & \mathbf{P}(\mathcal{E}) \\ \downarrow & & \downarrow \\ Y & \xrightarrow{f} & X \end{array}$$

such that $f'^*\mathcal{O}_{\mathbf{P}(\mathcal{E})}(1) \cong \mathcal{O}_{\mathbf{P}(f^*\mathcal{E})}(1)$. Both statements follow from Lemma 1.3.3 applied to $\mathcal{O}_{\mathbf{P}(\mathcal{E})}(1)$, where, for (ii), note that f' is surjective, being the base change of f; see [30, Lemma 03MH]. □

Nef locally free sheaves are also well behaved under field extensions.

Lemma 1.4.4 *Let X be a proper algebraic space over k. Let $\mathcal{E}$ be a finite locally free $\mathcal{O}_X$-module. Then $\mathcal{E}$ is nef if and only if for every field extension $k \subseteq k'$, the pullback of $\mathcal{E}$ to $X \otimes_k k'$ is nef.*

Proof It suffices to apply Lemma 1.3.4 to $\mathcal{O}_{\mathbf{P}(\mathcal{E})}(1)$ on $\mathbf{P}(\mathcal{E})$. □

To show some other important properties of nef locally free sheaves, we prove the following characterization of nefness. The statement for schemes is known as the Barton–Kleiman criterion; see [3, p. 437], [27, Proposition 6.1.18], and [21, Definition–Proposition 3.3].

Proposition 1.4.5 *Let X be a proper algebraic space over k. Let $\mathcal{E}$ be a finite locally free $\mathcal{O}_X$-module. Then the following are equivalent:*

(i) *$\mathcal{E}$ is nef.*

(ii) *For every k-morphism $f\colon C \to X$ from a projective k-scheme C of dimension 1, and for every surjection $f^*\mathcal{E} \to \mathcal{L}$ where $\mathcal{L}$ is invertible, we have $\deg_C(\mathcal{L}) \geq 0$.*

(iii) *For every k-morphism $f\colon C \to X$ from a regular projective curve C over k, and for every surjection $f^*\mathcal{E} \to \mathcal{L}$ where $\mathcal{L}$ is invertible, we have $\deg_C(\mathcal{L}) \geq 0$.*

If k is algebraically closed, then these conditions are also equivalent to:

(iv) *For every k-morphism $f\colon C \to X$ from a regular projective curve C over k, and for every ample invertible sheaf $\mathcal{H}$ on C, the locally free sheaf $\mathcal{H} \otimes_{\mathcal{O}_C} f^*\mathcal{E}$ is ample.*

Proof (i) $\Rightarrow$ (ii). Let $f\colon C \to X$ be a morphism as in (ii), and let $\mathcal{L}$ be an invertible quotient of $f^*\mathcal{E}$ on C. By the universal property of $\mathbf{P}(\mathcal{E})$, we obtain a morphism $r\colon C \to \mathbf{P}(\mathcal{E})$ such that $\mathcal{L} \cong r^*\mathcal{O}_{\mathbf{P}(\mathcal{E})}(1)$. By Lemma 1.3.3(i), $\mathcal{L}$ is nef. We then have $\deg_C(\mathcal{L}) \geq 0$ by Lemma 1.3.5.

(ii) $\Rightarrow$ (iii). This holds since the morphisms appearing in (iii) are special cases of those appearing in (ii).

(iii) $\Rightarrow$ (i). Let $g\colon C' \hookrightarrow \mathbf{P}(\mathcal{E})$ be an integral closed subspace of dimension 1. By the weak version of Chow's lemma, [30, Lemma 089J], there exists a proper surjective morphism $f\colon C \to C'$ from a scheme C projective over k, and by Lemma 1.3.2, we may replace C by a closed integral subscheme mapping onto C' to assume that $\dim(C) = 1$. Replacing C by a suitable irreducible component of its normalization, we may also assume that C is regular and integral. Let $\pi\colon \mathbf{P}(\mathcal{E}) \to X$ be the projection morphism. By the universal property of $\mathbf{P}(\mathcal{E})$, we have a surjection

$$(\pi \circ g \circ f)^*\mathcal{E} \to (g \circ f)^*\mathcal{O}_{\mathbf{P}(\mathcal{E})}(1)$$

on C. By (iii) and [30, Lemma 0BEY], the pullback $(g \circ f)^*\mathcal{O}_{\mathbf{P}(\mathcal{E})}(1)$ is nef. Thus $g^*\mathcal{O}_{\mathbf{P}(\mathcal{E})}(1)$ is also nef by Lemma 1.3.3(ii), and $(\mathcal{O}_{\mathbf{P}(\mathcal{E})}(1) \cdot C) \geq 0$.

We now show (i) $\Rightarrow$ (iv), assuming that k is algebraically closed. Let $\pi\colon \mathbf{P}(f^*\mathcal{E}) \to C$ be the projection morphism. We want to show that

$$\mathcal{O}_{\mathbf{P}(\mathcal{H}\otimes_{\mathcal{O}_C} f^*\mathcal{E})}(1) \cong \mathcal{O}_{\mathbf{P}(f^*\mathcal{E})}(1) \otimes_{\mathcal{O}_{\mathbf{P}(f^*\mathcal{E})}} \pi^*\mathcal{H}$$

is ample, where the isomorphism shown holds by the definition of relative Proj under the identification $\mathbf{P}(\mathcal{H} \otimes_{\mathcal{O}_C} f^*\mathcal{E}) \cong \mathbf{P}(f^*\mathcal{E})$. Let $Y \subset \mathbf{P}(f^*\mathcal{E})$ be an integral closed subscheme. By the Nakai–Moishezon criterion (see Proposition 1.2.4), it suffices to show that

$$\left((\mathcal{O}_{\mathbf{P}(f^*\mathcal{E})}(1) \otimes_{\mathcal{O}_{\mathbf{P}(f^*\mathcal{E})}} \pi^*\mathcal{H})^d \cdot Y\right) > 0$$

where $d = \dim(Y)$. If Y is contained in a closed fibre over C, then this positivity holds since $\mathcal{O}_{\mathbf{P}(f^*\mathcal{E})}(1)$ restricts to $\mathcal{O}_{\mathbf{P}^n}(1)$ on the closed fibre, where $n = \operatorname{rank}(f^*\mathcal{E}) - 1$. Otherwise, it suffices to show that

$$\left((\mathcal{O}_{\mathbf{P}(f^*\mathcal{E})}(1) \otimes_{\mathcal{O}_{\mathbf{P}(f^*\mathcal{E})}} \pi^*\mathcal{H})^d \cdot Y\right) \geq \left((\mathcal{O}_{\mathbf{P}(f^*\mathcal{E})}(1))^{d-1} \cdot \pi^*\mathcal{H} \cdot Y\right)$$

since the right-hand side is positive by the fact that $\pi^*\mathcal{H} \cdot Y$ corresponds to a closed subscheme of dimension $d-1$ contained in a union of closed fibres over C, so that we can apply the case above. This inequality holds since we can expand the left-hand side by additivity ([30, Lemma 0BER]) and then observe that since $f^*\mathcal{E}$ is nef by Lemma 1.4.3(i), every term involving $\mathcal{O}_{\mathbf{P}(f^*\mathcal{E})}(1)$ is nonnegative by [15, Lemma 2.12], and every term with more than one power of $\pi^*\mathcal{H}$ is zero.

Finally, we show (iv) $\Rightarrow$ (iii), assuming that k is algebraically closed. Let $f^*\mathcal{E} \to \mathcal{L}$ be a surjection where $\mathcal{L}$ is invertible. Choose an ample invertible sheaf $\mathcal{H}$ on C of degree 1, which exists since k is algebraically closed. Twist this surjection by $\mathcal{H}$. Since the quotient of an ample locally free sheaf is ample by Lemma 1.4.2, and ample invertible sheaves have positive degree by [30, Lemma 0B5X], we have

$$1 + \deg_C(\mathcal{L}) = \deg_C(\mathcal{H} \otimes_{\mathcal{O}_C} \mathcal{L}) \geq 1$$

where the equality holds by [30, Lemma 0AYX], and the inequality holds by (iv). This shows that $\deg_C(\mathcal{L}) \geq 0$. □

We can now show that nefness is preserved under extensions.

Lemma 1.4.6 *Let X be a proper algebraic space over k. Let*

$$0 \to \mathcal{E}' \to \mathcal{E} \to \mathcal{E}'' \to 0$$

be a short exact sequence of finite locally free $\mathcal{O}_X$-modules. If $\mathcal{E}'$ and $\mathcal{E}''$ are both nef, then $\mathcal{E}$ is nef.

Proof Let $f: C \to X$ be a k-morphism from a regular projective curve C over k, and let $f^*\mathcal{E} \to \mathcal{L}$ be an invertible quotient. By Proposition 1.4.5, it suffices to show that $\deg_C(\mathcal{L}) \geq 0$.

Denote by $\mathcal{L}'$ the image of $f^*\mathcal{E}'$ in $\mathcal{L}$ and by $\mathcal{L}''$ the quotient sheaf $\mathcal{L}/\mathcal{L}'$. We then have a commutative diagram

$$\begin{array}{ccccccccc} 0 & \longrightarrow & f^*\mathcal{E}' & \longrightarrow & f^*\mathcal{E} & \longrightarrow & f^*\mathcal{E}'' & \longrightarrow & 0 \\ & & \downarrow & & \downarrow & & \downarrow & & \\ 0 & \longrightarrow & \mathcal{L}' & \longrightarrow & \mathcal{L} & \longrightarrow & \mathcal{L}'' & \longrightarrow & 0 \end{array}$$

where the top row is exact since $\mathcal{E}''$ is locally free, and the bottom row is exact by definition. The sheaf $\mathcal{L}'$ is torsion-free since it is a subsheaf of $\mathcal{L}$ and is therefore locally free since C is regular of dimension 1; see [30, Lemma 0AUW].

First consider the case where $\operatorname{rank}(\mathcal{L}') = 0$, in which case $\mathcal{L}' = 0$ and $\mathcal{L} \to \mathcal{L}''$ is an isomorphism. We then have $\deg_C(\mathcal{L}) = \deg_C(\mathcal{L}'') \geq 0$ by Proposition 1.4.5 since $\mathcal{E}''$ is nef.

It remains to consider the case where $\operatorname{rank}(\mathcal{L}') = 1$, in which case $\operatorname{rank}(\mathcal{L}'') = 0$. The additivity of Euler characteristics, [30, Lemma 08AA], and the definition of degree, [30, Definition 0AYR], give the first three of the following equalities:

$$\begin{aligned} \deg_C(\mathcal{L}) &= \chi(C, \mathcal{L}) - \chi(C, \mathcal{O}_C) \\ &= \chi(C, \mathcal{L}') - \chi(C, \mathcal{O}_C) + \chi(C, \mathcal{L}'') \\ &= \deg_C(\mathcal{L}') + \chi(C, \mathcal{L}'') = \deg_C(\mathcal{L}') + h^0(C, \mathcal{L}'') \geq 0. \end{aligned}$$

The fourth equation follows from [30, Lemma 0AYT] as $\mathcal{L}''$ is rank 0, and the final inequality is Proposition 1.4.5 as $\mathcal{E}'$ is nef. □

Our next goal is to prove that nefness is preserved under various tensor operations. The idea is to use the Barton–Kleiman criterion, Proposition 1.4.5, to reduce to the curve case, in which case we will use the following:

Lemma 1.4.7 *Let C be a regular projective curve over an algebraically closed field k, $\mathcal{E}$ a nef finite locally free $\mathcal{O}_C$-module, and $\mathcal{H}$ an invertible $\mathcal{O}_C$-module of degree $\geq 2g$. Then $\mathcal{E} \otimes_{\mathcal{O}_C} \mathcal{H}$ is globally generated.*

Proof We first show that if $\mathcal{H}$ is an invertible $\mathcal{O}_C$-module such that

$$H^1(C, \mathcal{E} \otimes_{\mathcal{O}_C} \mathcal{H}) \neq 0,$$

then $\deg_C(\mathcal{H}) \leq 2g - 2$. By Serre Duality, [30, Lemma 0FVV], we have

$$H^1(C, \mathcal{E} \otimes_{\mathcal{O}_C} \mathcal{H}) \cong \operatorname{Hom}_{\mathcal{O}_C}(\mathcal{E} \otimes_{\mathcal{O}_C} \mathcal{H}, \omega_C) \neq 0,$$

and we therefore have a nonzero morphism $\mathcal{E} \otimes_{\mathcal{O}_C} \mathcal{H} \to \omega_C$. The image $\mathcal{M}$ of this morphism is torsion-free, hence invertible, since C is regular of dimension 1; see [30, Lemma 0AUW]. This invertible $\mathcal{O}_C$-module $\mathcal{M}$ satisfies

$$2g - 2 = -2\chi(C, \mathcal{O}_C) = \deg_C(\omega_C) \geq \deg_C(\mathcal{M})$$

since $\mathcal{M}$ is a subsheaf of ω_C. Twisting the surjection $\mathcal{E} \otimes_{\mathcal{O}_C} \mathcal{H} \to \mathcal{M}$ by $\mathcal{H}^{-1}$,

$$\begin{aligned} 2g - 2 - \deg_C(\mathcal{H}) &\geq \deg_C(\mathcal{M}) - \deg_C(\mathcal{H}) \\ &= \deg_C(\mathcal{M} \otimes_{\mathcal{O}_C} \mathcal{H}^{-1}) \geq 0 \end{aligned}$$

where the equality holds by [30, Lemma 0AYX], and the last inequality holds by the nefness of $\mathcal{E}$ and Proposition 1.4.5.

We now show the statement of the lemma. Let $x \in C$ be a closed point with ideal sheaf $\mathcal{O}(-x)$. We have a short exact sequence

$$0 \to \mathcal{E} \otimes_{\mathcal{O}_C} \mathcal{H}(-x) \to \mathcal{E} \otimes_{\mathcal{O}_C} \mathcal{H} \to \mathcal{E} \otimes_{\mathcal{O}_C} \mathcal{H}|_x \to 0.$$

Using [30, Lemma 0AYX] again, we have

$$\deg_C(\mathcal{H}(-x)) = \deg_C(\mathcal{H}) - \deg(\mathcal{O}_C(x)) = \deg_C(\mathcal{H}) - 1 \geq 2g - 1,$$

and hence $H^1(C, \mathcal{E} \otimes_{\mathcal{O}_C} \mathcal{H}(-x)) = 0$ by the previous paragraph. Thus, $\mathcal{E} \otimes_{\mathcal{O}_C} \mathcal{H}$ is globally generated. □

We will also need the following to reduce to the case when the ground field k is of positive characteristic.

Lemma 1.4.8 *Let Y be a Noetherian scheme, and let $f : X \to Y$ be a proper morphism from an algebraic space X. Let $\mathcal{E}$ be a finite locally free $\mathcal{O}_X$-module. Let $y \in Y$ be a point such that $\mathcal{E}_y$ is ample on the fibre X_y. Then there exists an open neighborhood $V \subseteq Y$ of y such that $\mathcal{E}_{y'}$ is ample on the fibre $X_{y'}$ for every point $y' \in V$.*

Proof Apply [30, Lemma 0D3A] to $\mathcal{O}_{\mathbf{P}(\mathcal{E})}(1)$ on $\mathbf{P}(\mathcal{E})$. □

Note that the statement analogous to Lemma 1.4.8 for nefness does not hold, as shown by Langer [24, 25], due to examples of Monsky, Brenner, and Trivedi [24, Example 5.3], of Ekedahl, Shepherd-Barron, and Taylor [24, Example 5.6], and of Moret-Bailly [25, Section 8].

We now prove the following result, originally due to Barton for schemes [3, Proposition 3.5(i)].

Proposition 1.4.9 *Let X be a proper algebraic space over k. Let $\mathcal{E}$ and $\mathcal{E}'$ be nef finite locally free $\mathcal{O}_X$-modules. Then $\mathcal{E} \otimes_{\mathcal{O}_X} \mathcal{E}'$ is nef, as are $\mathcal{E}^{\otimes n}$, $\mathrm{Sym}^n(\mathcal{E})$, $\Gamma^n(\mathcal{E}) := (\mathrm{Sym}^n(\mathcal{E}^\vee))^\vee$, and $\bigwedge^n(\mathcal{E})$ for all $n \geq 0$.*

Proof If $\mathcal{E}$ and $\mathcal{E}'$ are nef, then $\mathcal{G} := \mathcal{E} \oplus \mathcal{E}'$ is nef by Lemma 1.4.6, and $\mathcal{E} \otimes_{\mathcal{O}_X} \mathcal{E}'$ is a locally free quotient of the locally free sheaf $\mathcal{G}^{\otimes 2}$. By Lemma 1.4.2, it therefore suffices to show that $\mathcal{E}^{\otimes n}$, $\mathrm{Sym}^n(\mathcal{E})$, $\Gamma^n(\mathcal{E})$, and $\bigwedge^n(\mathcal{E})$ are nef. We will denote any such sheaf by $\rho^n(\mathcal{E})$. By Lemma 1.4.4, we may assume that k is algebraically closed.

Step 1 Proof when $\mathrm{char}(k) > 0$.

Fix a k-morphism $f\colon C \to X$ from a regular projective curve C over k. Let $\mathcal{L}$ be a quotient invertible sheaf of $\rho^n(\mathcal{E})$, and set $d := \deg_C(\mathcal{L})$. By Proposition 1.4.5, it suffices to show that $d \geq 0$.

Let $\mathcal{H}$ be an invertible $\mathcal{O}_C$-module of degree $2g$, where g is the genus of C. For every $e > 0$, consider the eth iterate of the absolute Frobenius morphism $F^e\colon C \to C$, which is a finite morphism of degree p^e. We claim that, for every $e > 0$, there is a generic isomorphism

$$(\mathcal{H}^{-n})^{\oplus r} \to F^{e*}\rho^n(f^*\mathcal{E}), \tag{$\star$}$$

where $r := \mathrm{rank}(\rho^n(f^*\mathcal{E}))$. Since $F^{e*}f^*\mathcal{E}$ is nef by Lemma 1.4.3, the sheaf $F^{e*}f^*\mathcal{E} \otimes_{\mathcal{O}_C} \mathcal{H}$ is globally generated by Lemma 1.4.7. By choosing $s := \mathrm{rank}(f^*\mathcal{E})$ global sections that form a basis after localizing at the generic point of C, we obtain a morphism $(\mathcal{H}^{-1})^{\oplus s} \to F^{e*}f^*\mathcal{E}$ that induces an isomorphism at the generic point of C. Applying the functor $\rho^n(-)$, we obtain the generic isomorphism

$$\rho^n\big((\mathcal{H}^{-1})^{\oplus s}\big) \to F^{e*}\rho^n(f^*\mathcal{E}).$$

The left-hand side is a direct sum of the sheaves $\mathcal{H}^{-n}$, so, passing to a direct summand, we obtain a generic isomorphism of the form in ($\star$).

We now show that $d = \deg_C(\mathcal{L}) \geq 0$. Note that $F^{e*}\mathcal{L} \cong \mathcal{L}^{\otimes p^e}$ is a quotient invertible $\mathcal{O}_C$-module of $F^{e*}\rho^n(f^*\mathcal{E})$ and that $\deg_C(F^{e*}\mathcal{L}) = p^e d$ by [30, Lemma 0AYZ]. By the previous paragraph, $(\mathcal{H}^{-n})^{\oplus r}$ surjects onto a subsheaf $\mathcal{M}$ of $F^{e*}\mathcal{L}$ that is torsion-free of rank 1, hence invertible since C is regular of dimension 1; see [30, Lemma 0AUW]. Twisting the surjection $(\mathcal{H}^{-n})^{\oplus r} \to \mathcal{M}$ by $\mathcal{H}^{\otimes n}$, we see that $\mathcal{M} \otimes_{\mathcal{O}_C} \mathcal{H}^{\otimes n}$ is nef since it is globally generated, and hence

$$\deg_C(\mathcal{M}) = \deg_C(\mathcal{M} \otimes_{\mathcal{O}_C} \mathcal{H}^{\otimes n}) + \deg_C(\mathcal{H}^{-n}) \geq -2gn$$

by [30, Lemma 0AYX] and Proposition 1.4.5. We then have

$$\begin{aligned} p^e d &= \deg_C(F^{e*}\mathcal{L}) = \chi(C, F^{e*}\mathcal{L}) - \chi(C, \mathcal{O}_C) \\ &= \chi(C, \mathcal{M}) - \chi(C, \mathcal{O}_C) + \chi(C, \mathcal{L}/\mathcal{M}) \\ &= \deg_C(\mathcal{M}) + h^0(C, \mathcal{L}/\mathcal{M}) \geq -2gn \end{aligned}$$

where the equalities hold by the additivity of Euler characteristics and the definition of degree; see [30, Lemma 08AA and Definition 0AYR]. Since this inequality must hold for all $e > 0$, we see that $d \geq 0$.

Step 2 Proof when $\operatorname{char}(k) = 0$.

It suffices to show that for every k-morphism $f\colon C \to X$ from a regular projective curve C over k, and every invertible quotient $\mathcal{L}$ of $\rho^n(f^*\mathcal{E})$, we have $\deg_C(\mathcal{L}) \geq -n$. Indeed, if $g\colon C' \to C$ is a finite surjective morphism of degree $e > 0$, then

$$e \cdot \deg_C(\mathcal{L}) = \deg_{C'}(g^*\mathcal{L}) \geq -n$$

holds by [30, Lemma 0AYZ]. Since this inequality must hold for all $e > 0$, we see that $\deg_C(\mathcal{L}) \geq 0$, and hence $\rho^n(f^*\mathcal{E})$ is nef by Proposition 1.4.5.

We now show that $\deg_C(\mathcal{L}) \geq -n$ for every morphism $f\colon C \to X$ and every quotient invertible sheaf $\mathcal{L}$ of $\rho^n(f^*\mathcal{E})$ as above. Since C is projective over k, there exists a finitely generated $\mathbf{Z}$-algebra $A \subset k$ and a projective morphism $C_A \to \operatorname{Spec}(A)$ such that the diagram

$$\begin{array}{ccc} C & \longrightarrow & C_A \\ {\scriptstyle f}\downarrow & & \downarrow{\scriptstyle f_A} \\ \operatorname{Spec}(k) & \longrightarrow & \operatorname{Spec}(A) \end{array}$$

is cartesian. Let $\mathcal{H}$ be an invertible sheaf on C of degree 1. By [30, Lemma 0B8W], after possibly enlarging A, we may assume that there exist invertible $\mathcal{O}_{C_A}$-modules $\mathcal{H}_A$ and $\mathcal{L}_A$ and a finite locally free $\mathcal{O}_{C_A}$-module $\mathcal{F}_A$ that pull back to $\mathcal{H}$, $\mathcal{L}$, and $f^*\mathcal{E}$, on C. By [30, Lemma 01ZR] and [12, Corollaire 8.5.7], we may also assume that there exists a surjection

$$\rho^n(\mathcal{F}_A) \to \mathcal{L}_A \tag{$\star\star$}$$

that pulls back to $\rho^n(f^*\mathcal{E}) \to \mathcal{L}$ on C. Now by Proposition 1.4.5, the $\mathcal{O}_C$-module $\mathcal{H} \otimes_{\mathcal{O}_C} f^*\mathcal{E}$ is ample. By Lemma 1.4.8, after possibly replacing A by a principal localization, we may assume that $\mathcal{H}_A \otimes_{\mathcal{O}_{C_A}} \mathcal{F}_A$ is ample on every fibre of f_A, since it is ample after pulling back to the generic fibre of f_A by applying [30, Lemma 0D2P] on $\mathbf{P}(\mathcal{H}_A \otimes_{\mathcal{O}_{C_A}} \mathcal{F}_A)$.

Moreover, by generic flatness, [30, Proposition 052A, and Lemma 05F7], we may assume that f_A is flat with one-dimensional fibres.

Let $y \in \mathrm{Spec}(A)$ be a closed point with residue field $\kappa(y)$, and set $C_y := f_A^{-1}(y)$. Since f_A is flat, the invertible $\mathcal{O}_{C_A}$-modules $\mathcal{L}_A$ and $\mathcal{O}_{C_A}$ are flat over A. So, writing η for the generic point of $\mathrm{Spec}(A)$, we have

$$\begin{aligned}\deg_C(\mathcal{L}) = \deg_{C_\eta}(\mathcal{L}_\eta) &= \chi(C_\eta, \mathcal{L}_\eta) - \chi(C_\eta, \mathcal{O}_{C_\eta}) \\ &= \chi(C_y, \mathcal{L}_y) - \chi(C_y, \mathcal{O}_{C_y}) = \deg_{C_y}(\mathcal{L}_y)\end{aligned}$$

where the first equality holds by [30, Lemma 0B59] applied to the field extension $\mathrm{Frac}(A) \subset k$ and the third equality follows from the constancy of Euler characteristics in proper flat families; see [30, Lemma 0B9T]. By the same argument, $\deg_{C_y}(\mathcal{H}_y) = 1$. Since $\mathcal{H}_y \otimes_{\mathcal{O}_{C_y}} \mathcal{F}_y$ is ample, it is nef, and hence $\mathcal{H}_y^{\otimes n} \otimes_{\mathcal{O}_{C_y}} \rho^n(\mathcal{F}_y)$ is nef by Step 1. Thus, the surjection ($\star\star$) twisted by $\mathcal{H}_A^{\otimes n}$ and then restricted to C_y implies that

$$\begin{aligned}\deg_{C_y}(\mathcal{L}_y) &= \deg_{C_y}(\mathcal{H}_y^{-n} \otimes_{\mathcal{O}_{C_y}} \mathcal{H}_y^{\otimes n} \otimes_{\mathcal{O}_{C_y}} \mathcal{L}_y) \\ &= -n + \deg_{C_y}(\mathcal{H}_y^{\otimes n} \otimes_{\mathcal{O}_{C_y}} \mathcal{L}_y) \geq -n\end{aligned}$$

by [30, Lemma 0AYX] and Proposition 1.4.5, as desired. □

We end this section with a criterion for bigness that will feature in the proof of Lemma 1.5.3:

Lemma 1.4.10 *Let X be a projective variety over k and let $\mathcal{L}$ be an invertible $\mathcal{O}_X$-module. Let $\mathcal{F}$ be a finite locally free $\mathcal{O}_X$-module with associated projective bundle $\pi\colon \mathbf{P} \to X$. Assume that*

(i) *$\mathcal{L}$ is nef,*
(ii) *$\mathcal{F}^\vee$ is nef, and*
(iii) *there exists $a \geq 1$ and an ample invertible sheaf $\mathcal{A}$ on X such that*

$$H^0(\mathbf{P}, \mathcal{O}_{\mathbf{P}}(a) \otimes_{\mathcal{O}_{\mathbf{P}}} \pi^*\mathcal{L} \otimes_{\mathcal{O}_{\mathbf{P}}} \pi^*\mathcal{A}^{-1}) \neq 0.$$

Then $\mathcal{L}$ is big and nef.

Proof Set $d := \dim(X)$. By (i) and the asymptotic Riemann–Roch theorem, [8, Proposition 1.31(b)], it suffices to show that the intersection number $(\mathcal{L}^d)$ is positive. By (iii), we may choose a nonzero morphism

$$\mathcal{O}_{\mathbf{P}} \to \mathcal{O}_{\mathbf{P}}(a) \otimes_{\mathcal{O}_{\mathbf{P}}} \pi^*\mathcal{L} \otimes_{\mathcal{O}_{\mathbf{P}}} \pi^*\mathcal{A}^{-1}.$$

Applying the projection formula and rearranging yields a nonzero morphism $\tau\colon \Gamma^a(\mathcal{E}^\vee) \to \mathcal{L} \otimes_{\mathcal{O}_X} \mathcal{A}^{-1}$. Since the sheaf on the right-hand side is locally trivial, the image of τ is of the form $\mathcal{I} \otimes_{\mathcal{O}_X} (\mathcal{L} \otimes_{\mathcal{O}_X} \mathcal{A}^{-1})$ for some coherent sheaf of ideals $\mathcal{I}$. Let $f: Y \to X$ be the blowup along $\mathcal{I}$, with exceptional divisor D. Then $f^*\tau$ gives a surjection

$$f^*\Gamma^a(\mathcal{E}^\vee) \twoheadrightarrow \mathcal{M} := f^*\mathcal{L} \otimes_{\mathcal{O}_Y} f^*\mathcal{A}^{-1} \otimes_{\mathcal{O}_Y} \mathcal{O}_Y(-D).$$

By (ii), Proposition 1.4.9, and Lemma 1.4.3(i), the sheaf on the left-hand side is nef, hence by Lemma 1.4.2, $\mathcal{M}$ is also nef. Rearranging gives

$$f^*\mathcal{L} \cong f^*\mathcal{A} \otimes_{\mathcal{O}_Y} \mathcal{M} \otimes_{\mathcal{O}_Y} \mathcal{O}_Y(D).$$

Since f is birational, $\dim(Y) = d = \dim(X)$ and, by [30, Lemma 0BET], $(f^*\mathcal{L}^d) = (\mathcal{L}^d)$ and $(f^*\mathcal{A}^d) = (\mathcal{A}^d)$. In particular, the latter quantity is positive since $\mathcal{A}$ is ample; see [30, Lemma 0BEV]. The additivity of intersection numbers, [30, Lemma 0BER], gives

$$(\mathcal{L}^d) = (f^*\mathcal{L}^d) = (f^*\mathcal{A}^d) + \sum_{i=1}^{d} (f^*\mathcal{A}^{d-i} \cdot f^*\mathcal{L}^{i-1} \cdot \mathcal{M}(D)).$$

The latter sum is nonnegative: by additivity and restriction, [30, Lemmas 0BER and 0BEU], the ith summand is the sum

$$(f^*\mathcal{A}^{d-i} \cdot f^*\mathcal{L}^{i-1} \cdot \mathcal{M}) + (f^*\mathcal{A}^{d-1}|_D \cdot f^*\mathcal{L}^{i-1}|_D) \geq 0$$

of intersection numbers of nef invertible sheaves, and hence each is nonnegative by [15, Lemma 2.12]. Therefore $(\mathcal{L}^d) \geq (\mathcal{A}^d) > 0$. □

1.5 Ampleness lemma

In this section, we formulate a method for proving the ampleness of line bundles of the form $\det(Q)$, where Q is a locally free quotient of a symmetric power of a nef finite locally free sheaf $\mathcal{E}$. The basic method is due to Kollár in [21, Lemmas 3.9 and 3.13], refining an idea of Viehweg [31]. We also include a refinement due to Kovács and Patakfalvi [22].

The idea is as follows: locally, Q is a quotient by a trivial vector bundle, so $\det(Q)$ is locally the pullback of the Plücker bundle under a classifying map to a Grassmannian. Globalize this by passing to its frame bundle to universally trivialize $\mathcal{E}$; the quotient bundle now gives a classifying map to a stack of the form $[\mathbf{G}(N, q)/\mathrm{PGL}_n]$. The ampleness lemma (see Proposition 1.5.4) is then a generalization of the familiar fact that the pullback of an ample sheaf under a finite map is ample.

We begin by constructing frame bundles. Let S be a scheme and let $\mathcal{E}$ be a finite locally free $\mathcal{O}_S$-module of rank n. Let T be a scheme and consider triples $(f: T \to S, \mathcal{L}, \psi)$ where

(i) $f\colon T \to S$ is a morphism of schemes,
(ii) $\mathcal{L}$ is an invertible $\mathcal{O}_T$-module, and
(iii) $\psi\colon \mathcal{O}_T^{\oplus n} \to f^*\mathcal{E} \otimes_{\mathcal{O}_T} \mathcal{L}$ is an isomorphism of $\mathcal{O}_T$-modules.

Call two triples $(f, \mathcal{L}, \psi)$ and $(f', \mathcal{L}', \psi')$ over T *equivalent* if $f = f'$ and if there exists an isomorphism $\beta\colon \mathcal{L} \to \mathcal{L}'$ such that $\beta \circ \psi = \psi'$. The *frame functor of* $\mathcal{E}$ is the functor

$$\begin{aligned} \mathrm{Fr}(\mathcal{E})\colon \mathrm{Sch}^{\mathrm{opp}} &\to \mathrm{Sets} \\ T &\mapsto \{\text{equivalence classes of } (f\colon T \to S, \mathcal{L}, \psi) \text{ as above}\} \end{aligned}$$

with pullbacks under $T' \to T$ defined as expected.

There are two important structures. First, the projection of $(f\colon T \to S, \mathcal{L}, \psi)$ onto the first factor yields a morphism of functors $\mathrm{Fr}(\mathcal{E}) \to S$. Second, given $f\colon T \to S$, the set of equivalence classes of $(f, \mathcal{L}, \psi)$ admits a simply transitive action of PGL_n via precomposition on ψ; note this is well defined since automorphisms of $\mathcal{L}$ are given by scalar multiplication. Therefore $\mathrm{Fr}(\mathcal{E})$ is a functor of PGL_n-sets over S.

Lemma 1.5.1 *Let S be a scheme. Let $\mathcal{E}$ be a finite locally free $\mathcal{O}_S$-module of rank n and set $\mathbf{P} := \mathbf{P}(\mathcal{H}om(\mathcal{E}, \mathcal{O}_S^{\oplus n}))$. There exists an effective Cartier divisor $\mathbf{D} \subset \mathbf{P}$ such that* $\mathrm{Fr}(\mathcal{E})$ *is represented by the open subscheme*

$$\mathbf{Fr}(\mathcal{E}) = \mathbf{P} \setminus \mathbf{D}.$$

The structure map $\mathbf{Fr}(\mathcal{E}) \to S$ exhibits this as a PGL_n*-torsor over S.*

Proof Consider a triple $(f\colon T \to S, \mathcal{L}, \psi)$ as above. By adjunction, the isomorphism $\psi\colon \mathcal{O}_T^{\oplus n} \to f^*\mathcal{E} \otimes_{\mathcal{O}_T} \mathcal{L}$ uniquely determines a surjection $\varphi\colon f^*\mathcal{H}om(\mathcal{E}, \mathcal{O}_S^{\oplus n}) \to \mathcal{L}$. This exhibits $\mathrm{Fr}(\mathcal{E})$ as the subfunctor of the projective bundle $\mathbf{P}$ on which ψ is an isomorphism.

On the other hand, let $\pi\colon \mathbf{P} \to S$ be the structure map and consider the universal quotient $\varphi_{\mathrm{univ}}\colon \pi^*\mathcal{H}om(\mathcal{E}, \mathcal{O}_S^{\oplus n}) \to \mathcal{O}_{\mathbf{P}}(1)$. By adjunction, this yields an injective map $\varphi^{\#}_{\mathrm{univ}}\colon \mathcal{O}_{\mathbf{P}}^{\oplus n} \to \mathcal{O}_{\mathbf{P}}(1) \otimes_{\mathcal{O}_{\mathbf{P}}} \pi^*\mathcal{E}$ and hence a universal determinant

$$\det(\varphi^{\#}_{\mathrm{univ}})\colon \mathcal{O}_{\mathbf{P}} \to \det(\mathcal{O}_{\mathbf{P}}(1) \otimes_{\mathcal{O}_{\mathbf{P}}} \pi^*\mathcal{E}).$$

Let $\mathbf{D}$ be the divisor determined by its vanishing. Then the open subscheme $\mathbf{Fr}(\mathcal{E}) := \mathbf{P} \setminus \mathbf{D}$ represents the functor $\mathrm{Fr}(\mathcal{E})$. □

We call the scheme $\mathbf{Fr}(\mathcal{E})$ the *frame bundle* of $\mathcal{E}$ over S. The torsor structure on the frame bundle induces a classifying map from S to the classifying stack $B\mathrm{PGL}_n$ fitting into a cartesian diagram

$$\begin{array}{ccc} \mathbf{Fr}(\mathcal{E}) & \longrightarrow & \mathrm{pt} \\ \pi \downarrow & & \downarrow \\ S & \longrightarrow & B\mathrm{PGL}_n \end{array}$$

We now construct lifts of this classifying map to quotient stacks of certain Grassmannians whenever we are given, additionally, $\alpha\colon \mathrm{Sym}^d(\mathcal{E}) \to Q$, a finite locally free quotient of rank q, with d some positive integer. The strategy is to pull the quotient back to the frame bundle and take symmetric powers of the *universal trivialization map*

$$\psi_{\mathrm{univ}} := \varphi^{\#}_{\mathrm{univ}}|_{\mathbf{Fr}(\mathcal{E})} : \mathcal{O}^{\oplus n}_{\mathbf{Fr}(\mathcal{E})} \to \mathcal{O}_{\mathbf{Fr}(\mathcal{E})}(1) \otimes_{\mathcal{O}_{\mathbf{Fr}(\mathcal{E})}} \pi^*\mathcal{E},$$

to give PGL_n-equivariant morphisms to $\mathbf{G} := \mathbf{G}(N, q)$, the Grassmannian parameterizing rank-q quotients of the module $\mathrm{Sym}^d(\mathbf{Z}^{\oplus n}) \cong \mathbf{Z}^{\oplus N}$ with $N = \binom{n+d-1}{d}$. Note that ψ_{univ} is equivariant for the action of PGL_n on $\mathbf{Fr}(\mathcal{E})$, where the action is tautological on the source and trivial on the target; likewise, PGL_n acts on $\mathbf{G}$ via the action on $\mathrm{Sym}^d(\mathbf{Z}^{\oplus n})$ induced by its tautological action.

Lemma 1.5.2 *Given the notation as above, there exists a commutative diagram*

$$\begin{array}{ccccc} \mathbf{Fr}(\mathcal{E}) & \xrightarrow[{[\pi^*\alpha]}]{} & \mathbf{G} & \longrightarrow & \mathrm{pt} \\ \pi \downarrow & & \downarrow & & \downarrow \\ S & \xrightarrow{[\alpha]} & [\mathbf{G}/\mathrm{PGL}_n] & \longrightarrow & B\mathrm{PGL}_n \end{array}$$

such that all squares are cartesian. Moreover, writing $\mathcal{O}_{\mathbf{G}}(1)$ for the Plücker line bundle on $\mathbf{G}$, we have

$$[\pi^*\alpha]^*\mathcal{O}_{\mathbf{G}}(1) \cong \mathcal{O}_{\mathbf{Fr}(\mathcal{E})}(qd) \otimes_{\mathcal{O}_{\mathbf{Fr}(\mathcal{E})}} \pi^* \det(Q).$$

Proof Pulling back α to $\mathbf{Fr}(\mathcal{E})$ and precomposing with the dth symmetric power of the universal trivialization ψ_{univ} gives a surjection

$$\mathrm{Sym}^d(\mathcal{O}^{\oplus n}_{\mathbf{Fr}(\mathcal{E})}) \to \mathcal{O}_{\mathbf{Fr}(\mathcal{E})}(d) \otimes \pi^* \mathrm{Sym}^d(\mathcal{E}) \to \mathcal{O}_{\mathbf{Fr}(\mathcal{E})}(d) \otimes \pi^*Q.$$

The universal property of $\mathbf{G}$ yields a morphism $[\pi^*\alpha]\colon \mathbf{Fr}(\mathcal{E}) \to \mathbf{G}$ which is PGL_n-equivariant by the description of the actions above, and such that the pullback of the universal quotient bundle on $\mathbf{G}$ is $\mathcal{O}_{\mathbf{Fr}(\mathcal{E})}(d) \otimes \pi^*Q$. Since the Plücker line bundle is the determinant of the universal quotient, this gives an identification of the line bundles. Finally, this data of a PGL_n-torsor over S together with a PGL_n-equivariant morphism to

$\mathbf{G}$ is precisely the data of a morphism $[\alpha]: S \to [\mathbf{G}/\mathrm{PGL}_n]$ lifting the classifying map for $\mathbf{Fr}(\mathcal{E})$; see [30, Sections 04UI and 04UV]. □

The morphism $[\alpha]: S \to [\mathbf{G}/\mathrm{PGL}_n]$ is called the *classifying map* of α. The aim is to pull positivity back to $\det(Q)$ via $[\alpha]$ from $O_{\mathbf{G}}(1)$. This is achieved most directly by asking for $[\alpha]$ to be a quasi-finite morphism of stacks; see [30, Definition 0G2M] and compare with [21, Definition 3.8]. Concretely, since S is a scheme, $[\alpha]$ is a representable morphism, so, by [30, Lemma 04XD], $[\alpha]$ is quasi-finite if and only if $[\pi^*\alpha]: \mathbf{Fr}(\mathcal{E}) \to \mathbf{G}$ is a quasi-finite morphism of schemes. Kovács and Patakfalvi observed in [22, Theorem 5.5] that, when working over a field k, it is sufficient to ask for $[\alpha]$ to have finite fibres on $\bar{k}$-points.

The following statement is the heart of the ampleness lemma, and is an analogue of the fact that the pullback of an ample line bundle by a generically quasi-finite morphism is big.

Lemma 1.5.3 *In the situation of Lemma 1.5.2, assume that*

(i) *S is a normal projective variety over k,*
(ii) *$\mathcal{E}$ is nef, and*
(iii) *there exists a dense open subset $S_0 \subseteq S$ over which the classifying map $[\alpha]$ has finite fibres on $\bar{k}$-points.*

Then $\det(Q)$ *is big and nef. In particular,* $(\det(Q)^{\dim(S)}) > 0$.

Proof We aim to apply Lemma 1.4.10 with $\mathcal{F} := \mathcal{H}om(\mathcal{E}, O_S^{\oplus n})$ and $\mathcal{L} := \det(Q)^{\otimes m}$ for some appropriately chosen integer $m > 0$. The first two hypotheses are already satisfied: 1.4.10(i) holds because $\mathcal{L}$ is a tensor power of a determinant of a quotient of a nef sheaf; see Lemma 1.4.2 and Proposition 1.4.9; 1.4.10(ii) holds because $\mathcal{F}^\vee \cong \mathcal{E}^{\oplus n}$ is a sum of nef bundles and hence is itself nef by Lemma 1.4.6.

It remains to arrange for condition 1.4.10(iii). The construction of the classifying map in Lemma 1.5.2 gives a rational map $[\pi^*\alpha]: \mathbf{P} \dashrightarrow \mathbf{G}$. Blowing up the ideal sheaf in the image of

$$\bigwedge^q \mathrm{Sym}^d(O_{\mathbf{P}}^{\oplus n}) \to O_{\mathbf{P}}(qd) \otimes_{O_{\mathbf{P}}} \pi^* \det(Q)$$

induced by $\pi^*\alpha \circ \mathrm{Sym}^d(\psi_{\mathrm{univ}})$ yields a birational morphism $b: \mathbf{P}' \to \mathbf{P}$, a morphism $f: \mathbf{P}' \to \mathbf{G}$ resolving $[\pi^*\alpha]$, and an effective Cartier divisor D of $\mathbf{P}'$ such that

$$f^*O_{\mathbf{G}}(1) = b^*\big(O_{\mathbf{P}}(qd) \otimes_{O_{\mathbf{P}}} \pi^* \det(Q)\big) \otimes_{O_{\mathbf{P}'}} O_{\mathbf{P}'}(-D).$$

Let $\mathbf{T}$ be the schematic image of $(\pi \circ b, f): \mathbf{P}' \to S \times_k \mathbf{G}$, and let

$\rho\colon \mathbf{T} \to S$ and $g\colon \mathbf{T} \to \mathbf{G}$ be the induced morphisms. We claim that g is generically quasi-finite. Identify $\mathbf{Fr}(\mathcal{E})$ as a dense open subscheme of $\mathbf{P}'$ via the birational morphism b. Let $\mathbf{T}_0 \subseteq \mathbf{T}$ be the dense set obtained as the intersection of the image of $\mathbf{Fr}(\mathcal{E})$, which is a dense constructible subset by Chevalley's theorem [30, Theorem 054K], and the open set $S_0 \times_k \mathbf{G}$ with S_0 from hypothesis (iii). Then there is a commutative diagram

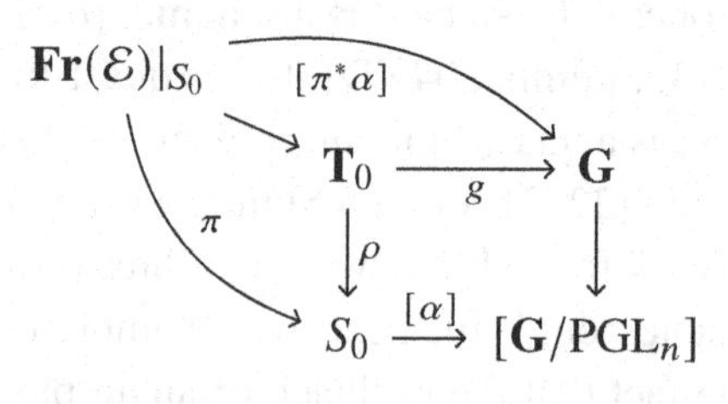

Let $\bar{x} \in \mathbf{G}(\bar{k})$ be a $\bar{k}$-point and let $\bar{y} \in [\mathbf{G}/\mathbf{PGL}_n](\bar{k})$ be its image along the quotient map. Since the outer square is cartesian by Lemma 1.5.2, the fibre $(\mathbf{Fr}(\mathcal{E})|_{S_0})_{\bar{x}}$ maps via π to $S_{0,\bar{y}}$. Since $\mathbf{T}_0$ is the image of $\mathbf{Fr}(\mathcal{E})|_{S_0}$ in $S_0 \times_k \mathbf{G}$, this implies that $\mathbf{T}_{0,\bar{x}}$ is contained in the finite set $S_{0,\bar{y}} \times \{\bar{x}\}$. Thus g has finite fibres on $\bar{k}$-points over $\mathbf{T}_0$, so $\mathbf{T}$ contains a dense set of closed points at which the fibre dimension of g is 0. This latter set is open, by [30, Lemma 02FZ]. Therefore g is generically quasi-finite.

We may now complete the proof of the lemma. Since $\mathcal{O}_{\mathbf{G}}(1)$ is ample and g is generically quasi-finite, $g^*\mathcal{O}_{\mathbf{G}}(1)$ is a big invertible sheaf on $\mathbf{T}$. Let $\mathcal{A}$ be any very ample invertible sheaf on S. Then Lemma 1.3.8 gives

$$H^0(\mathbf{T}, g^*\mathcal{O}_{\mathbf{G}}(m) \otimes_{\mathcal{O}_T} \rho^*\mathcal{A}^{-1}) \neq 0 \quad \text{for some integer } m > 0.$$

Pulling back to $\mathbf{P}'$, multiplying by an equation of the effective divisor D, and then applying the projection formula gives

$$\begin{aligned} 0 &\neq H^0(\mathbf{P}', f^*\mathcal{O}_{\mathbf{G}}(m) \otimes_{\mathcal{O}_{\mathbf{P}'}} b^*\pi^*\mathcal{A}^{-1}) \\ &\subset H^0(\mathbf{P}', b^*(\mathcal{O}_{\mathbf{P}}(qdm) \otimes_{\mathcal{O}_{\mathbf{P}}} \pi^* \det(Q)^{\otimes m} \otimes_{\mathcal{O}_{\mathbf{P}}} \pi^*\mathcal{A}^{-1})) \\ &\cong H^0(\mathbf{P}, (\mathcal{O}_{\mathbf{P}}(qdm) \otimes_{\mathcal{O}_{\mathbf{P}}} \pi^* \det(Q)^{\otimes m} \otimes_{\mathcal{O}_{\mathbf{P}}} \pi^*\mathcal{A}^{-1}) \otimes_{\mathcal{O}_{\mathbf{P}}} b_*\mathcal{O}_{\mathbf{P}'}). \end{aligned}$$

Now $\mathbf{P}$ is normal by hypothesis (i), so the Stein factorization of the birational map b is trivial; see [30, Theorem 03H0]. In particular, $b_*\mathcal{O}_{\mathbf{P}'} \cong \mathcal{O}_{\mathbf{P}}$. Setting $\mathcal{L} := \det(Q)^{\otimes m}$ and $a := qdm$, we conclude that

$$H^0(\mathbf{P}, \mathcal{O}_{\mathbf{P}}(a) \otimes_{\mathcal{O}_{\mathbf{P}}} \pi^*\mathcal{L} \otimes_{\mathcal{O}_{\mathbf{P}}} \pi^*\mathcal{A}^{-1}) \neq 0.$$

Thus hypothesis (iii) of Lemma 1.4.10 is satisfied and it can be applied to show that $\det(Q)^{\otimes m}$ is big and nef, and so $\det(Q)$ is itself big and nef. □

Proposition 1.5.4 (Ampleness lemma) *Let X be a proper algebraic space over k, let $\mathcal{E}$ be a locally free $\mathcal{O}_X$-module of rank n, let d be a positive integer, and let $\alpha\colon \operatorname{Sym}^d(\mathcal{E}) \to Q$ be a locally free quotient of rank q. Assume that*

(i) *$\mathcal{E}$ is nef, and*
(ii) *the classifying map $[\alpha]$ has finite fibres on $\bar{k}$-points.*

Then $\det(Q)$ *is ample on X.*

Proof We aim to apply the Nakai–Moishezon criterion, Proposition 1.2.4. Thus we need to show that $\det(Q)$ has positive degree on each integral closed subspace $\iota\colon Y \hookrightarrow X$. Applying Chow's lemma [30, Lemma 088U] and normalizing gives a modification $f\colon Y' \to Y$ from a normal projective variety Y'. The compatibility of intersection numbers with pullbacks, [30, Lemma 0EDJ], gives

$$(\det(Q)^{\dim(Y)} \cdot Y) = (\iota^* \det(Q)^{\dim(Y)}) = (f^* \iota^* \det(Q)^{\dim(Y')}).$$

This final quantity is positive by Lemma 1.5.3: the pullback of $\mathcal{E}$ to Y' is nef by Lemma 1.4.3(i), and the classifying map on Y' associated with the pullback of α is the composite

$$[f^*\iota^*\alpha]\colon Y' \xrightarrow{f} Y \xrightarrow{\iota} X \xrightarrow{[\alpha]} [\mathbf{G}/\mathrm{PGL}_n],$$

which generically has finite fibres on $\bar{k}$-points, as each of f, i, and $[\alpha]$ do. □

1.6 Nefness for families of nodal curves

In this section, we prove that $f_*\omega_{S/C}^{\otimes m}$ is nef for all $m \geq 2$ and any family $f\colon S \to C$ of stable curves over a smooth projective curve C over k; see Theorem 1.6.10. In other words, we show that the corresponding vector bundle on the stack $\overline{\mathcal{M}}_g$ is nef.

Since nefness is insensitive to field extensions, by Lemmas 1.3.4 and 1.4.4, throughout we assume that our base field k is algebraically closed. Furthermore, all schemes and morphisms appearing will be over k. We will make constant use of the following transitivity property of relative dualizing sheaves: by [30, Lemma 0E30], there is an isomorphism

$$\omega_{S/C} \cong \omega_S \otimes_{\mathcal{O}_S} f^*\omega_C^{-1}.$$

The first positivity result is Proposition 1.6.3 and it concerns families in which the generic fibre is smooth. This is generalized in Proposition 1.6.7 to positivity when $\omega_{S/C}$ is twisted up by some sections. Finally, as a general family of stable curves is essentially obtained by glueing generically smooth families along horizontal curves, this gives us the main positivity result in Proposition 1.6.9.

To begin, we discuss the local structure of nodal families of curves. So let $f: S \to C$ be a nodal family of curves over a smooth projective curve C. Consider the closed subset $\mathrm{Sing}(f) \subset S$ of points at which f is not smooth. This has a canonical scheme structure given by the first Fitting ideal of $\Omega^1_{S/C}$; see [30, Section 0C3H].

Lemma 1.6.1 *Let $f: S \to C$ be a family of nodal curves over a smooth projective curve C.*

(i) *If s is an isolated point of* $\mathrm{Sing}(f)$, *then*

$$\mathcal{O}^{\wedge}_{S,s} \cong \mathcal{O}^{\wedge}_{C,f(s)}[[x,y]]/(xy - \pi^n)$$

where π is a uniformizer of $\mathcal{O}^{\wedge}_{C,f(s)}$ and $n \geq 1$.

(ii) *If s is not isolated in* $\mathrm{Sing}(f)$, *then there exists a commutative diagram*

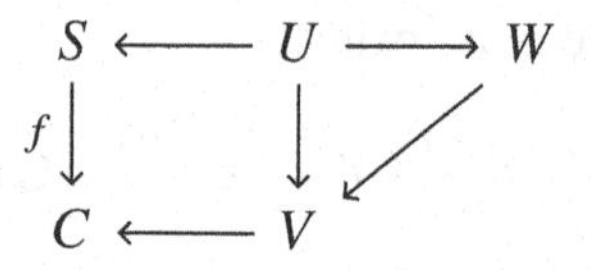

where $W := V \otimes_k k[u,v]/(uv)$, the morphisms $S \leftarrow U \to W$ and $C \leftarrow V$ are étale, and there is a point $u \in U$ mapping to $s \in S$.

Proof In the isolated case, this follows from [30, Lemma 0CBX], noting that all nodes are split since we assume k is algebraically closed. In the non-isolated case, this follows from [30, Lemma 0CBY]. See also [30, Lemma 0CDD]. □

The isolated points in $\mathrm{Sing}(f)$ as in Lemma 1.6.1(i) are rational double points and can be resolved by repeated blowup. See [30, Section 0BGB] and also [2]. Since the singularity is rational, we may harmlessly pass to a resolution of such singularities:

Lemma 1.6.2 *Let $f: S \to C$ be a family of nodal curves over a smooth projective curve C. Let $b: S' \to S$ be the minimal resolution of the isolated singularities of S. Then $b_*\omega_{S'/C} \cong \omega_{S/C}$.*

Proof There is a canonical morphism $b_*\omega_{S'/C} \to \omega_{S/C}$ obtained by dualizing the map $b^{\#}: \mathcal{O}_S \to b_*\mathcal{O}_{S'}$. This map is an isomorphism: it is clear on the locus where b is an isomorphism; around the singular points, this follows from [30, Lemma 0BBU]. □

We are now ready for the first positivity result, concerning families of nodal curves in which the generic fibre is smooth. In this case, the total space is normal as only the isolated singularities of Lemma 1.6.1(i) may appear. Compare with [21, Proposition 4.5].

Proposition 1.6.3 *Let $f\colon S \to C$ be a family of nodal curves over a smooth projective curve C. If the generic fibre of f is smooth of genus $g \geq 2$, then $f_*\omega_{S/C}^{\otimes m}$ is nef for any $m \geq 2$.*

We first prove Proposition 1.6.3 under a series of simplifying assumptions in Lemma 1.6.5, then explain afterward how these assumptions may be removed. The crucial input is the following consequence of Ekedahl's vanishing theorems for surfaces of general type.

Lemma 1.6.4 *Suppose that* $\mathrm{char}(k) = p > 0$. *Let S be a smooth projective minimal surface of general type, and D a reduced effective Cartier divisor with smooth connected components of genus at least* 2 *and $\mathcal{O}_S(D)|_D \cong \mathcal{O}_D$. Then for any $m \geq 2$,*

$$h^1(S, \omega_S^{\otimes m}(D)) \begin{cases} = 0 & \text{if } \mathrm{char}(k) \neq 2 \text{ or } m \neq 2, \\ \leq 1 & \text{if } \mathrm{char}(k) = 2 \text{ and } m = 2, \end{cases}$$

Proof From the cohomology of the exact sequence

$$0 \to \omega_S^{\otimes m} \to \omega_S^{\otimes m}(D) \to \omega_S^{\otimes m}(D)|_D \to 0,$$

it suffices to show $h^1(S, \omega_S^{\otimes m}(D)|_D) = 0$ and to bound $h^1(S, \omega_S^{\otimes m})$. For the former, the adjunction formula, [30, Lemma 0B4B], gives

$$\omega_S^{\otimes m}(D)|_D \cong \omega_S^{\otimes m}(mD)|_D \cong \omega_D^{\otimes m}.$$

Since the genus of each connected component of D is at least 2,

$$H^1(S, \omega_S^{\otimes m}(D)|_D) \cong H^1(D, \omega_D^{\otimes m}) = 0$$

when $m \geq 2$, for degree reasons; see [30, Lemma 0B90]. The bound on $h^1(S, \omega_S^{\otimes m})$ follows from the vanishing theorem of Ekedahl [10, Main theorem (i)]. □

Lemma 1.6.5 *Proposition 1.6.3 holds with additional assumptions that*

(i) *the characteristic of k is $p > 0$,*
(ii) *S is minimal, and*
(iii) *the genus of C is at least 2.*

Proof If $f_*\omega_{S/C}^{\otimes m}$ were not nef, then the Barton–Kleiman criterion, Proposition 1.4.5, gives an invertible quotient $\alpha\colon f_*\omega_{S/C}^{\otimes m} \to \mathcal{M}^{-1}$ such that $d := \deg(\mathcal{M}) > 0$. The assumption on the genus of the fibres of f together with (iii) imply that a resolution of S is of general type; see [5, Theorem 1.3]. We now seek a contradiction to Lemma 1.6.4.

Let $F_C\colon C \to C$ be the absolute Frobenius of C and consider the base change $f'\colon S' \to C$ of f along F_C. This is still a family of nodal curves by [30, Lemma 0C5B]. Since smoothness is stable under base change, by [30, Lemma 01VB], the generic fibre of f' is also smooth. Since the formation of dualizing sheaves commutes with base change; see [30, Lemmas 0B91 and 0E6R],

$$F_C^* f_*\omega_{S/C}^{\otimes m} \cong f'_* g^* \omega_{S/C}^{\otimes m} \cong f'_* \omega_{S'/C}^{\otimes m}$$

where $g\colon S' \to S$ is the projection. Pulling α back by F_C yields a negative quotient $f'_*\omega_{S'/C}^{\otimes m} \to F_C^*\mathcal{M}^{-1}$ of degree $-dp$. Replacing f by f', we can take $d = \deg(\mathcal{M})$ to be arbitrarily large. Thus we may assume $\mathcal{M} \cong \mathcal{L} \otimes_{\mathcal{O}_C} \omega_C^{\otimes m}$ for some very ample invertible $\mathcal{O}_C$-module $\mathcal{L}$.

Since the generic fibre of $f\colon S \to C$ is assumed to be smooth, S has only isolated rational double points as in Lemma 1.6.1(i). A minimal resolution of singularities is obtained by repeated blowups and the resulting exceptional divisor is a chain of projective lines joined along nodes. Thus the minimal resolution of singularities of S will remain a family of nodal curves over C. Therefore, using Lemma 1.6.2, we may replace S by the minimal resolution of its singularities and assume that S is a smooth minimal surface of general type.

Upon rearranging the terms of α, we obtain a surjection of sheaves

$$\mathcal{L} \otimes_{\mathcal{O}_C} \omega_C^{\otimes m} \otimes_{\mathcal{O}_C} f_*\omega_{S/C}^{\otimes m} \twoheadrightarrow \mathcal{O}_C.$$

Since C is of dimension 1, we obtain the inequality

$$h^1(C, \mathcal{L} \otimes_{\mathcal{O}_C} \omega_C^{\otimes m} \otimes_{\mathcal{O}_C} f_*\omega_{S/C}^{\otimes m}) \geq h^1(C, \mathcal{O}_C) = g.$$

On the other hand, consider the invertible $\mathcal{O}_S$-module

$$\mathcal{F} := f^*\mathcal{L} \otimes_{\mathcal{O}_S} (f^*\omega_C^{\otimes m} \otimes_{\mathcal{O}_S} \omega_{S/C}^{\otimes m}) \cong f^*\mathcal{L} \otimes_{\mathcal{O}_S} \omega_S^{\otimes m}$$

where we have used the transitivity of dualizing sheaves. Since f has relative dimension 1, the Leray spectral sequence, [30, Lemma 01F2], for f and $\mathcal{F}$ degenerates on the E_2-page and yields a short exact sequence

$$0 \to H^1(C, f_*\mathcal{F}) \to H^1(S, \mathcal{F}) \to H^0(C, R^1 f_*\mathcal{F}) \to 0.$$

The projection formula gives $f_*\mathcal{F} \cong \mathcal{L} \otimes_{O_C} \omega_C^{\otimes m} \otimes_{O_C} f_*\omega_{S/C}^{\otimes m}$, so this sequence together with the inequality above gives

$$h^1(S, f^*\mathcal{L} \otimes_{O_S} \omega_S^{\otimes m}) = h^1(S, \mathcal{F}) \geq h^1(C, f_*\mathcal{F}) \geq g \geq 2.$$

Since $\mathcal{L}$ is very ample, we may choose an effective Cartier divisor D in $|f^*\mathcal{L}|$ which is the union of smooth fibres of f. Then $f^*\mathcal{L} \cong O_S(D)$ yields a contradiction to Lemma 1.6.4. Therefore $f_*\omega_{S/C}^{\otimes m}$ is nef. □

Proof of Proposition 1.6.3 We now explain how to remove the assumptions (i), (ii), and (iii) of Lemma 1.6.5.

We may reduce to characteristic $p > 0$ as in Step 2 in the proof of Proposition 1.4.9. That is, if k is of characteristic 0 and $f_*\omega_{S/C}^{\otimes m}$ has a negative quotient, then choose a finitely generated $\mathbf{Z}$-algebra over which everything is defined. We may then reduce modulo some prime p to yield a contradiction to Lemma 1.6.5. Thus we may drop assumption (i).

If S is not minimal, consider any (-1)-curve E. Then E is contained in fibres of f since, otherwise, $f|_E \colon E \to C$ would be a dominant morphism from a curve of genus 0 to a curve of genus $g \geq 2$, which is impossible. So contracting E as in [30, Lemma 0C2N] yields a normal projective surface S', a morphism $f' \colon S' \to C$ such that $f = f' \circ b$, where $b \colon S \to S'$ is the contraction map. Since $b_*\omega_{S'} \cong \omega_S$, the transitivity of relative dualizing sheaves implies that $f'_*\omega_{S'/C}^{\otimes m} \cong f_*\omega_{S/C}^{\otimes m}$. Successively contracting (-1)-curves will produce a minimal model $f_{\min} \colon S_{\min} \to C$ of $f \colon S \to C$. Induction on the number of contractions gives

$$f_*\omega_{S/C}^{\otimes m} \cong f_{\min,*}\omega_{S_{\min}/C}^{\otimes m}$$

and nefness of the former follows from the nefness of the latter. Thus we may drop both assumptions (i) and (ii) in Lemma 1.6.5.

Finally, supposse that the genus of C is less than 2. Let $g \colon C' \to C$ be any finite cover from a smooth projective curve C' of genus at least 2 and let $f' \colon S' \to C$ be the base change of f. Then, as before, f' is a family of nodal curves with smooth generic fibre and $g^* f_*\omega_{S/C}^{\otimes m} = f'_*\omega_{S'/C}^{\otimes m}$. This is nef by Lemma 1.6.5. Hence $f_*\omega_{S/C}^{\otimes m}$ is also nef by Lemma 1.4.3(ii). This completes the proof. □

As a consequence, we obtain the following weak positivity result for $\omega_{S/C}$ on S. See also [21, Corollary 4.6].

Corollary 1.6.6 *In the situation of Proposition 1.6.3, let C_t be a section of f. Then $(\omega_{S/C} \cdot C_t) \geq 0$.*

Proof Consider the pushforward along f of the sequence

$$0 \to \omega_{S/C}^{\otimes m}(-C_t) \to \omega_{S/C}^{\otimes m} \to \omega_{S/C}^{\otimes m}|_{C_t} \to 0.$$

We have $R^1 f_*(\omega_{S/C}^{\otimes m}(-C_t)) = 0$ by looking at degrees along fibres; see [30, Lemma 0B90]. So this gives a surjection $f_*\omega_{S/C}^{\otimes m} \twoheadrightarrow f_*\omega_{S/C}^{\otimes m}|_{C_t}$. But $f|_{C_t} : C_t \to C$ is an isomorphism, so this is an invertible quotient of degree $(\omega_{S/C} \cdot C_t)$. By the nefness of Proposition 1.6.3, we conclude that $(\omega_{S/C} \cdot C_t) \geq 0$. □

To establish positivity for general families of stable curves, we need the following generalization of Proposition 1.6.3, in which the relative dualizing sheaf is twisted up by sections and where the fibres of f may be of genus 0 or 1. Compare with [21, Proposition 4.7].

Proposition 1.6.7 *Let $f : S \to C$ be a family of nodal curves over a smooth projective curve C. If the generic fibre of f is smooth, then, for any set of pairwise distinct sections $C_1, \ldots, C_n$ of f contained in the smooth locus of S,*

$$f_*(\omega_{S/C}^{\otimes m}(a_1 C_1 + \cdots + a_n C_n))$$

is nef for any $m \geq 2$ and any $0 \leq a_1, \ldots, a_n \leq m$.

Proof Since the C_i avoid the singularities of S, we may reduce to the case in which S is smooth by passing to a minimal resolution of singularities of S using Lemma 1.6.2. We split the proof into three cases, depending on whether the genus of the generic fibre of f is ≥ 2, 0, or 1. Each case will proceed by induction on $j := \sum a_i$.

Case 1 The generic fibre of f is of genus $g \geq 2$.

The base case where each $a_i = 0$ is Proposition 1.6.3. Assume the claim is proven for $D_j := \sum a_i C_i$; we will prove it for $D_{j+1} := D_j + C_t$ for any index t such that $a_t + 1 \leq m$. Consider the exact sequence

$$0 \to \omega_{S/C}^{\otimes m}(D_j) \to \omega_{S/C}^{\otimes m}(D_{j+1}) \to \omega_{S/C}^{\otimes m}(D_{j+1})|_{C_t} \to 0$$

obtained by twisting the sequence for C_t by $\omega_{S/C}^{\otimes m}(D_{j+1})$. Since the C_i are pairwise disjoint, together with the transitivity of relative dualizing sheaves and the adjunction formula, we have

$$\begin{aligned} \omega_{S/C}^{\otimes m}(D_{j+1})|_{C_t} &\cong \omega_{S/C}^{\otimes m-a_t-1}|_{C_t} \otimes (\omega_S^{\otimes a_t+1}((a_t+1)C_t)|_{C_t} \otimes \omega_{C_t}^{\otimes -a_t-1}) \\ &\cong \omega_{S/C}^{\otimes m-a_t-1}|_{C_t}. \end{aligned}$$

Because $a_t + 1 \leq m$, Corollary 1.6.6 together with [30, Lemma 0BEY] shows that this invertible sheaf has nonnegative degree on C_t. Also note that $R^1 f_*(\omega_{S/C}^{\otimes m}(D_j)) = 0$ because of the degree on the fibres; see [30, Lemma 0B90].

Thus applying f_* to the above exact sequence yields an exact sequence

$$0 \to f_*(\omega_{S/C}^{\otimes m}(D_j)) \to f_*(\omega_{S/C}^{\otimes m}(D_{j+1})) \to f_*(\omega_{S/C}^{\otimes m - a_t - 1}|_{C_t}) \to 0.$$

The subsheaf is nef by the induction hypothesis, and the quotient sheaf is a nonnegative invertible sheaf on C, as C_t is a section. Thus the extension is nef by Lemma 1.4.6, completing the induction in this case.

Case 2 The generic fibre of f is of genus $g = 0$.

When $j = \sum a_i \leq 2m - 1$, the sheaf $\omega_{S/C}^{\otimes m}(\sum a_i C_i)$ is negative on fibres of f and hence has vanishing, whence nef, pushforward; these are the base cases. Let $j \geq 2m - 1$ and assume that the claim is true for all divisors of the form $\sum a_i C_i$ with $\sum a_i = j$; we will prove it for $D = C_t + \sum a_i C_i$ for any index t such that $a_t + 1 \leq m$.

We can assume that $(C_t^2) \leq 0$. Indeed, by the Hodge index theorem, we may assume that among $C_1, \ldots, C_n$ the only section with positive self-intersection is C_1. So, if $D = C_1 + \sum a_i C_i$, as $a_1 \leq m < 2m - 1$, there is some index $t \neq 1$ such that $a_t \neq 0$. Thus we may write

$$D = C_1 + \sum a_i C_i = C_t + \sum a_i' C_i$$

with $a_1' := a_1 + 1$, $a_t' := a_t - 1$, and $a_i' := a_i$ for $i \neq 1, t$. Then $\sum a_i = \sum a_i' = j$ and induction will apply to $\sum a_i' C_i$. With this, we see by the adjunction formula as in Case 1 that

$$\omega_{S/C}(C_t)|_{C_t} \cong \mathcal{O}_{C_t} \quad \text{so} \quad (\omega_{S/C} \cdot C_t) = -(C_t^2) \geq 0.$$

From here, induction proceeds as in Case 1.

Case 3 The generic fibre of f is of genus $g = 1$.

To establish the base case and the nonnegativity $(\omega_{S/C} \cdot C_t) \geq 0$, we claim that it suffices to show that $\chi(S, \mathcal{O}_S) \geq 0$. Indeed, the canonical bundle formula for elliptic surfaces in [4, Theorem 2] gives

$$\omega_{S/C} \cong f^* \mathcal{M} \otimes_{\mathcal{O}_S} \mathcal{O}_S(F)$$

where F is an effective Cartier divisor supported along fibres of f and $\mathcal{M}$ is an invertible $\mathcal{O}_C$-module with degree $\geq \chi(S, \mathcal{O}_S)$. Thus

$$f_*(\omega_{S/C}^{\otimes m}) \cong \mathcal{M}^{\otimes m} \quad \text{and} \quad (\omega_{S/C} \cdot C_t) \geq (f^* \mathcal{M} \cdot C_t) \geq \chi(S, \mathcal{O}_S),$$

so the nefness of $f_*(\omega_{S/C}^{\otimes m})$ and the nonnegativity will follow from $\chi(S, \mathcal{O}_S) \geq 0$.

Since $\chi(S, \mathcal{O}_S)$ is a birational invariant, we may in fact assume S is minimal over C. In this case, the effective Cartier divisor F is actually a sum of fibre classes, at least viewed as a $\mathbf{Q}$-Cartier divisor: see [4, the text at the bottom of page 28]. Thus ω_S is a sum of fibre classes and so $(\omega_S^2) = 0$. Noether's formula then gives

$$12\chi(S, \mathcal{O}_S) = (\omega_S^2) + e(S) = e(S) \geq e(C)e(S_{\bar{\eta}}) = 0,$$

where e denotes ℓ-adic topological Euler characteristic, ℓ any prime different from p, and $S_{\bar{\eta}}$ is the geometric generic fibre of $f\colon S \to C$. The inequality follows from [23, Lemma 1], and $e(S_{\bar{\eta}}) = 0$ since the generic fibre of f is a smooth curve of genus 1. With this, induction may proceed as in Case 1, and the proof of the proposition is complete. □

To obtain a positivity result for a general family $f\colon S \to C$ of stable curves, it remains to consider the non-isolated singularities of Lemma 1.6.1(ii). Let D be the subscheme of one-dimensional components of $\mathrm{Sing}(f)$, and call it the *double locus* of S. The following explains how a general family of nodal curves is obtained by glueing nodal families with only double points along the double locus:

Lemma 1.6.8 *Let $f\colon S \to C$ be a family of nodal curves over a smooth projective curve C. Let $\nu\colon S^\nu \to S$ be the normalization. Then S^ν is a disjoint union of nodal families of curves over C with smooth generic fibre and $\omega_{S^\nu/S} \cong \mathcal{O}_{S^\nu}(-D^\nu)$ where $D^\nu := \nu^{-1}(D)$.*

Proof Let $s \in D$ be a point of the double locus of S and consider the diagram of Lemma 1.6.1(ii):

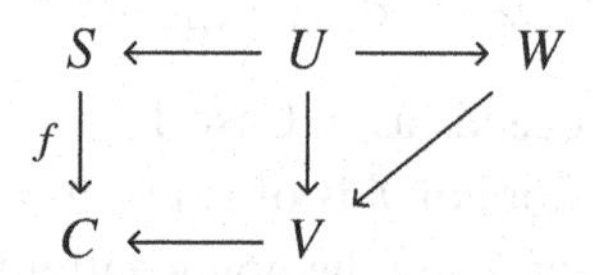

Since the morphisms $S \leftarrow U \to W$ are étale and normalization commutes with smooth base change, by [30, Lemma 03GV], there are étale morphisms $S^\nu \leftarrow U^\nu \to W^\nu$. Since $C \leftarrow V$ is also étale, V is smooth, so the same lemma gives

$$W^\nu = V \otimes_k (k[u] \times k[v]) \to V \otimes_k k[u, v]/(uv) = W.$$

In particular, W^ν is smooth. As the morphisms from U^ν are étale, we conclude that S^ν, locally around s, is the disjoint union of two families of nodal curves over C with smooth generic fibre. Since this is true for all $s \in D$, S^ν itself is a disjoint union of families of nodal curves over C with smooth generic fibre.

For $\omega_{S^\nu/S}$, since ν is a finite morphism, its relative dualizing sheaf is characterized by the formula

$$\nu_*\omega_{S^\nu/S} = \mathcal{H}om_{O_S}(\nu_*O_{S^\nu}, O_S);$$

see [30, Section 0FKW]. Evaluation at 1 yields an injection $\nu_*\omega_{S^\nu/S} \to O_S$ whose image is an ideal sheaf $\mathcal{I}$ of a subscheme supported on D. In fact, this is the ideal sheaf of D. To see this, since the formation of the evaluation map commutes with flat pullback (see [30, Lemmas 0C6I and 02KH]), using the local structure of S around $s \in D$ above it suffices to show that, for

$$R^\nu := k[u] \times k[v] \leftarrow k[u,v]/(uv) =: R,$$

we have $I := \mathrm{Hom}_R(R^\nu, R) = (u, v)$. Indeed, R^ν is generated as an R-module by $(1,0)$ and $(0,1)$, and they are annihilated by v and u, respectively, so any R-module map $\varphi\colon R^\nu \to R$ must be of the form

$$\varphi((1,0)) = \alpha u \quad \text{and} \quad \varphi((0,1)) = \beta v \quad \text{for some } \alpha, \beta \in R.$$

Furthermore, this shows that the image of I under the ring extension $R \to R^\nu$ is the ideal of the two preimages of the node. Hence we conclude that $\nu_*\omega_{S^\nu/S} \cong \mathcal{I}$ is the ideal sheaf of D in S, and so $\omega_{S^\nu/S} \cong O_{S^\nu}(-D^\nu)$ is the ideal sheaf of D^ν in S^ν. □

With the notation above, we have the following intermediate result:

Proposition 1.6.9 *Let $f\colon S \to C$ be a family of stable curves over a smooth projective curve C. Assume that*

(i) *the double curve D is a union of sections of f, and*
(ii) *its preimage $D^\nu := \nu^{-1}(D)$ is a union of sections of $f^\nu\colon S^\nu \to C$.*

Then $f_*\omega_{S/C}^{\otimes m}$ is nef for any $m \geq 2$.

Proof By the transitivity of relative dualizing sheaves and Lemma 1.6.8, $\nu^*\omega_{S/C} \cong \omega_{S^\nu/C}(D^\nu)$. Thus pulling $\omega_{S/C}^{\otimes m}$ back to S^ν and tensoring with the subscheme sequence for D^ν yields

$$0 \to \omega_{S^\nu/C}^{\otimes m}((m-1)D^\nu) \to \nu^*(\omega_{S/C}^{\otimes m}) \to \omega_{S^\nu/C}^{\otimes m}(mD^\nu)|_{D^\nu} \to 0.$$

Since D^ν is an effective Cartier divisor of S^ν, the adjunction formula, [30, Lemma 0AA4], together with hypothesis (ii) gives

$$\omega_{S^\nu/C}(D^\nu)|_{D^\nu} \cong \omega_{D^\nu/C} \cong O_{D^\nu}.$$

Applying ν_* to the above short exact sequence yields an exact sequence on S:

$$0 \to \nu_*(\omega_{S^\nu/C}^{\otimes m}((m-1)D^\nu)) \to \omega_{S/C}^{\otimes m} \otimes_{\mathcal{O}_S} \nu_*\mathcal{O}_{S^\nu} \to \mathcal{O}_D^{\oplus 2} \to 0.$$

Since the preimage of the antidiagonal $\mathcal{O}_D$ along the map $\nu_*\mathcal{O}_{S^\nu} \to \mathcal{O}_D^{\oplus 2}$ is $\mathcal{O}_S$, there is a short exact sequence

$$0 \to \nu_*(\omega_{S^\nu/C}^{\otimes m}((m-1)D^\nu)) \to \omega_{S/C}^{\otimes m} \to \mathcal{O}_D \to 0.$$

Now push down to C. Write $f^\nu := f \circ \nu \colon S^\nu \to C$. Since the fibres of f are stable curves, the fibres of f^ν are stable pointed curves, so $R^1 f_*^\nu(\omega_{S^\nu/C}^{\otimes m}((m-1)D^\nu)) = 0$ for all $m \geq 2$. The relative Leray spectral sequence for f^ν, [30, Lemma 0734], shows that $R^1 f_*\nu_*(-)$ is a subsheaf of $R^1 f_*^\nu(-)$. Thus applying f_* to the preceding short exact sequence yields

$$0 \to f_*^\nu(\omega_{S^\nu/C}^{\otimes m}((m-1)D^\nu))) \to f_*\omega_{S/C}^{\otimes m} \to f_*\mathcal{O}_D \to 0.$$

The term on the left is nef by Lemma 1.6.8 together with (ii) and Proposition 1.6.7; by (i), the sheaf $f_*\mathcal{O}_D$ is isomorphic to the sum of copies of $\mathcal{O}_C$. Thus $f_*\omega_{S/C}^{\otimes m}$ is an extension of a direct sum of nonnegative line bundles by a nef bundle, and hence is nef by Lemma 1.4.6. □

Putting everything together now gives the main positivity result.

Theorem 1.6.10 *Let $f \colon S \to C$ be a family of stable curves over a smooth projective curve C. Then $f_*\omega_{S/C}^{\otimes m}$ is nef for any $m \geq 2$.*

Proof In order to apply Proposition 1.6.9 to f, we need to arrange for the components of the double curve D and its preimage D^ν in the normalization $\nu \colon S^\nu \to S$ to be sections over C. So let C' be any such component and form the cartesian diagram

$$\begin{array}{ccc} S' & \longrightarrow & S \\ {\scriptstyle f'}\downarrow & & \downarrow{\scriptstyle f} \\ C' & \xrightarrow{g} & C \end{array}$$

By [30, Lemma 0E76], f' is still a family of stable curves. Moreover, the inverse image of C' is now a section over C'. Since $g^* f_*\omega_{S/C}^{\otimes m} \cong f'_*\omega_{S'/C'}^{\otimes m}$, by Lemma 1.4.3(ii), we may replace f by f'. Repeating this for every component of D and D^ν, we can arrange for hypotheses (i) and (ii) of Proposition 1.6.9 to be verified, upon which we may conclude. □

1.7 Projectivity of the moduli of curves

Finally, we put everything together to show that the Deligne–Mumford moduli space $\overline{M}_g$ of stable curves is projective over Spec($\mathbf{Z}$).

The first step is to show that for a family of curves $f: X \to S$ over an algebraically closed field k whose moduli map has finite fibres, there is some m such that λ_m pulls back to an ample invertible sheaf on S. In fact, $m = 6$ works by using the fact that tri-canonically embedded stable curves are projectively normal and are determined by their quadratic equations; see [28, the corollary on page 58]. In the following, we argue directly and show only that $m = 3d$ works for all sufficiently large d, perhaps depending on the family f.

Lemma 1.7.1 *Let $f: X \to S$ be a family of stable curves of genus $g \geq 2$ over an algebraically closed field k. If the moduli map $[f]: S \to \overline{M}_g$ is of finite type and has finite fibres on k-points, then $[f]^*\lambda_{3d} = \det(f_*\omega_{X/S}^{\otimes 3d})$ is ample on S for all $d \gg 0$.*

Proof We apply the ampleness lemma (see Proposition 1.5.4) to the multiplication map

$$\mu_d \colon \mathrm{Sym}^d(f_*\omega_{X/S}^{\otimes 3}) \to f_*\omega_{X/S}^{\otimes 3d}.$$

We choose d sufficiently large that

(i) the fibres of f are determined by the degree-d equations in their tri-canonical embedding, and

(ii) μ_d is surjective.

To see that this is possible, note that by [30, Lemma 0E8X], $\omega_{X/S}^{\otimes 3}$ is f-very ample, so there is a commutative diagram

$$\begin{array}{ccc} X & \xrightarrow{\iota} & \mathbf{P} \\ & {\scriptstyle f}\searrow & \downarrow{\scriptstyle \pi} \\ & & S \end{array} \qquad \text{where } \mathbf{P} := \mathbf{P}(f_*\omega_{X/S}^{\otimes 3})$$

and ι is a closed immersion in which the fibres of $f: X \to S$ are embedded as tri-canonical curves of degree $6g - 6$. Thus (i) is satisfied for any $d \geq 6g - 6$, as can be seen by taking joins with disjoint codimension-3 linear spaces; see for example [28, Theorem 1].

As for (ii), let $\mathcal{I}$ be the ideal sheaf of X in $\mathbf{P}$ and consider the sequence

$$0 \to \mathcal{I} \otimes_{\mathcal{O}_\mathbf{P}} \mathcal{O}_\mathbf{P}(d) \to \mathcal{O}_\mathbf{P}(d) \to \iota_*\omega_{X/S}^{\otimes 3d} \to 0.$$

Then μ_d is the direct image under π of the surjection. Now S is Noetherian as it is of finite type over $\overline{\mathcal{M}}_g$, which is of finite presentation over Spec($\mathbf{Z}$) (see [30, Lemmas 0DSS and 0E9B]). Thus relative Serre vanishing, [30, Lemma 02O1], applies to give a d_0 such that

$$R^1\pi_*(\mathcal{I} \otimes_{\mathcal{O}_\mathbf{P}} \mathcal{O}_\mathbf{P}(d)) = 0 \quad \text{for all } d \geq d_0,$$

whence (ii) is satisfied for any $d \geq d_0$.

Choose any $d \geq \max(6g-6, d_0)$ and set $\mu := \mu_d$. We now verify the hypotheses of the ampleness lemma (see Proposition 1.5.4). The basic positivity is given by Theorem 1.6.10, ensuring that $f_*\omega_{X/S}^{\otimes 3}$ is nef. To understand the classifying map, fix a closed point $0 \in S$ and set $V := H^0(X_0, \omega_{X_0/k}^{\otimes 3})$. For each closed point $s \in S$, choose an isomorphism

$$\varphi_s \colon V \xrightarrow{\cong} H^0(X_s, \omega_{X_s/k}^{\otimes 3})$$

to view X_s as being embedded in $\mathbf{P}V$. We obtain maps

$$\mu_{0,s} \colon \mathrm{Sym}^d(V) \xrightarrow{\mathrm{Sym}^d(\varphi_s)} \mathrm{Sym}^d(H^0(X_s, \omega_{X_s/k}^{\otimes 3})) \xrightarrow{\mu|_s} H^0(X_s, \omega_{X_s/k}^{\otimes 3d})$$

whose kernel is the space of degree-d equations defining X_s in $\mathbf{P}V$. Up to the action of PGL(V) on the source, $\mu_{0,s}$ is independent of the choice of isomorphism φ_s. Since $R^1f_*\omega_{X/S}^{\otimes 3} = 0$ by [30, Lemma 0E8X], the base change maps on direct images are isomorphisms by [30, Lemma 0D2M]. Therefore the classifying map of Lemma 1.5.2 is identified with the map

$$[\mu] \colon S \to [\mathbf{G}(\mathrm{Sym}^d(V), q)/\mathrm{PGL}(V)] \quad \text{where } q := (6d-1)(g-1),$$

which sends a closed point s of S to the PGL(V) equivalence class K_s of $\ker(\mu_{0,s})$. Now condition (i) implies from our choice of d that, for any two closed points $s, s' \in S$,

$$K_s = K_{s'} \quad \text{if and only if} \quad X_s \cong X_{s'}$$

meaning that $[\mu]$ has finite fibres on k-points if and only if $[f] \colon S \to \overline{\mathcal{M}}_g$ does. Thus the ampleness lemma applies to show that $f_*\omega_{X/S}^{\otimes 3d}$ is ample. □

Theorem 1.7.2 *The moduli space $\overline{M}_g$ of stable curves of genus $g \geq 2$ is projective over* Spec($\mathbf{Z}$).

Proof Since $\overline{\mathcal{M}}_g$ is quasi-compact, by [30, Lemma 0E9B], Lemma 1.1.5 allows us to choose an integer n such that the invertible sheaf $\lambda_m^{\otimes n}$ descends to an invertible sheaf $\mathcal{L}_m$ on $\overline{M}_g$ for all m. We show that there exists some m such that $\mathcal{L}_m$ is ample over Spec($\mathbf{Z}$).

By [30, Lemma 0E7A] and Lemma 1.1.3, $\overline{\mathcal{M}}_g$ is a Deligne–Mumford stack with a moduli space, so [32, Proposition 2.6] shows that there exists a scheme S and a finite surjective morphism $\varphi\colon S \to \overline{\mathcal{M}}_g$. We have a diagram

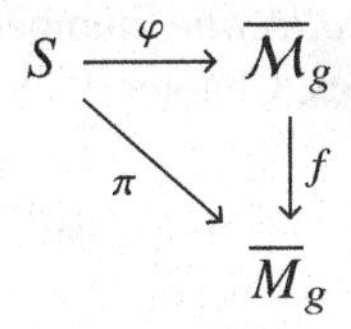

We claim that $\pi := f \circ \varphi$ is a finite surjective morphism of algebraic spaces. Indeed, π is the composition of a finite surjective map φ with a universal homeomorphism f (see [30, Theorem 0DUT]), so π is surjective with discrete fibres. By [30, Lemma 0A4X], the finiteness of π now follows from the properness of π. Since $\overline{\mathcal{M}}_g$ is proper over Spec$(\mathbf{Z})$, by [30, Theorem 0E9C], the same is true for both S and $\overline{M}_g$, by [30, Lemmas 0CL7 and 0DUZ], respectively. Hence π is proper by [30, Lemma 04NX].

By [30, Lemma 0GFB], $\mathcal{L}_m$ is ample on $\overline{M}_g$ over Spec$(\mathbf{Z})$ if and only if $\pi^*\mathcal{L}_m$ is ample on S over Spec$(\mathbf{Z})$. Thus it suffices to show that there exists some m such that $\pi^*\mathcal{L}_m = \varphi^*\lambda_m^{\otimes n}$ is ample over Spec$(\mathbf{Z})$. Let p be a prime number and let S_p be the base change of S along Spec$(\overline{\mathbf{F}}_p) \to$ Spec$(\mathbf{Z})$. The restriction $\varphi_p\colon S_p \to \overline{\mathcal{M}}_g$ of φ to S_p is finite and satisfies

$$\varphi_p^*\lambda_m^{\otimes n} = \varphi^*\lambda_m^{\otimes n}|_{S_p} = \pi^*\mathcal{L}_m|_{S_p}.$$

By Lemma 1.7.1, we may choose d_p such that $\varphi_p^*\lambda_{3d}^{\otimes n}$ is ample for all $d \geq d_p$. Now [30, Lemma 0D2N] gives an open neighbourhood U_p of p in Spec$(\mathbf{Z})$ over which $\varphi^*\lambda_{3d}^{\otimes n}$ is ample. By quasi-compactness, there exists a finite set of primes P such that Spec$(\mathbf{Z}) = \bigcup_{p\in P} U_p$. Then $\pi^*\mathcal{L}_m$ is ample over Spec$(\mathbf{Z})$ for any $m = 3d$ with $d \geq \max(d_p \mid p \in P)$. □

Acknowledgements This project began at the Stacks Project Workshop held at the University of Michigan during the first week of August in 2017. The current document was prepared by RC, CL, and TM, and is based on a draft written with Yordanka Kovacheva and Monica Marinescu during the workshop; we are most grateful for their contributions. We thank our workshop mentor, Davesh Maulik, for suggesting the topic and many hours of helpful conversation in the classrooms of Ann Arbor. We would also like to thank the referee for their very thorough comments which greatly improved this chapter. CL was supported by the National Science Foundation under Grant No. DMS-2001976. TM was supported by the National Science Foundation under Grant No. DMS-1902616.

References

[1] Jarod Alper, Pieter Belmans, Daniel Bragg, Jason Liang, and Tuomas Tajakka. Projectivity of the moduli space of vector bundles on a curve. In *Stacks project expository collection*, number 480 in London Mathematical Society Lecture Note Series, Chapter 3, chapter 3. Cambridge University Press, 2022.

[2] M. Artin. On isolated rational singularities of surfaces. *Amer. J. Math.*, 88:129–136, 1966. doi:10.2307/2373050.

[3] C. M. Barton. Tensor products of ample vector bundles in characteristic p. *Amer. J. Math.*, 93:429–438, 1971. doi:10.2307/2373385.

[4] E. Bombieri and D. Mumford. Enriques' classification of surfaces in char. p. II. In *Complex analysis and algebraic geometry*, pages 23–42. Iwanami Shoten, Tokyo, 1977.

[5] Y. Chen and L. Zhang. The subadditivity of the Kodaira dimension for fibrations of relative dimension one in positive characteristics. *Math. Res. Lett.*, 22(3):675–696, 2015. doi:10.4310/MRL.2015.v22.n3.a3.

[6] G. Codogni and Zs. Patakfalvi. Positivity of the CM line bundle for families of K-stable klt Fano varieties. *Invent. Math.*, 223(3):811–894, 2021. doi:10.1007/s00222-020-00999-y.

[7] M. D. T. Cornalba. On the projectivity of the moduli spaces of curves. *J. Reine Angew. Math.*, 443:11–20, 1993. doi:10.1515/crll.1993.443.11.

[8] O. Debarre. *Higher-dimensional algebraic geometry*. Universitext. Springer, 2001. doi:10.1007/978-1-4757-5406-3.

[9] P. Deligne and D. Mumford. The irreducibility of the space of curves of given genus. *Inst. Hautes Études Sci. Publ. Math.*, 36:75–109, 1969.

[10] T. Ekedahl. Canonical models of surfaces of general type in positive characteristic. *Inst. Hautes Études Sci. Publ. Math.*, 67:97–144, 1988.

[11] D. Gieseker. *Lectures on moduli of curves*, volume 69 of *Tata Inst. Fund. Res. Lectures on Math. and Phys.* Springer, 1982.

[12] A. Grothendieck. Éléments de géométrie algébrique. IV. Étude locale des schémas et des morphismes de schémas. III. *Inst. Hautes Études Sci. Publ. Math.*, (28):255, 1966.

[13] J. Harris and D. Mumford. On the Kodaira dimension of the moduli space of curves. *Invent. Math.*, 67(1):23–88, 1982. With an appendix by W. Fulton. doi:10.1007/BF01393371.

[14] B. Hassett. Moduli spaces of weighted pointed stable curves. *Adv. Math.*, 173(2):316–352, 2003. doi:10.1016/S0001-8708(02)00058-0.

[15] D. S. Keeler. Ample filters of invertible sheaves. *J. Algebra*, 259(1):243–283, 2003. doi:10.1016/S0021-8693(02)00557-4.

[16] S. L. Kleiman. Toward a numerical theory of ampleness. *Ann. Math. (2)*, 84:293–344, 1966. doi:10.2307/1970447.

[17] F. F. Knudsen. The projectivity of the moduli space of stable curves. II. The stacks $M_{g,n}$. *Math. Scand.*, 52(2):161–199, 1983. doi:10.7146/math.scand.a-12001.

[18] F. F. Knudsen. The projectivity of the moduli space of stable curves. III. The line bundles on $M_{g,n}$, and a proof of the projectivity of $\overline{M}_{g,n}$ in characteristic 0. *Math. Scand.*, 52(2):200–212, 1983. doi:10.7146/math.scand.a-12002.

[19] F. F. Knudsen and D. Mumford. The projectivity of the moduli space of stable curves. I. Preliminaries on "det" and "Div". *Math. Scand.*, 39(1):19–55, 1976. doi:10.7146/math.scand.a-11642.

[20] K. Kodaira. Holomorphic mappings of polydiscs into compact complex manifolds. *J. Differential Geom.*, 6:33–46, 1971/72.

[21] J. Kollár. Projectivity of complete moduli. *J. Differential Geom.*, 32(1):235–268, 1990.

[22] S. J. Kovács and Z. Patakfalvi. Projectivity of the moduli space of stable log-varieties and subadditivity of log-Kodaira dimension. *J. Amer. Math. Soc.*, 30(4):959–1021, 2017. doi:10.1090/jams/871.

[23] W. E. Lang. On the Euler number of algebraic surfaces in characteristic p. *Amer. J. Math.*, 102(3):511–516, 1980. doi:10.2307/2374113.

[24] A. Langer. On positivity and semistability of vector bundles in finite and mixed characteristics. *J. Ramanujan Math. Soc.*, 28A:287–309, 2013.

[25] A. Langer. Generic positivity and foliations in positive characteristic. *Adv. Math.*, 277:1–23, 2015. doi:10.1016/j.aim.2015.02.015.

[26] R. Lazarsfeld. *Positivity in algebraic geometry. I. Classical setting: line bundles and linear series*, volume 48 of *Ergeb. Math. Grenzgeb. (3)*. Springer, 2004. doi:10.1007/978-3-642-18808-4.

[27] R. Lazarsfeld. *Positivity in algebraic geometry. II. Positivity for vector bundles, and multiplier ideals*, volume 49 of *Ergeb. Math. Grenzgeb. (3)*. Springer, 2004. doi:10.1007/978-3-642-18808-4.

[28] D. Mumford. Varieties defined by quadratic equations. In *Questions on Algebraic Varieties*, pages 29–100. Edizioni Cremonese, 1970.

[29] D. Mumford. Stability of projective varieties. *Enseignement Math. (2)*, 23(1-2):39–110, 1977.

[30] The Stacks Project Authors. *The Stacks project*. https://stacks.math.columbia.edu.

[31] E. Viehweg. Weak positivity and the stability of certain Hilbert points. *Invent. Math.*, 96(3):639–667, 1989. doi:10.1007/BF01393700.

[32] A. Vistoli. Intersection theory on algebraic stacks and on their moduli spaces. *Invent. Math.*, 97(3):613–670, 1989. doi:10.1007/BF01388892.

[33] C. Xu and Z. Zhuang. On positivity of the CM line bundle on K-moduli spaces. *Ann. Math. (2)*, 192(3):1005–1068, 2020. doi:10.4007/annals.2020.192.3.7.

2

The stack of admissible covers is algebraic

Elsa Corniani
University of Ferrara

Neeraj Deshmukh
Universität Zürich

Brett Nasserden
Western University

Emanuel Reinecke
Max-Planck-Institut für Mathematik

Nawaz Sultani
University of Michigan

Rachel Webb
University of California, Berkeley

Abstract We show that the moduli stack of admissible G-covers of prestable curves is an algebraic stack, loosely following [1, Appendix B]. As preparation, we discuss finite group actions on algebraic spaces.

2.1 Introduction

One of the most studied objects in algebraic geometry is the moduli space $\mathcal{M}_g$ parametrizing smooth curves of genus g. From early on, there has been considerable interest in describing the finite étale covers of $\mathcal{M}_g$. Their understanding is not only of intrinsic interest, but also offers technical advantages. For example, while $\mathcal{M}_g$ is only smooth when considered as an algebraic stack, many of its finite étale covers are given by smooth *schemes*; in some cases, statements for $\mathcal{M}_g$ can then be reduced to scheme-theoretic statements for these covers.

Another important tool in the study of $\mathcal{M}_g$ is its compactification by the moduli stack $\overline{\mathcal{M}}_g$ of stable curves of genus g. Admissible G-covers were introduced in [1] to give modular interpretations for compactifications of many finite étale covers of $\mathcal{M}_g$. The authors showed that the stacks $\mathcal{A}dm_g(G)$ of admissible G-covers of curves of genus g are still smooth and the maps $\mathcal{A}dm_g(G) \to \overline{\mathcal{M}}_g$ are still proper and quasi-finite. Moreover, they constructed groups G for which $\mathscr{A}dm_g(G)$ is a smooth projective scheme which is finite over $\overline{\mathscr{M}}_g$. In this chapter, we give a detailed proof of one of the fundamental theorems of [1] in the more general setting of prestable curves, relying solely on results from the Stacks project.

Theorem A *$\mathscr{A}dm(G) = \bigsqcup \mathscr{A}dm_g(G)$ is an algebraic stack over* $\operatorname{Spec}(\mathbf{Z}[1/|G|])$.

As explained in [1, Section 4.2], the stack of admissible G-covers is also the normalization of (a generalization of) the stack of admissible covers studied initially in [7] and also in [11, 8]. A general reference for these covers and their associated moduli stacks is [5]. In the following sections, we define $\mathscr{A}dm(G)$ in full detail and then outline our proof of Theorem A.

Definition of admissible G-covers

An admissible G-cover is a finite cover $X \to Y$ of prestable curves that is generically a G-torsor and has prescribed behavior on the nodes and markings. Note that we will allow a family of prestable curves to have disconnected geometric fibers (Definition 2.2.4).

Definition 2.1.1 Let G be a finite group and let S be a scheme over $\operatorname{Spec}(\mathbf{Z}[1/|G|])$. Let $(Y \to S, \{q_i\}_{i=1}^n)$ be a marked connected prestable curve over S. An *admissible G-cover* is a finite morphism $f: X \to Y$ from a prestable curve X, together with a G-action on X which leaves f invariant, such that:

(i) Away from the nodes and markings of Y, f is a G-torsor.
(ii) The nodes of X map to nodes of Y.
(iii) For any geometric point $\bar{s}$: $\operatorname{Spec} k \to S$ and any node $x \in X_{\bar{s}}$, the map $f^{\mathrm{sh}}: \mathcal{O}^{\mathrm{sh}}_{Y_{\bar{s}}, f(x)} \to \mathcal{O}^{\mathrm{sh}}_{X_{\bar{s}}, x}$ can be identified with the strict henselization of the map

$$\left(k[u', v']/(u'v')\right)_{(u',v')} \to \left(k[u, v]/(uv)\right)_{(u,v)},$$
$$u' \mapsto u^e, \qquad v' \mapsto v^e$$

for some $e \in \mathbf{Z}_{\geq 1}$. Under this identification, the action of the stabilizer G_x on $\mathcal{O}^{\mathrm{sh}}_{X_{\bar{s}}, x}$ is *balanced*, i.e., given by $u \mapsto \xi(g)u$, $v \mapsto \xi^{-1}(g)v$ for some character $\xi: G_x \to k^\times$.
(iv) For any geometric point $\bar{s}$: $\operatorname{Spec} k \to S$ and any $x \in X_{\bar{s}}$ lying over a

marked point in $Y_{\bar{s}}$, the map $f^{\mathrm{sh}}\colon \mathscr{O}^{\mathrm{sh}}_{Y_{\bar{s}},\phi(x)} \to \mathscr{O}^{\mathrm{sh}}_{X_{\bar{s}},x}$ can be identified with the strict henselization of

$$k[v]_{(v)} \to k[u]_{(u)}, \quad v \mapsto u^e$$

for some $e \in \mathbf{Z}_{\geq 1}$. Under this identification, the action of the stabilizer G_x on $\mathscr{O}^{\mathrm{sh}}_{X_{\bar{s}},x}$ is given by $u \mapsto \xi(g)u$ for some character $\xi\colon G_x \to k^{\times}$.

Admissible G-covers form a category fibered in groupoids $\mathscr{A}dm(G)$; for a rigorous description, see Definition 2.3.30.

Remark 2.1.2 Our definition of an admissible G-cover differs slightly from that of [1, Definition 4.3.1] in that we require the descriptions (iii) and (iv) only in geometric fibers.

Remark 2.1.3 We follow [1, Definition 4.3.1] in allowing a prestable curve $X \to S$ to have disconnected geometric fibers. If one modifies Definition 2.1.1 to require $X \to S$ to be a prestable curve with geometrically connected fibers, the discussion in this chapter shows (with very minor changes) that the resulting moduli stack is algebraic.

Remark 2.1.4 In part (iii) of Definition 2.1.1 we have required G_x to act on u and v via inverse characters; that is, we required the action to be *balanced*. To motivate this requirement, recall from [10, Tag 0CBX] that if $f\colon X \to S$ is a morphism of schemes and a nodal curve and if $x \in X$ is a split node, then there is an isomorphism

$$\mathscr{O}^{\wedge}_{X,x} = \mathscr{O}^{\wedge}_{S,f(x)}[\![u,v]\!]/(uv - h)$$

for some $h \in \mathfrak{m}_{f(x)}$. If moreover G acts on X leaving $X \to S$ invariant, then since h comes from S it is G-invariant, so if u and v are weight vectors for the G_x-action they must have inverse weights. In other words, a smoothable split node must be balanced.

Remark 2.1.5 The analytically minded reader may replace conditions (iii) and (iv) on strict henselizations in Definition 2.1.1 by the corresponding conditions on completions. This will follow from Lemma 2.3.20 below.

Roadmap of the proof

One standard approach to proving a theorem like Theorem A is to use Artin's axioms [10, Tag 07SZ]. Hower, we will provide a proof that is logically independent of Artin's axioms by constructing $\mathscr{A}dm(G)$ as a locally closed substack of a certain Hom-stack $\mathfrak{H}_0$.

To construct $\mathfrak{H}_0$, we begin with the stacks $\mathfrak{M}$ and $\mathfrak{N}$ of connected prestable curves and (possibly disconnected) prestable curves, respectively. The algebraicity of these stacks was proved by Hall in [9, Appendix B] by finding explicit presentations in terms of Hilbert schemes of projective spaces. From these we construct universal curves $\mathfrak{C}_{\mathfrak{M}}$ and $\mathfrak{C}_{\mathfrak{N}}$, as well as the stack of marked connected prestable curves $\mathfrak{M}_\star$ (Definition 2.2.12), and we explain why they are algebraic (Lemmas 2.2.11 and 2.2.13).

To finish the construction of $\mathfrak{H}_0$, we construct the stack of prestable curves with balanced G-action $\mathfrak{N}^{\mathrm{bal}}$. We first define the stack $\mathfrak{N}(G)$ of prestable curves with G-action and exhibit it as a locally closed substack of the $|G|$-fold product $\prod_G \mathscr{I}$ of the inertia stack $\mathscr{I}$ of $\mathfrak{N}$ (Lemma 2.3.22). An analysis of the local picture at the nodes reveals that the balancedness of the G-action is an open condition on the base, so $\mathfrak{N}^{\mathrm{bal}}$ is an open substack of $\mathfrak{N}(G)$ and hence also algebraic (Lemma 2.3.25). Finally we set $\mathfrak{H}_0 := \underline{\mathrm{Hom}}_{\mathfrak{N}^{\mathrm{bal}} \times \mathfrak{M}_\star}(\mathfrak{C}_{\mathfrak{N}^{\mathrm{bal}}}, \mathfrak{C}_{\mathfrak{M}_\star})$. It is algebraic by a general argument about Hom-stacks (Lemma 2.2.14).

After constructing $\mathfrak{H}_0$, we realize $\mathscr{A}dm(G)$ as a locally closed substack in the following way. A general element of $\mathfrak{H}_0$ over a scheme S defines a morphism of prestable curves $f\colon X \to Y$ over S such that G acts on X and Y has markings. We prove that the family $f\colon X \to Y$ is in $\mathscr{A}dm(G)$ if and only if Y is the (coarse) quotient of X by G and f maps smooth G-fixed points to marked points (Theorem 2.3.40). Since these conditions are locally closed and closed (Lemmas 2.3.31, 2.3.32, and 2.3.33), respectively, Theorem A is proven.

Outline of the contents

Section 2.2 provides some of the prerequisites that are necessary in the proof of Theorem A and are interesting in their own right. Namely, in Section 2.2.1, we review and extend results of [10] on prestable curves. Then, we recall some general algebraicity criteria for stacks in Section 2.2.2. Finally, we explain many basic facts about finite group actions on algebraic spaces (Section 2.2.3), including coarse quotients by those actions (Section 2.2.4).

Section 2.3 contains a discussion of the objects of Theorem A and its proof. We begin in Section 2.3.1 by introducing balanced group actions on prestable curves and their complete local pictures. In Section 2.3.2, we study the deformation theory of these objects, which we use in Section 2.3.3 to show that balanced prestable G-curves form the algebraic stack $\mathfrak{N}^{\mathrm{bal}}$. We conclude our analysis of the substack $\mathcal{A}dm(G) \subset \mathfrak{H}_0$ as outlined above in Section 2.3.4.

Notation and conventions

We adopt the following notation and conventions, listed in order of appearance:

- For $x \in X$ a point of a scheme, $\mathcal{O}_{X,x}$ is the local ring of X at x, $\mathfrak{m}_x$ is the maximal ideal of $\mathcal{O}_{X,x}$ and $\kappa(x) := \mathcal{O}_{X,x}/\mathfrak{m}_x$ is the residue field of X at x.
- $\bar{k}$ is the algebraic closure of a field k.
- A geometric point of a scheme X is a map $\mathrm{Spec}(k) \to X$ from an algebraically closed field k ([10, Tag 0486]).
- $|X|$ is the topological space of an algebraic space ([10, Tag 03BY])
- $\mathrm{Hom}_T(X, Y)$ is the set of morphisms of algebraic spaces over T.
- f^{-1} will refer to the scheme-theoretic inverse image ([10, Tag 01JV]).
- Δ is the diagonal morphism.
- 1 is the identity element of a group. ([10, Tag 03BU]).
- $[X/G]$ denotes the stack quotient ([10, Tag 044Q]), while X/G denotes the coarse quotient (Definition 2.2.26).
- $\mathbf{Z}[1/|G|]$ is the smallest subring of $\mathbf{Q}$ containing $\mathbf{Z}$ and $1/|G|$.

2.2 Setup

2.2.1 Prestable curves

We recall some definitions leading up to that of a family of prestable curves. The following is from [10, Tags 0C47 and 0CBV].

Definition 2.2.1 Let X be a one-dimensional scheme locally of finite type over a field k.

(i) A closed point $x \in X$ is a *split node* if there exists an isomorphism $\mathcal{O}^{\wedge}_{X,x} \simeq k[\![u, v]\!]/(uv)$.

(ii) A closed point $x \in X$ is a *node* if there is a split node $\bar{x} \in X_{\bar{k}}$ mapping to x.

(iii) The scheme X has *at-worst-nodal singularities* if every closed point of X is either contained in the smooth locus of $X \to \mathrm{Spec}(k)$ or is a node of X.

Observe that by definition a split node of $X \to \mathrm{Spec}(k)$ has residue field k. By [10, Tag 0C4D], one may equivalently demand in Definition 2.2.1(ii) that all $\bar{x} \in X_{\bar{k}}$ mapping to x are split nodes.

We will need some properties of (split) nodes beyond what is currently in the Stacks project, so we write these out in Lemmas 2.2.2 and 2.2.3 below. The proofs of these lemmas both use the very general statement of Lemma 2.2.9. Since Lemma 2.2.9 applies to more schemes than just nodal curves we defer it to the end of this subsection.

Lemma 2.2.2 *Let X be a one-dimensional scheme locally of finite type over a field k, let $x \in X$ be a split node, and let K/k be a field extension. Then there is a unique $y \in X_K$ mapping to x and y is a split node.*

Proof We note that the lemma does not immediately follow from Lemma 2.2.9 because our field extension K/k may not be finite. However, since $x \in X$ is k-rational, we can at least use Lemma 2.2.9 to say that there is a unique point $y \in X_K$ mapping to x and that y has residue field K. By [10, Tag 0CBW], the fiber of the normalization $X^\nu \to X$ over x has precisely two points. By [10, Tag 0C3N] and surjectivity of the map $\mathrm{Spec}(K) \to \mathrm{Spec}(k)$, the fiber of $X^\nu_K \to X_K$ over y has at least two points, and by [10, Tag 0CBT] and [10, Tag 0C56] it has exactly two points. By another application of [10, Tag 0CBW], the node y is split. □

General nodes have a complete local description similar to that of split nodes.

Lemma 2.2.3 *Let X be a one-dimensional scheme locally of finite type over a field k and let $x \in X$ be a closed point. The following are equivalent:*

(i) *The point x is a node.*
(ii) *The extension $\kappa(x)/k$ is separable, and $\mathcal{O}^\wedge_{X,x} \simeq \kappa(x)[[u, v]]/(q)$ for some nondegenerate quadratic form $q = au^2 + buv + cv^2$ over $\kappa(x)$.*
(iii) *There exists a finite separable field extension κ' of k and a split node $y \in X_{\kappa'}$ mapping to x.*

Proof The equivalence of (i) and (ii) is shown in [10, Tag 0C4D], and that (iii) implies (i) is immediate from the definitions and Lemma 2.2.2.

Now assume (ii). Set $\kappa := \kappa(x)$ and set

$$\kappa' := \begin{cases} \kappa & \text{if } q \text{ is a split quadratic form,} \\ \kappa[t]/(a + bt + ct^2) & \text{if } q \text{ is not a split quadratic form.} \end{cases}$$

The latter is the splitting field of q; when q is not split over κ, it factors over κ' as $q = (v - tu)(cv + (b + ct)u)$. Since κ is finite separable over k and q is nondegenerate, $\kappa'/\kappa/k$ must be a tower of finite separable field extensions.

First, we construct a closed point $\tilde{x} \in X_\kappa(\kappa)$ lying over x. By the primitive element theorem [10, Tag 030N], there is a polynomial $P \in k[t]$ such that $\kappa = k[t]/(P)$. Then $\kappa \otimes_k \kappa \simeq \kappa[t]/(P)$ has a factor κ because P has a zero in κ. The prime ideal given as the kernel of the associated surjection $\kappa \otimes_k \kappa \twoheadrightarrow \kappa$ determines the point $\tilde{x}$ over x with residue field κ [10, Tag 01JT]. By [10, Tag 0C50, Tag 0AGX], we have an isomorphism $\mathcal{O}^\wedge_{X,x} \simeq \mathcal{O}^\wedge_{X_\kappa,\tilde{x}}$. Finally, by Lemma 2.2.9, there is a unique $y \in X_{\kappa'}$ with residue field κ' and complete local ring

$$\begin{aligned}\mathcal{O}^\wedge_{X_{\kappa'},y} &\simeq \mathcal{O}^\wedge_{X_\kappa,\tilde{x}} \otimes_\kappa \kappa' \simeq \mathcal{O}^\wedge_{X,x} \otimes_\kappa \kappa' \simeq \kappa[\![u,v]\!]/(q) \otimes_\kappa \kappa' \\ &\simeq \kappa'[\![u,v]\!]/(q) \simeq \kappa'[\![u,v]\!]/(uv)\end{aligned}$$

of the desired form. □

Definition 2.2.4 A morphism $\pi\colon X \to S$ from an algebraic space X to a scheme S is a *nodal curve* if π is flat and of finite presentation and every geometric fiber has pure dimension 1 and at-worst-nodal singularities. We say π is a *connected prestable curve* if in addition π is proper and every geometric fiber is connected. A *prestable curve* is a finite disjoint union $\bigsqcup_i X_i \to S$ where each $X_i \to S$ is a connected prestable curve.

Remark 2.2.5 Definition 2.2.4 is a compilation of [10, Tags 0D4Z, 0DSE, 0C54, 0C56, and 0E6S] with some important differences in nomenclature:

- A nodal curve in our sense is a morphism that is *at-worst-nodal of relative dimension* 1 in the sense of the Stacks project [10, Tag 0C5A]; this recovers the notion of nodal curves from [10, Tag 0C46] when S is the spectrum of a field.
- A prestable curve in our sense is a *nodal family of curves* in the sense of the Stacks project [10, Tag 0DSX] (noting that families of curves are by definition always proper in the Stacks project [10, Tag 0D4Z])
- A connected prestable curve in our sense is a prestable curve in the sense of the Stacks project [10, Tag 0E6T].

In the next definition, recall from [10, Tag 03BU] that $|X|$ is the set of points of an algebraic space, and elements of this set are equivalence classes of morphisms from fields.

Definition 2.2.6 Let $\pi: X \to S$ be a nodal curve. A point $x \in |X|$ is a *node in its fiber* if it is the image of a node in $X_{\pi(x)}$.

Definition 2.2.7 If $\pi: X \to S$ is a nodal curve, the *smooth locus* $X^{\mathrm{sm}} \subset X$ is the maximal open subspace such that $\pi|_{X^{\mathrm{sm}}}: X^{\mathrm{sm}} \to S$ is smooth (see [10, Tag 0DZI]).

Remark 2.2.8 If $X \to S$ is a nodal curve, the smooth locus $X^{\mathrm{sm}} \to S$ is precisely the complement of the points in X that are nodes in their fibers. In particular, X^{sm} commutes with arbitrary base change by [10, Tag 0C56].

Define $\mathfrak{M}$ and $\mathfrak{N}$ to be the moduli stacks over Sch_{fppf} (see [10, Tag 021R]) whose groupoid fibers over a scheme S are given by

$$\mathfrak{M}(S) = \{\text{connected prestable curves } X \to S\},$$
$$\mathfrak{N}(S) = \{\text{prestable curves } X \to S\}.$$

By [10, Tag 0D5A] and [10, Tag 0DSQ] there is a larger moduli stack $Curves$ over Sch_{fppf} with a diagonal that is separated and of finite presentation. By [10, Tag 0E6U] and [10, Tag 0DSY], respectively, the stacks $\mathfrak{M}$ and $\mathfrak{N}$ are open substacks of $Curves$, and hence are algebraic stacks with diagonals that are separated and of finite presentation. The proof of algebraicity in [10, Tag 0D5A] uses Artin's axioms; for a proof using only Hilbert schemes of projective spaces, see [9, Appendix B].

We use the following general lemma to prove Lemmas 2.2.2 and 2.2.3 earlier in this section.

Lemma 2.2.9 *Let X be a k-scheme of locally finite type. Let K/k be a field extension and let $x \in X$ be a rational point. Then there is a unique point $y \in X_K$ lying over x with residue field $\kappa(y) = K$. Moreover, if K/k is finite, then $\mathcal{O}_{X,x} \otimes_k K \simeq \mathcal{O}_{X_K,y}$ and $\mathcal{O}^{\wedge}_{X,x} \otimes_k K \simeq \mathcal{O}^{\wedge}_{X_K,y}$.*

Proof By [10, 01JT], y corresponds to the unique prime ideal in $k \otimes_k K \simeq K$ which has residue field equal to K.

Now assume K/k is a finite extension. By [10, Tag 0C4Y], we have $\mathcal{O}_{X_K,y} = (\mathcal{O}_{X,x} \otimes_k K)_{\mathfrak{p}}$, where $\mathfrak{p}$ is the kernel of $\mathcal{O}_{X,x} \otimes_k K \to \kappa(x) \otimes_k K \simeq K$. Since K is flat over k, there is an exact sequence

$$0 \to \mathfrak{m} \otimes_k K \to \mathcal{O}_{X,x} \otimes_k K \to K \to 0.$$

This shows that $\mathfrak{p} = \mathfrak{m} \otimes_k K$ and that this ideal is maximal.

In fact, we claim that it is the unique maximal ideal, so $(\mathcal{O}_{X,x} \otimes_k K)_{\mathfrak{p}} \simeq \mathcal{O}_{X,x} \otimes_k K$ as desired. To see this, let $\mathfrak{n} \subset \mathcal{O}_{X,x} \otimes_k K$ be a maximal ideal. Since K/k is finite, $\mathcal{O}_{X,x} \otimes_k k \subset \mathcal{O}_{X,x} \otimes_k K$ is integral, so $(\mathcal{O}_{X,x} \otimes_k k)/(\mathfrak{n} \cap \mathcal{O}_{X,x} \otimes_k k) \to (\mathcal{O}_{X,x} \otimes_k K)/\mathfrak{n}$ is also integral. However, $(\mathcal{O}_{X,x} \otimes_k K)/\mathfrak{n}$ is a field, so [10, Tag 00GR] implies that $(\mathcal{O}_{X,x} \otimes_k k)/(\mathfrak{n} \cap \mathcal{O}_{X,x} \otimes_k k)$ is a field as well. Hence $\mathfrak{n} \cap \mathcal{O}_{X,x} \otimes_k k$ is maximal and thus equals $\mathfrak{m} \otimes_k k$. In particular, $\mathfrak{n}$ contains $\mathfrak{m} \otimes_k k$, so $\mathfrak{n}$ contains $\mathfrak{m} \otimes_k K$, so $\mathfrak{n}$ must equal $\mathfrak{m} \otimes_k K$ because we already know the latter is maximal.

Finally, the natural map $\mathcal{O}^{\wedge}_{X,x} \otimes_k K \to (\mathcal{O}_{X,x} \otimes_k K)^{\wedge}$ is an isomorphism by [10, Tag 00MA(3)]. □

2.2.2 Some algebraic stacks

We will use the following lemma to construct algebraic stacks. Let S be a scheme in Sch_{fppf} (see [10, Tag 021R]).

Lemma 2.2.10 *Let $\mathcal{Y} \to (Sch/S)_{\mathrm{fppf}}$ and $\mathcal{X} \to \mathcal{Y}$ be categories fibered in groupoids, and assume $\mathcal{X} \to \mathcal{Y}$ is representable by algebraic spaces. If $\mathcal{Y}$ is an algebraic stack then so is $\mathcal{X}$.*

Proof This is found in [10, Tag 09WW] and [10, Tag 05UN]. □

Define the *universal curve* $\mathfrak{C} \to \mathfrak{M}$ to be the category whose groupoid fiber over an object $S \in Sch_{\mathrm{fppf}}$ is

$$\mathfrak{C}(S) := \{(X \to S, q) \mid (X \to S) \in \mathfrak{M}(S) \text{ and } q \in \mathrm{Hom}_S(S, X)\}$$

together with the apparent forgetful morphism to $\mathfrak{M}$. The universal curve $\mathfrak{C} \to \mathfrak{N}$ is defined similarly.

Lemma 2.2.11 *The morphisms $\mathfrak{C} \to \mathfrak{M}$ and $\mathfrak{C} \to \mathfrak{N}$ are representable morphisms of algebraic stacks.*

Proof We prove the statement for $\mathfrak{C} \to \mathfrak{M}$; the argument for $\mathfrak{C} \to \mathfrak{N}$ is identical. Let $T \to \mathfrak{M}$ be a map from a scheme T corresponding to a connected prestable curve $X \to T$. By Lemma 2.2.10 it suffices to show that $T \times_{\mathfrak{M}} \mathfrak{C}$ is representable. Using the construction of [10, Tag 0040], objects of $T \times_{\mathfrak{M}} \mathfrak{C}$ over a scheme S are given by quadruples

$$(S, S \to T, (Y \to S, f \in \mathrm{Hom}_S(S, Y)), \phi\colon Y \simeq S \times_T X)$$

where $Y \to S$ is a connected prestable curve and arrows are unique when they exist. The data of such a quadruple is equivalent to an element of $\mathrm{Hom}_T(S, X)$, so $X \to T$ represents $T \times_{\mathfrak{M}} \mathfrak{C}$. □

If $\pi \colon X \to S$ is a connected prestable curve, let $X^{\mathrm{sm}} \subset X$ be the open subspace where π is smooth. We will use this notation even when S is an algebraic stack (see [10, Tag 0DZR]). Define $\mathfrak{M}_n \to Sch_{\mathrm{fppf}}$ to be the category fibered in groupoids whose objects over S are

$$\mathfrak{M}_n(S) = \left\{ (X \to S, q_1, \dots, q_n) \;\middle|\; \begin{array}{l} (X \to S) \in \mathfrak{M}, \\ q_i \in \mathrm{Hom}_S(S, X^{\mathrm{sm}}) \\ \text{are disjoint sections} \end{array} \right\}$$

Definition 2.2.12 A *marked connected prestable curve* is an object of $\mathfrak{M}_n$ for some $n \geq 0$. We set $\mathfrak{M}_\star = \bigsqcup_{n \in \mathbf{Z}_{\geq 0}} \mathfrak{M}_n$.

Lemma 2.2.13 *The category $\mathfrak{M}_\star$ is an algebraic stack.*

Proof We prove that $\mathfrak{M}_n$ is an algebraic stack. Let $\mathfrak{C}_n$ denote the n-fold fiber product of $\mathfrak{C}$ over $\mathfrak{M}$. The category $\mathfrak{M}_n$ naturally a full subcategory of $\mathfrak{C}_n$. We show that the inclusion $\mathfrak{M}_n \to \mathfrak{C}_n$ is representable and apply Lemma 2.2.10.

Let $T \to \mathfrak{C}_n$ be an arbitrary map from a scheme, and let $(X \to T, q_i)$ be the corresponding prestable curve and sections. The fiber product $\mathfrak{M}_n \times_{\mathfrak{C}_n} T$ is the full subcategory of T whose objects over S are maps $f \colon S \to T$ such that the pullback $(X_S \to S, q_i : S \to X_S)$ of $(X \to T, q_i)$ is an object of $\mathfrak{M}_n$; that is, the q_i are pairwise disjoint and land in $(X_S)^{\mathrm{sm}}$. Let $U_1 \subset T$ be the intersection of the finitely many open sets $q_i^{-1}(X^{\mathrm{sm}})$. Since $X \to T$ is separated, the equalizer of any two of the q_i is a closed subspace (see e.g. [10, Tag 01KM]); let U_2 be the open subspace of T equal to the complement of the union of the equalizers. Now $U_1 \cap U_2$ is an open subscheme of T representing $\mathfrak{M}_n \times_{\mathfrak{C}_n} T$. □

Let $\mathscr{X} \to \mathscr{Z}$ and $\mathscr{Y} \to \mathscr{Z}$ be representable morphisms of algebraic stacks. The Hom stack $\underline{\mathrm{Hom}}_{\mathscr{Z}}(\mathscr{X}, \mathscr{Y})$ is the category fibered in groupoids over $\mathscr{Z}$ whose objects over a scheme $T \to \mathscr{Z}$ are

$$\underline{\mathrm{Hom}}_{\mathscr{Z}}(\mathscr{X}, \mathscr{Y})(T) = \mathrm{Hom}_T(\mathscr{X}_T, \mathscr{Y}_T).$$

Lemma 2.2.14 *If $\mathscr{X} \to \mathscr{Z}$ is representable, of finite presentation, flat, and proper, and $\mathscr{Y} \to \mathscr{Z}$ is representable, of finite presentation, and separated, then $\underline{\mathrm{Hom}}_{\mathscr{Z}}(\mathscr{X}, \mathscr{Y})$ is an algebraic stack.*

Proof Let $T \to \mathscr{Z}$ be a scheme. By [10, Tag 0D1C], the fiber product $T \times_{\mathscr{Z}} \underline{\mathrm{Hom}}_{\mathscr{Z}}(\mathscr{X}, \mathscr{Y})$ is representable by an algebraic space, so the lemma follows from Lemma 2.2.10. □

Remark 2.2.15 The proof of representability in [10, Tag 0D1C] uses

Artin's axioms. If $\mathscr{X} \to \mathscr{Y}$ and $\mathscr{Y} \to \mathscr{Z}$ are families of curves (as they will be in our applications), one may alternatively use an étale base change of T to reduce to the case where the families are H-projective ([10, Tag 0E6F]), and then use the more classical [3, Theorem 2.6].

2.2.3 Finite group actions on algebraic spaces

Let G be a finite group and let S be a scheme. A *G-space (resp. scheme) over S* is an algebraic space (resp. scheme) $X \to S$ equipped with a group homomorphism $G \to \mathrm{Aut}_S(X)$. If $X \to S$ is a morphism of affine schemes $\mathrm{Spec}(A) \to \mathrm{Spec}(B)$, then it is equivalent to giving a homomorphism from G to the automorphisms of A as a B-algebra.

Definition 2.2.16 For $g \in G$ the *fixed locus* X^g is the fiber product of the diagram

$$\begin{array}{ccc} X^g & \longrightarrow & X \\ \downarrow & & \downarrow \Delta \\ X & \xrightarrow{(g,id)} & X \times_S X \end{array} \tag{1}$$

If X is separated, then X^g is a closed subspace of X. We define the fixed locus of G to be $X^G = \bigcap_{g \in G} X^g$.

For any scheme T, G acts on $X(T)$ and we have $X^g(T) = (X(T))^g$, where the right-hand side is the subset fixed by g.

Remark 2.2.17 If $X \to S$ and $Y \to S$ are G-spaces over S, then the fiber product $X \times_S Y$ has a unique structure of a G-space over S such that the projection maps to X and Y are G-equivariant. In terms of functors of points, this action is given by the diagonal action of G on $(X \times_S Y)(T) = X(T) \times_{S(T)} Y(T)$.

Lemma 2.2.18 *Let X be a G-space over S. If $S' \to S$ is an S-scheme with trivial G-action then $X^g \times_S S' \simeq (X \times_S S')^g$ for all $g \in G$.*

Proof We argue using functors of points. If $T \to S'$ is a scheme, then an element of $(X \times_S Y)^g$ is given by a pair $(\phi_x, \phi_y) \in X(T) \times_{S(T)} S'(T)$ such that $g \circ \phi_x = \phi_x$ and $g \circ \phi_y = \phi_y$. An element of $X^g \times_S S'$ is a pair $(\phi_x, \phi_y) \in X(T) \times_{S(T)} S'(T)$ such that $g \circ \phi_x = \phi_x$. Since G acts trivially on S' these are the same set. □

In the next lemma, recall that $|X|$ is the set of points of an algebraic space [10, Tag 03BU], and elements of this set are equivalence classes of morphisms from fields.

Lemma 2.2.19 *Let X be a G-space over S. Let $x \in |X|$ and $g \in G$. The following are equivalent:*

(i) *For every field K and morphism $\alpha\colon \operatorname{Spec}(K) \to X$ representing x, we have $g \circ \alpha = \alpha$.*
(ii) *For some field K and morphism $\alpha\colon \operatorname{Spec}(K) \to X$ representing x, we have $g \circ \alpha = \alpha$.*
(iii) *We have $x \in |X^g|$.*

The subset of elements $g \in G$ satisfying conditions (i)–(iii) *is a subgroup of G.*

Proof For the equivalence of (i) and (ii), observe that if Ω/L is a field extension then the induced map $\beta\colon \operatorname{Spec}(\Omega) \to \operatorname{Spec}(L)$ is an epimorphism. So if $\gamma\colon \operatorname{Spec}(L) \to X$ represents x, we have $g \circ \gamma = \gamma$ if and only if $g \circ \gamma \circ \beta = \gamma \circ \beta$. This means that the representative γ is fixed by g if and only if the representative $\gamma \circ \beta$ is fixed by g. For the equivalence of (ii) and (iii), use (1). For the last assertion, it is clear that the set of elements satisfying (ii) is a subgroup. □

Definition 2.2.20 If X is a G-space over S, the *stabilizer* of a point $x \in |X|$ is the subgroup $G_x \subset G$ of elements satisfying the equivalent conditions of Lemma 2.2.19.

The next definition is a generalization of the fact that, for any open subspace U of a G-space X, there is a natural G-action on $\cap_{g\in G}(g \cdot U)$.

Definition 2.2.21 Let X be a G-space over S and let $\mu\colon U \to X$ be a morphism of algebraic spaces. Define μ_g to be the composition $g \circ \mu\colon U \to X$. The *equivariantization* of U, denoted $\mu^{\mathrm{e}}\colon U^{\mathrm{e}} \to X$, is the fiber product $U \times_X \cdots \times_X U \to X$ of all the μ_g (see [10, Tag 02XC]).

Lemma 2.2.22 *Let X be a G-space over S and let $\mu\colon U \to X$ be a morphism of algebraic spaces. Then U^{e} has a natural G-action making μ^e equivariant. Furthermore, if μ is an open immersion or étale, then so is μ^{e}.*

Note that the equivariantization U^{e} is nonempty if the image of μ has a fixed point of X.

Proof Let $T \to X$ be an X-scheme. An element $u = (u_g)_{g\in G} \in U^{\mathrm{e}}(T) \subseteq \prod_{g\in G} U(T)$ is given by morphisms of X-schemes $u_g\colon T \to U$ such that $g \circ \mu \circ u_g = \mu \circ u_1$ for all $g \in G$, where 1 denotes the identity of G. Given $h \in G$, set $h \cdot (u_g)_{g\in G} := (u_{h^{-1}g})_{g\in G}$. One can check that this defines the desired action. The last assertion of the lemma follows from the stability of open immersions and étale morphisms under pullback. □

Example 2.2.23 Recall from [10, Tag 02LE] that if X is a scheme, an elementary étale neighborhood $\mu\colon (U,u) \to (X,x)$ is an étale morphism $\mu\colon U \to X$ of schemes such that $\mu(u) = x$ and the induced morphism $\kappa(x) \to \kappa(u)$ on residue fields is an isomorphism. If X is moreover a G-space over S and $x \in X^G$ with an elementary étale neighborhood $\mu\colon (U,u) \to (X,x)$, then define

$$u^{\mathrm{e}} \in U^{\mathrm{e}} \quad \text{by} \quad (u^{\mathrm{e}})_g := u. \tag{2}$$

Note that $u^{\mathrm{e}} \in (U^{\mathrm{e}})^G$ and that $\mu^{\mathrm{e}}\colon (U^{\mathrm{e}}, u^{\mathrm{e}}) \to (X,x)$ is an elementary étale neighborhood by Lemma 2.2.22 and [10, Tag 01JT].

Lemma 2.2.24 *Let X be a G-space over S and let $\mu\colon U \to X$ be an open morphism of algebraic spaces. Any $x \in |X^G \cap \mu(U)|$ has an open neighborhood V such that for any $h \in G$ the restriction of μ^{e} to $\left((\mu^{\mathrm{e}})^{-1}(V)\right)^h \to V^h$ is surjective.*

Proof Let $V := \bigcap_{g\in G}(g \circ \mu)(U)$; this is an open neighborhood of x. The equivariant map μ^{e} necessarily sends $\left((\mu^{\mathrm{e}})^{-1}(V)\right)^h$ to V^h. Let $v \in V^h$, and let $g_1, \dots, g_r \in G$ be a set of right coset representatives for the subgroup $\langle h \rangle$ of G and write $[g_i]$ for the coset of g_i. Choose $u_{[g_i]} \in U$ such that $\mu(u_{[g_i]}) = g_i^{-1}v$ and define $(u_g)_{g\in G}$ by $u_g := u_{[g]}$. Since $v \in V^h$, one can check that $g\mu(u_{[g]}) = \mu(u_{[1]}) = v$ and hence $(u_g)_{g\in G}$ defines a point of U^{e} mapping to v. Likewise we compute

$$(h \cdot (u_g)_{g\in G})_g = u_{h^{-1}g} = u_{[h^{-1}g]} = u_{[g]} = u_g. \qquad \square$$

Lemma 2.2.25 *Let $\mu\colon U \to X$ be an equivariant morphism of G-algebraic spaces over S. If μ is a monomorphism in the category of algebraic spaces over S, then the equivariantization U^{e} of U is canonically isomorphic to U.*

Proof For any $g \in G$, let $\mathrm{pr}_g\colon U^{\mathrm{e}} \to U$ be the projection onto the copy of U whose given map to X is μ_g; note that $\mu_g \circ \mathrm{pr}_g = \mu \circ \mathrm{pr}_1$ where $1 \in G$ is the identity. Let $\widetilde{\Delta}\colon U \to U^{\mathrm{e}}$ be the unique map such that $\mathrm{pr}_g \circ \widetilde{\Delta} = g^{-1}$ for all $g \in G$. Clearly, $\mathrm{pr}_1 \circ \widetilde{\Delta}$ is the identity on U. Conversely, $\widetilde{\Delta} \circ \mathrm{pr}_1$ is the identity on U^{e} if and only if $\mathrm{pr}_g \circ \widetilde{\Delta} \circ \mathrm{pr}_1 = \mathrm{pr}_g$ for each $g \in G$, and this holds if and only if

$$\mu_g \circ \mathrm{pr}_g \circ \widetilde{\Delta} \circ \mathrm{pr}_1 = \mu_g \circ \mathrm{pr}_g \qquad \text{for each } g \in G \tag{3}$$

since μ (and hence μ_g) is a monomorphism. Using the identities $\mu_g \circ \mathrm{pr}_g = \mu \circ \mathrm{pr}_1$ and $\mathrm{pr}_1 \circ \widetilde{\Delta} = id_U$, we see that (3) holds. □

2.2.4 Quotients by finite group actions on algebraic spaces

Let G be a finite group acting on an algebraic space X over a base scheme S.

Definition 2.2.26 The *coarse quotient* of X by G is a morphism $X \to X/G$ to an algebraic space X/G with the universal property that every G-invariant morphism $X \to Y$ of algebraic spaces factors uniquely through $X \to X/G$.

The universal property of X/G implies that it is canonically an algebraic space over S and that if X/G exists, it is unique up to unique isomorphism.

Lemma 2.2.27 *If X is a G-space over S such that $X \to S$ is separated and locally of finite type, then X/G exists and its formation commutes with flat base change on S.*

The construction of X/G and the proof of Lemma 2.2.27 use the quotient stack $[X/G]$. This stack is defined in [10, Tag 044Q(2)] to be the quotient stack associated to a certain groupoid [10, Tag 0444] in algebraic spaces, which in particular has $U = X$ and relations $R = G \times X$. Note that the definition of $[X/G]$ is independent of the base S (notated B in [10, Tag 044Q(2)]). One may also describe $[X/G]$ as the moduli stack parametrizing G-torsors together with maps to X (see [10, Tag 04UV]).

Remark 2.2.28 The formation of $[X/G]$ commutes with arbitrary base change on S. More precisely, let $S' \to S$ be a map of schemes and $X' := X \times_S S'$; we claim that the natural 1-morphism $[X'/G] \to [X/G] \times_S S'$ is an equivalence. Indeed, $X' = X \times_S S' \to [X/G] \times_S S'$ is a smooth cover, and the associated relations R (in the sense of [10, Tag 04T4]) are precisely the pullback of the relations associated with the smooth cover $X \to [X/G]$, which in turn are given by $G \times X$ with the maps defining the group action (using [10, Tag 04M9]). So a computation shows that R is $G \times X'$ with the maps defining the G-action on X'. Now use [10, Tag 04T4, Tag 04T5] to conclude that $[X'/G] \to [X/G] \times_S S'$ is an equivalence.

Proof of Lemma 2.2.27 We consider the stack quotient $\mathscr{X} := [X/G]$ as in [10, Tag 044Q(2)], noting that this operation commutes with arbitrary base change (as follows from Remark 2.2.28) and then observe that our definition of X/G agrees with the definition of a categorical moduli space for $\mathcal{X}$ in the sense of [10, Tag 0DUG] (this need not exist, in general). So it suffices to show that we can apply the Keel–Mori theorem [10, Tag 0DUT]; i.e., we need to show that $\mathcal{X}$ has finite inertia.[1] From the description of the inertia stack as the fiber product of diagonal morphisms [10, Tag 034H], it suffices to show that the diagonal $\Delta\colon \mathcal{X} \to \mathcal{X} \times_S \mathcal{X}$ is finite.

By [10, Tag 02XE] and [10, Tag 04M9] there is a fiber square

$$\begin{array}{ccc} G \times X & \longrightarrow & X \times_S X \\ \downarrow & & \downarrow \\ \mathcal{X} & \xrightarrow{\Delta} & \mathcal{X} \times_S \mathcal{X} \end{array}$$

where the top horizontal morphism is given by the action map $(g,x) \mapsto (g \cdot x, x)$. By [10, Tag 04X0] the projection $X \to \mathcal{X}$ is smooth, hence $X \times_S X \to \mathcal{X} \times_S \mathcal{X}$ is also smooth since it factors as a composition of smooth maps $X \times_S X \to X \times_S \mathcal{X} \to \mathcal{X} \times_S \mathcal{X}$. By [10, Tag 04XD], we have reduced the proof to showing that $G \times X \to X \times_S X$ is finite, or equivalently that it is proper and locally quasi-finite ([10, Tag 0A4X]). This follows if the composition

$$G \times X \to X \times_S X \xrightarrow{\mathrm{pr}_2} X$$

is proper and locally quasi-finite (this uses [10, Tags 04NX and 03XN] and the fact that $X \to S$ is separated). These properties hold since the composition is just $\mathrm{pr}_2\colon G \times X \to X$ and G is finite. □

Lemma 2.2.29 *If X is a G-space over S such that $X \to S$ is separated and locally of finite type, then $X \to X/G$ is finite and surjective.*

Proof We recall from the proof of Lemma 2.2.27 the factorization

$$X \to [X/G] \to X/G \tag{4}$$

of the map $X \to X/G$. We will show that each of the maps in (4) is proper, locally quasi-finite, and surjective, hence the composition is as well, and by [10, Tag 0A4X] it is finite and surjective.

[1] Note that the final assertion of Lemma 2.2.27, namely that X/G commutes with flat base change, also follows from the Keel–Mori theorem [10, Tag 0DUT]. This is so because the cited tag asserts that X/G exists and is a *uniform* categorical moduli space, meaning that its formation commutes with flat base change.

By [10, Tag 04M9] there is a fiber square

$$\begin{array}{ccc} G \times X & \xrightarrow{\mathrm{pr}_2} & X \\ \downarrow & & \downarrow \\ X & \longrightarrow & [X/G], \end{array}$$

where the left vertical arrow is given by the action of G on X. By [10, Tag 04X0], the (vertical) map $X \to [X/G]$ is smooth, so by [10, Tag 04XD] the (horizontal) map $X \to [X/G]$ is proper, locally quasi-finite, and surjective since $\mathrm{pr}_2 \colon G \times X \to X$ has those properties.

Since $X \to S$ is locally of finite type, the morphism $[X/G] \to S$ is also locally of finite type, and hence $[X/G] \to X/G$ is as well (using [10, Tag 06FM] and [10, Tag 06U9], respectively). By the Keel–Mori theorem [10, Tag 0DUT], the morphism $[X/G] \to X/G$ is also separated, quasi-compact, and a universal homeomorphism, hence proper and surjective. Since $[X/G] \to X/G$ is a universal homeomorphism it has discrete fibers, so by [10, Tag 06RW] it is locally quasi-finite. □

Lemma 2.2.30 *Let X be a G-space over S such that $X \to S$ is separated and locally of finite type. Let* $\operatorname{Spec} C \to X/G$ *be an étale morphism from an affine scheme. Then* $X \times_{X/G} \operatorname{Spec} C \simeq \operatorname{Spec} B$ *is an affine G-scheme étale over X, and the projection* $\operatorname{Spec} B \to \operatorname{Spec} C$ *is induced by the inclusion of the invariant ring* $C \simeq B^G \hookrightarrow B$.

Proof By Lemma 2.2.29 the map $X \to X/G$ is finite, hence affine, so the fiber product $X \times_{X/G} \operatorname{Spec} C$ is isomorphic to an affine scheme $\operatorname{Spec} B$. By Remark 2.2.17, the ring B has a G-action and $C \to B$ factors through the invariant subring B^G. In fact, by Remark 2.2.28 with $S = X/G$ and $S' = \operatorname{Spec} C$, we have

$$[X/G] \times_{X/G} \operatorname{Spec} C = [(\operatorname{Spec} B)/G]. \tag{5}$$

Since the formation of the coarse space commutes with flat base change ([10, Tag 0DUT]), the map $[(\operatorname{Spec} B)/G] \to \operatorname{Spec} C$ is a coarse moduli space. In the parlance of [10, Tag 0DUK], the stack $[(\operatorname{Spec} B)/G]$ is well-nigh affine, so its coarse moduli space is constructed in [10, Tag 0DUP]; as explained there, $C \hookrightarrow B^G$ is an isomorphism. □

The following is a special case of [2, Corollary 3.3].

Lemma 2.2.31 *Let X be a G-space over S such that $X \to S$ is separated and locally of finite type, where S is a scheme over* $\operatorname{Spec}(\mathbf{Z}[1/|G|])$.

(i) *The formation of X/G commutes with arbitrary base change on S. That is, for any morphism of schemes $T \to S$ the natural map $(X \times_S T)/G \to (X/G) \times_S T$ is an isomorphism.*
(ii) *If $X \to S$ is flat, then $X/G \to S$ is flat.*

Proof If A is a G-module on which $|G| \in \mathbf{Z}$ acts as a unit (so A is also a $\mathbf{Z}[1/|G|]$-module), then we may define the Reynolds operator $R_A := (1/|G|) \sum_{g \in G} g \colon A \to A^G$, a projection onto the submodule $A^G \subset A$. The existence of this operator has two important applications: (a) taking G-invariants is an exact functor from $\mathbf{Z}[1/|G|]$-modules with G-action to $\mathbf{Z}[1/|G|]$-modules (use the functoriality of R_A), and (b) if R is an algebra with a homomorphism $R \to A^G$, then for any R-module N the inclusion $A^G \otimes_R N \subset (A \otimes_R N)^G$ is an equality, where G acts on $A \otimes_R N$ by the rule $g \cdot (a \otimes n) = (g \cdot a) \otimes n$ (show that $R_A \otimes_R \mathrm{id}_N = R_{A \otimes_R N}$).

To prove (i) we may take $S = X/G$. Since the property of being an isomorphism is étale local on the target [10, Tag 041Y] and the formation of X/G commutes with flat base change, we reduce to the case when $S = \operatorname{Spec} B$, $T = \operatorname{Spec} B'$, and hence by Lemma 2.2.30 we have $X = \operatorname{Spec} A$ with $B = A^G$. Now $A^G \otimes_B B' \to (A \otimes_B B')^G$ is an isomorphism by (b) above.

For (ii) we may check flatness locally, so assume that $S = \operatorname{Spec} B$ and $X = \operatorname{Spec} A$ are affine. By (b) above, the functor $A^G \otimes_B (-)$ is isomorphic to $(A \otimes_R (-))^G$, but $A \otimes_B (-)$ and taking G-invariants are both exact functors, by assumption and (a) above. □

The next lemma is similar to [10, Tag 0DUZ] but we remove the hypothesis that the base is locally Noetherian.

Lemma 2.2.32 *Let X be a G-space over a $\mathbf{Z}[1/|G|]$-scheme S. If $f \colon X \to S$ is proper and of finite presentation, then so is $X/G \to S$.*

Proof Since the properties of being finitely presented and proper are local on the base ([10, Tag 041V] and [10, Tag 0422]), we may assume $S = \operatorname{Spec} A$ is affine. Write $A = \operatorname{colim}_{i \in I} A_i$ as the filtered colimit of its finitely generated (hence Noetherian) $\mathbf{Z}[1/|G|]$-subalgebras and set $S_i = \operatorname{Spec} A_i$. By [10, Tag 07SK], we can find $i \in I$ and a G-space X_i over S_i with $f_i \colon X_i \to S_i$ of finite presentation, such that f is the base change of f_i. By [10, Tag 08K1] we can increase i and assume f_i is still proper. By Lemma 2.2.27 the coarse quotient $X_i/G \to S_i$ exists. Since S_i is Noetherian, $X_i/G \to S_i$ is proper (using [10, Tag 0DUZ]) and of finite presentation (using [10, Tag 06G4(2)]). Lastly, $X/G \simeq (X_i/G) \times_{S_i} S$ by Lemma 2.2.31, so $X/G \to S$ is proper and of finite presentation. □

2.3 Admissible G-covers

2.3.1 Balanced group actions on nodal curves

Let G be a finite group.

Definition 2.3.1 A *nodal G-curve* is a G-space X over a scheme S where

(i) S is a $\mathrm{Spec}(\mathbf{Z}[1/|G|])$-scheme,
(ii) $X \to S$ is a nodal curve and,
(iii) for every geometric fiber $X_{\bar{k}}$ and $g \in G$ with $g \neq 1$, the locus $X_{\bar{k}}^g$ does not contain any irreducible component of $X_{\bar{k}}$.

A *prestable G-curve* is a nodal G-curve where we replace (ii) with the condition that $X \to S$ is a prestable curve.

We observe that the notation $X_{\bar{k}}^g$ is well defined by Lemma 2.2.18.

Remark 2.3.2 Unlike our definition of a G-space X over S, our definitions of nodal and prestable G-curves $X \to S$ require S to be a $\mathrm{Spec}(\mathbf{Z}[1/|G|])$-scheme.

Remark 2.3.3 The pullback of a nodal (resp. prestable) G-curve is a nodal (resp. prestable) G-curve.

Lemma 2.3.4 *Let X be a nodal G-curve over a field k and $x \in X$ be a smooth point with residue field k and stabilizer G_x. Then G_x is cyclic, and there is a faithful character $\xi\colon G_x \to k^\times$ such that we have a G_x-equivariant isomorphism*

$$\mathscr{O}_{X,x}^\wedge \simeq k[\![u]\!], \qquad g \cdot u = \xi(g)u \text{ for } g \in G_x$$

sending u to an element of $\mathfrak{m}_x \subset \mathscr{O}_{X,x}^\wedge$.

Proof Let $\mathfrak{m} \subset \mathscr{O}_{X,x}^\wedge$ be the maximal ideal and let $r := |G_x|$. Note that $\mathrm{char}(k) \nmid r$. Since x is a smooth point of X, we have $\mathscr{O}_{X,x}^\wedge \simeq k[\![u]\!]$ for any $u \in \mathfrak{m} \setminus \mathfrak{m}^2$ by the proof of [10, Tag 0C0S].

Since x is a G_x-fixed point, the action of G_x on X induces an action on $\mathscr{O}_{X,x}$ and thus a character $\xi\colon G_x \to \mathrm{Aut}_k(\mathfrak{m}/\mathfrak{m}^2) \simeq k^\times$. We show that ξ is faithful. Let $g \in G_x$ such that $\xi(g) = 1$. Then g also acts trivially on $\mathscr{O}_{X,x}^\wedge \simeq k[\![u]\!]$. Indeed, otherwise we can choose $\ell \geq 2$ minimal and

$a \in k^\times$ such that $g \cdot u \equiv u + au^\ell \mod \mathfrak{m}^{\ell+1}$. An inductive argument then shows that

$$u = g^r \cdot u \equiv v + rau^\ell \not\equiv v \mod \mathfrak{m}^{\ell+1}$$

because $r \neq 0$ in k, a contradiction. Further, if Y denotes the irreducible component of X containing x, then g must act trivially on $k(Y) \subset \operatorname{Frac} \mathcal{O}^\wedge_{X,x}$ and thus on Y by [10, Tag 0BY1]. Hence $g = 1$ as desired.

Next, the finiteness of G_x guarantees that ξ identifies G_x with the rth roots of unity $\mu_r(k)$. Thus, G_x is cyclic ([10, Tag 09HX]). To finish the proof, it suffices to exhibit a uniformizer $t \in \mathfrak{m}$ with $g \cdot t = \xi(g)t$, because the natural map

$$k[\![u]\!] \to \mathcal{O}^\wedge_{X,x}, \quad u \mapsto t,$$

will then be a G_x-equivariant isomorphism with the required properties. Pick $v \in \mathfrak{m} \setminus \mathfrak{m}^2$ arbitrary and set $t := \sum_{g \in G} \xi(g)^{-1}(g \cdot v)$. Since $g \cdot v \equiv \xi(g)v \mod \mathfrak{m}^2$, we have $t \equiv rv \neq 0 \mod \mathfrak{m}^2$ (again using that $r \in k^\times$) and thus t is not in $\mathfrak{m}^2$; i.e., it is a uniformizer of $\mathfrak{m}$. On the other hand, $g \cdot t = \xi(g)t$, so t does the job. □

Lemma 2.3.5 *Let X be a nodal G-curve over a field k and $x \in X$ be a split node with stabilizer G_x. If $X^\nu \to X$ is the normalization, let x_1, x_2 be the two preimages of x in X^ν (see [10, Tag 0CBW]) and assume that $G_{x_1} = G_x$. Then G_x is cyclic and there exist faithful characters $\xi_1, \xi_2 \colon G_x \to k^\times$ such that we have a G_x-equivariant isomorphism*

$$\mathcal{O}^\wedge_{X,x} = k[\![u,v]\!]/(uv), \quad g \cdot u = \xi_1(g)u, \quad g \cdot v = \xi_2(g)v \ \text{for } g \in G_x, \quad (1)$$

identifying u and v with elements of $\mathfrak{m}_x \subset \mathcal{O}^\wedge_{X,x}$.

Proof Let $A = \mathcal{O}^\wedge_{X,x}$ and let A' be the integral closure of A in its total ring of fractions. By [10, Tag 0C3V], we have $\operatorname{Spec}(A') = X^\nu \times_X \operatorname{Spec}(A)$. Since the node is split, the δ-invariant of A (as defined in [10, Tag 0C3T]) is 1 and there are two maximal ideals $\mathfrak{m}_1, \mathfrak{m}_2$ in A' lying over $\mathfrak{m}_x \subset A$: indeed, these properties hold by definition and by [10, Tag 0C4A] when A is replaced with $\mathcal{O}_{X,x}$, and they pass to the completion by [10, Tag 0C3W] and the fact that $\operatorname{Spec}(A') = X^\nu \times_X \operatorname{Spec}(A)$.

We observe that $(A')_{\mathfrak{m}_i}$ is by definition the local ring of $X^\nu \times_X \operatorname{Spec}(A)$ at x_i, and this is in turn equal to $\mathcal{O}^\wedge_{X^\nu,x_i}$ by [10, Tag 07N9]. Our assumptions guarantee that G_x acts on the branch of X^ν containing x_i and hence $\mathcal{O}^\wedge_{X^\nu,x_i}$ has a description as in Lemma 2.3.4 for some character ξ_i. (In particular, $G_x = G_{x_1}$ is cyclic.)

We are in the situation of [10, Tag 0C4A(2)], in fact of Case I of the proof. There it is shown that $\kappa(\mathfrak{m}_1) = \kappa(\mathfrak{m}_2) = \kappa(x)$ and A is equal to the subring of A' given by the *wedge* ([10, Tag 0C41]) of $(A')_{\mathfrak{m}_1}$ and $(A')_{\mathfrak{m}_2}$; i.e. (using the explicit description of $(A')_{\mathfrak{m}_i} = \mathcal{O}^{\wedge}_{X^\nu, x_i} \simeq k[\![u]\!]$ in the previous paragraph),

$$\mathcal{O}^{\wedge}_{X,x} \simeq \{(f_1(u), f_2(v)) \in k[\![u]\!] \times k[\![v]\!] \mid f_1(0) = f_2(0)\}. \qquad (2)$$

We observe that G_x acts on A and on A' and that the identification (2) as given in [10, Tag 0C4A(2)] is completely G_x-equivariant. Thus, we get a G_x-equivariant isomorphism $\mathcal{O}^{\wedge}_{X,x} = k[\![u, v]\!]/(uv)$ with the desired properties. □

Definition 2.3.6 Let X be a nodal G-curve over a field k.

(i) If k is algebraically closed, a split node $x \in X$ with stabilizer G_x is *balanced* if G_x is cyclic and, for some faithful character ξ of G_x, there is a G_x-equivariant isomorphism as in (1) with $\xi_2 = \xi_1^{-1}$.

(ii) For general k, a node $x \in X$ is *balanced* if there is a balanced split node in $X_{\bar{k}}$ that maps to x.

We say X is *balanced* if every node is balanced, and a nodal G-curve $X \to S$ is balanced if every geometric fiber is balanced.

Remark 2.3.7 Let x be a split node of a nodal G-curve X over a field k with cyclic stabilizer $G_x = \langle g \rangle$ and G_x-equivariant trivialization $\mathcal{O}^{\wedge}_{X,x} = k[\![u, v]\!]/(uv)$ as in (1). The induced action of g on the tangent space $T_x := (\mathfrak{m}/\mathfrak{m}^2)^\vee$ of x is identified with an endomorphism of $k \cdot u \oplus k \cdot v$. Then $\xi_2 = \xi_1^{-1}$ if and only if $\det g = 1$. In particular, $\xi_2 = \xi_1^{-1}$ is independent of the choice of trivialization in (1) because the condition $\det g = 1$ is independent of the choice of basis of the vector space $(\mathfrak{m}/\mathfrak{m}^2)^\vee$.

Lemma 2.3.8 *Let X be a nodal G-curve over a field k and $x \in X$ a node.*

(i) *If x is a split node, then x is balanced if and only if $\mathcal{O}^{\wedge}_{X,x}$ may be described as in* (1) *with $\xi_2 = \xi_1^{-1}$.*

(ii) *If K/k is a field extension and $y \in X_K$ has image x, then x is a balanced node if and only if y is a balanced node.*

(iii) *If $Y \to X$ is an G-equivariant étale morphism of nodal G-curves over k, and $y \in Y$ has image x with $G_x = G_y$, then x is a balanced node if and only if y is a balanced node.*

Proof Let x be a split node and K/k be a field extension. If $y \in X_K$ maps to x, then y is a split node by Lemma 2.2.2. If x_1 (resp. y_1) is a preimage of x (resp. y) in the normalization of X (resp. X_K), then $G_x = G_y$ and $G_{x_1} = G_{y_1}$ by Lemma 2.2.18. Hence $G_x = G_{x_1}$ if and only if $G_y = G_{y_1}$, and if this is the case then by Lemma 2.3.5 the rings $\mathcal{O}^\wedge_{X,x}$ and $\mathcal{O}^\wedge_{X_K,y}$ admit trivializations as in (1) with characters ξ_1, ξ_2 and $\tilde{\xi}_1, \tilde{\xi}_2$, respectively. Moreover, the G_x-equivariant map $\mathcal{O}^\wedge_{X,x} \to \mathcal{O}^\wedge_{X_K,y}$ induces a G_x-equivariant morphism of tangent spaces $T_y \xrightarrow{\sim} T_x \otimes_k K$. Now Remark 2.3.7 shows that $\xi_2 = \xi_1^{-1}$ if and only if $\tilde{\xi}_2 = \tilde{\xi}_1^{-1}$. In the special case $K = \bar{k}$, this implies (i).

Next, we prove (ii). If y is balanced, then by definition we have a balanced split node $y' \in X_{\bar{K}}$ mapping to y. The inclusion $k \hookrightarrow \bar{K}$ factors through $k \hookrightarrow \bar{k}$, so we get a map $X_{\bar{K}} \to X_{\bar{k}}$; let $x' \in X_{\bar{k}}$ be the image of y' under this map. It is a (necessarily split) node by [10, Tag 0C56]. Moreover it is balanced, by the first paragraph of this proof. Since x' maps to x we see that x is balanced.

Conversely, assume x is balanced. By [10, Tag 0C56], y is a node. Let $\bar{k}^{\text{sep}} \subset \bar{k}$ denote the separable algebraic closure of k. To prove that y is balanced, we may assume K is algebraically closed. Since such a K contains $\bar{k} \supset \bar{k}^{\text{sep}}$, by the first paragraph of this proof it suffices to prove that if $y \in X_{\bar{k}^{\text{sep}}}$ has image x, then y is a balanced split node. Let $y \in X_{\bar{k}^{\text{sep}}}$ have image $x \in X$. By Lemmas 2.2.3(iii) and 2.2.2 we know there is a split node $y' \in X_{\bar{k}^{\text{sep}}}$ mapping to x. By [10, Tag 04KP] there is an automorphism in $\mathrm{Gal}(\bar{k}^{\text{sep}}/k) \subset \mathrm{Aut}_k(X_{\bar{k}^{\text{sep}}})$ that maps y to y'; hence y is split (as is every node in the preimage of x). Since x is balanced the discussion in the first paragraph shows that there is some balanced split node $y'' \in X_{\bar{k}^{\text{sep}}}$ mapping to x. There is an element of $\mathrm{Gal}(\bar{k}^{\text{sep}}/k) \subset \mathrm{Aut}_k(X_{\bar{k}^{\text{sep}}})$ sending y to y'', and moreover this automorphism is G-equivariant (consider the action in terms of the functor of points). From the characterization of balanced split nodes in Lemma 2.3.5 it follows that y is balanced.

To prove (iii), let $y' \in Y_{\bar{k}}$ be a node mapping to y and let $x' \in X_{\bar{k}}$ be its image. By (ii), x (resp. y) is balanced if and only if x' (resp. y') is balanced. Both x' and y' are split nodes; let x'_1, x'_2 and y'_1, y'_2 be the two preimages in the respective normalizations, indexed so that the G-equivariant map of normalizations sends x_i to y_i. Note that $G_{y'_1} \subset G_{x'_1} \subset G_{x'} = G_{y'}$ (the equality is our hypothesis). So $G_{y'_1} = G_{y'}$ implies $G_{x'_1} = G_{x'}$, and conversely if $g \in G_{y'} \setminus G_{y'_1}$ then $gy'_1 = y'_2$ implies $gx'_1 = x'_2$, so in fact $G_{y'_1} = G_{y'}$ if and only if $G_{x'_1} = G_{x'}$. In the situation $G_{x'_1} = G_{y'_1} = G_{x'} = G_{y'}$, the local rings of x' and y' can be trivialized as in (1) and the nodes are balanced if and only if the characters ξ_1 and ξ_2 are inverse. On the other hand since $\mathcal{O}^{\wedge}_{X_{\bar{k}},x'} \to \mathcal{O}^{\wedge}_{Y_{\bar{k}},y'}$ is G-equivariant and k-linear, the characters for $\mathcal{O}^{\wedge}_{X_{\bar{k}},x'}$ and $\mathcal{O}^{\wedge}_{Y_{\bar{k}},y'}$ must be the same. □

2.3.2 Families of balanced nodal curves

In this subsection we study the deformation theory of balanced nodal G-curves. This problem is simplest when we work over a base ring that contains the $|G|$th roots of unity because the group homomorphisms $G_x \to k(s)^\times$ of Lemmas 2.3.4 and 2.3.5 can be lifted to morphisms of group schemes (see Remark 2.3.12 below).

Remark 2.3.9 If X is a balanced nodal G-curve over k and $x \in X$ has residue field k, it follows from Lemmas 2.3.4 and 2.3.5 that k contains all the $|G_x|$th roots of unity.

Definition 2.3.10 Let $r \geq 1$ be an integer, and define the *ring of cyclotomic integers* as $\mathbf{Z}[\zeta_r] := \mathbf{Z}[x]/(\Phi_r(x))$, where $\Phi_r(x)$ denotes the rth cyclotomic polynomial. This is the integral closure of $\mathbf{Z}$ in the rth cyclotomic field $\mathbf{Q}(\zeta_r)$, and it is also equal to the smallest subring of $\mathbf{C}$ containing $\mathbf{Z}$ and ζ_r.

Remark 2.3.11 If $S = \operatorname{Spec}(\mathbf{Z}[1/r, \zeta_r])$, then there is an isomorphism of group schemes $(\mathbf{Z}/r\mathbf{Z})_S \to \mu_{r,S}$. In terms of functors of points, for any S-scheme T this morphism is given by group homomorphisms $(\mathbf{Z}/r\mathbf{Z})(T) \to \mu_r(T)$ sending the generator $1 \in \mathbf{Z}/r\mathbf{Z}$ to $\zeta_r \in \Gamma(T, \mathcal{O}_T)$. One can check it is an isomorphism by noting that as a morphism of affine schemes it is given by the ring map

$$\mathbf{Z}[1/r, \zeta_r, x]/(x^r - 1) \to \prod_{i=0}^{r-1} \mathbf{Z}[1/r, \zeta_r] \tag{3}$$

sending x to $(1, \zeta_r, \ldots, \zeta_r^{r-1})$. The homomorphism (3) is an isomorphism by the Chinese remainder theorem ([10, Tag 00DT]) if each pair of ideals $(x - \zeta_r^i)$ and $(x - \zeta_r^j)$ is coprime (for $i \neq j$). To show that these ideals are coprime it is enough to show that $\zeta_r^i - \zeta_r^j$ is invertible in $\mathbf{Z}[1/r, \zeta_r]$, or equivalently that $1 - \zeta_r^k$ is invertible in $\mathbf{Z}[1/r, \zeta_r]$ for each integer k.

To see this, observe that

$$\prod_{k=1}^{r-1}(x-\zeta_r^k) = 1+x+\cdots+x^{r-1}$$

and then set $x = 1$.

Remark 2.3.12 Let S be a scheme over $\mathrm{Spec}(\mathbf{Z}[1/r,\zeta_r])$, let $s \in S$, and let G be a group of order r with $\xi\colon G \to k(s)^{x}$ a group homomorphism. Then there is a morphism of group schemes $G_S \longrightarrow \mathbf{G}_{m,S}$ such that the map on sections over $\mathrm{Spec}(k(s))$ is given by ξ: one may see this as follows. Since G has order r, the homomorphism ξ factors through $\mu_r(k(s)) \subset k(s)^{\times}$. By Remark 2.3.11 we have a group homomorphism $G \to \mathbf{Z}/r\mathbf{Z}$. This defines a morphism $G_S \longrightarrow (\mathbf{Z}/r\mathbf{Z})_S$ of constant group schemes, and again by Remark 2.3.11 this yields the desired $G_S \to \mu_{r,S}$. (Note that $G_S \longrightarrow \mathbf{G}_{m,S}$ is uniquely determined by ξ on the connected component of s.)

The following lemma is the analog of [10, Tag 0CBX].

Lemma 2.3.13 *Let X and S be schemes over $\mathbf{Z}[1/|G|,\zeta_{|G|}]$ with $X \to S$ a nodal G-curve, and choose $x \in X$ with $G_x = G$. Let $s \in S$ be the image of x and assume that $\kappa(x) = \kappa(s)$ and that $\mathscr{O}_{S,s}$ is Noetherian. Let $\xi\colon G \to \kappa(s)^{\times}$ be defined as in Lemmas 2.3.4 or 2.3.5. Then the morphism of group schemes in Remark 2.3.12 defines a group homomorphism $\xi\colon G \to (\mathscr{O}^{\wedge}_{S,s})^{\times}$ such that*

(i) *If x is a smooth point of X_s, then there is a G-equivariant isomorphism of $\mathscr{O}^{\wedge}_{S,s}$-algebras*

$$\begin{aligned} \mathscr{O}^{\wedge}_{X,x} &\simeq \mathscr{O}^{\wedge}_{S,s}[\![u]\!], \\ g\cdot u &= \xi(g)u \quad \textit{for } g \in G. \end{aligned} \tag{4}$$

(ii) *If x is a balanced split node of X_s, then there exists $h \in \mathfrak{m}_s \subset \mathscr{O}^{\wedge}_{S,s}$ and a G-equivariant isomorphism of $\mathscr{O}^{\wedge}_{S,s}$-algebras*

$$\begin{aligned} \mathscr{O}^{\wedge}_{X,x} &\simeq \mathscr{O}^{\wedge}_{S,s}[\![u,v]\!]/(uv-h), \\ g\cdot u = \xi(g)u,&\quad g\cdot v = \xi^{-1}(g)v, \quad \textit{for } g \in G. \end{aligned} \tag{5}$$

Proof We replace S by $\mathrm{Spec}(\mathscr{O}_{S,s})$ and X by $\mathrm{Spec}(\mathscr{O}_{S,s}) \times_S X$. Let $A = \mathscr{O}^{\wedge}_{S,s}$ and $B = \mathscr{O}^{\wedge}_{X,x}$ and let $\mathfrak{m} := \mathfrak{m}_A$ be the maximal ideal of A. Then $\mathscr{O}_{S,s} \to \mathscr{O}_{X,x}$ is a flat homomorphism of Noetherian rings and by [10, 0C4G] the map $A \to B$ is also flat (we will use this at the end of the proof). Note that $\mathscr{O}^{\wedge}_{X_s,x} = B/\mathfrak{m}B$. By Remark 2.3.12, ξ defines a faithful character $G \to \mathbf{G}_{m,S}$ that identifies G_S with $\mu_{|G|,S}$. Hence, any $\mathcal{O}_{S,s}$-module R with a G-action may be written as a direct sum

$$R = \bigoplus_{i=1}^{|G|} R_{\xi^i} \tag{6}$$

of $\mathcal{O}_{S,s}$-modules such that for $g \in G$ and $r \in R_{\xi^i}$ we have $g \cdot r = \xi(g)^i r$. We say that $r \in R_{\xi^i}$ has weight ξ^i. To see (6), set $r_i := \frac{1}{|G_x|} \sum_{g \in G} \xi^{-i}(g) g \cdot r$ and check that $\sum_i r_i = r$ and that r_i has weight ξ^i.

To prove (i), observe that by Lemma 2.3.4 there is an isomorphism

$$(A/\mathfrak{m})[\![u]\!] \to B/\mathfrak{m}B,$$

where u maps to some element of $\mathfrak{m}_B/\mathfrak{m}B$ of weight ξ. Since the map $B \to B/\mathfrak{m}$ respects the decompositions (6), we can choose a lift $\tilde{t} \in \mathfrak{m}_B$ of weight ξ. This defines a morphism $f \colon A[\![u]\!] \to B$ of Noetherian rings that is an isomorphism modulo $\mathfrak{m}$. We conclude by [10, Tags 0315 and 00ME], using the flatness of $A \to B$, that f is an isomorphism.

To prove (ii) observe that by Lemma 2.3.8(i) there is an isomorphism

$$(A/\mathfrak{m})[\![u, v]\!]/(uv) \to B/\mathfrak{m}B, \qquad u \mapsto \bar{x},\ v \mapsto \bar{y},$$

where $\bar{x}$ (resp. $\bar{y}$) has weight ξ (resp. ξ^{-1}) and $A/\mathfrak{m}$ maps to $(B/\mathfrak{m}B)_{\xi^0}$. Let x_1 and y_1 be lifts of $\bar{x}$ and $\bar{y}$ of weights ξ and ξ^{-1}, respectively, and observe that $x_1 y_1 \in \mathfrak{m}B$ has weight ξ^0.

Set $h_1 = 0$ and $\delta_1 = x_1 y_1$. For $n > 1$ we recursively construct elements $x_n \in B_\xi, y_n \in B_{\xi^{-1}}, h_n \in A$, and $\delta_n \in (\mathfrak{m}^n B)_{\xi^0}$ such that $x_n - x_{n+1}, y_n - y_{n+1} \in \mathfrak{m}^n B$, $h_n - h_{n+1} \in \mathfrak{m}^n$, and $x_n y_n = h_n + \delta_n$. To define the $(n+1)$th collection of elements, write $\delta_n = \sum f_i b_i$ with $f_i \in \mathfrak{m}^n$ and $b_i \in B$. Since δ_n and f_i have weight ξ^0, we can replace each summand of $\sum f_i b_i$ with its weight-ξ^0 part and thus assume b_i has weight ξ^0. Since $A/\mathfrak{m} \simeq B/\mathfrak{m}_B$ and $\mathfrak{m}_B = x_1 B + y_1 B + \mathfrak{m}B = x_n B + y_n B + \mathfrak{m}B$, we can write $b_i = a_i + x_n b_{i,1} + y_n b_{i,2} + \delta_{i,n}$ where $a_i \in A,\ b_{i,1}, b_{i,2} \in B$, and $\delta_{i,n} \in \mathfrak{m}B$. The element b_i already has weight ξ^0, so by replacing every summand on the right-hand side with its weight-ξ^0 part we can choose $\delta_{i,n}$ of weight ξ^0 and $b_{i,1}, b_{i,2}$ of weights ξ^{-1} and ξ, respectively.

Define $x_{n+1} := x_n - \sum b_{i,2} f_i$ and $y_{n+1} := y_n - \sum b_{i,1} f_i$; note that these have the desired weights. One checks that

$$x_{n+1} y_{n+1} = h_n + \sum f_i a_i + \sum f_i \delta_{i,n} + \sum c_{ij} f_i f_j$$

for some $c_{ij} \in B_{\xi^0}$. Thus, $h_{n+1} := h_n + \sum f_i a_i$ and $\delta_{n+1} := \sum f_i \delta_{i,n} + \sum c_{ij} f_i f_j$ satisfy the recursive assumptions. Since A and B are complete, we can define $x_\infty := \lim x_n$, $y_\infty := \lim y_n$, and $h_\infty := \lim h_n$, noting that $x_\infty y_\infty = h_\infty$ and that x_∞ and y_∞ have weights ξ and ξ^{-1}, respectively. We define $f\colon A[\![u, v]\!]/(uv - h_\infty) \to B$ to send u to x_∞ and v to y_∞. Since f is an isomorphism modulo $\mathfrak{m}$, it is an isomorphism as in the proof of (i). □

Lemma 2.3.14 *Let $X \to S$ be a nodal G-curve and a morphism of schemes, and choose $x \in X$ with image s such that $\kappa(x) = \kappa(s)$ and $\mathcal{O}_{S,s}$ is Noetherian. Assume that x is either a smooth point or a balanced split node of its fiber. After passing to an elementary étale neighborhood of $s \in S$ we can arrange that the hypotheses of Lemma 2.3.13 are satisfied.*

Proof By restricting the G-action to G_x, we obtain the hypothesis that $G = G_x$. We may assume $S = \operatorname{Spec}(A)$. Let $r := |G|$ and set $S' := \operatorname{Spec}(A[t]/(\Phi_r(t))$. Since r is invertible in A by Definition 2.3.1, the map $S' \to S$ is étale and S' is defined over $\operatorname{Spec}(\mathbf{Z}[1/r, \zeta_r])$. Let $x' \in X_{S'}$ be any point mapping to x and let s' be its image in S'. As s' lies in the fiber $\kappa(s) \times_S S' \simeq \operatorname{Spec}(\kappa(s)[t]/\Phi_r(t))$ and r is invertible in $\kappa(s)$, we have $\kappa(s) = \kappa(s')$, so $(S', s') \to (S, s)$ is actually an elementary étale neighborhood. Moreover, if x is a smooth rational point (resp. balanced split node) of its fiber, then so is x' by Lemmas 2.2.2 and 2.3.8(ii). By Lemmas 2.2.18 and 2.2.19, we have $G_{x'} = G$. □

The remainder of this section is dedicated to "spreading out" the description of $\mathcal{O}^\wedge_{X,x}$ in Lemma 2.3.13(ii) to an étale neighborhood of x. That is, we prove the following equivariant analog of [10, Tag 0CBY] (recall from [10, Tag 03BU] that $|X|$ is the topological space of an algebraic space X).

Proposition 2.3.15 *Let $X \to S$ be a nodal G-curve over S. Let $x \in |X|$ be a point which is a balanced node of the fiber X_s. Then the stabilizer G_x of $x \in X_s$ is a cyclic group of order r, and there exists a commutative diagram of G_x-spaces and equivariant morphisms*

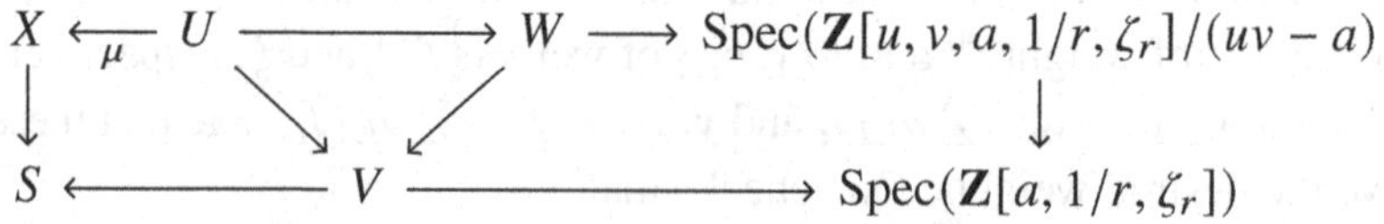

$$\begin{array}{ccccccc}
X & \xleftarrow{\mu} & U & \longrightarrow & W & \longrightarrow & \operatorname{Spec}(\mathbf{Z}[u, v, a, 1/r, \zeta_r]/(uv - a) \\
\downarrow & & & \searrow \quad \swarrow & & & \downarrow \\
S & \longleftarrow & & V & & \longrightarrow & \operatorname{Spec}(\mathbf{Z}[a, 1/r, \zeta_r])
\end{array}$$

such that the arrows $X \xleftarrow{\mu} U$, $S \leftarrow V$, and $U \to W$ are étale morphisms, the right-hand square is cartesian, the G_x-action at the top right is given by $g \cdot u = \zeta_r^i u$, $g \cdot v = \zeta_r^{-i} v$ for some integer i, the G_x-action on V is *trivial, and there exists a point* $u \in U$ *mapping to* $x \in X$ *with* $G_u = G_x$*. Moreover* U^g *surjects onto* $\mu(U) \cap X^g$ *for all* $g \in G_x$*.*

Remark 2.3.16 Since the description of complete local rings in Lemma 2.3.13 may be read as a kind of flat-local description of nodal G-curves, one may wonder why the étale-local description in Proposition 2.3.15 is necessary. The reason is that flat maps do not preserve the smooth and nodal points of curves. A naive observation is that flat maps need not preserve dimension, so the notion "$x \in X$ is a node in its fiber" does not make sense after replacing X with an arbitrary flat cover. However, even flat maps of relative dimension zero need not preserve nodes and smooth points: for example, the ring map

$$k[\![u,v]\!]/(u-v^2) \to k[\![u^{1/2},v]\!]/(u-v^2) \simeq k[\![u^{1/2},v]\!]/((u^{1/2}-v)(u^{1/2}+v))$$

induces a flat morphism of schemes that sends a node to a smooth point. On the other hand, by [10, Tag 0C57] (resp. Lemma 2.3.8(iii)), étale covers *do* preserve nodes and smooth points (resp. balanced split nodes). Looking at the proof of [10, Tag 0C57], one sees that the additional condition that f is unramified (hence étale) is precisely what is needed to make the tag work.

Our proof of Proposition 2.3.15 uses Artin approximation and closely follows the proof of [10, Tag 0CBY]. Before proving Proposition 2.3.15, we state and prove a version of Artin approximation for limit-preserving functors from [4] as well as equivariant analogs of many of the tags used to prove [10, Tag 0CBY].

Following [10, Tag 049J], a contravariant functor $F\colon (Sch/X)^{\mathrm{opp}} \to Sets$ is limit-preserving if for every directed inverse system of affine schemes T_i in (Sch/X) with affine inverse limit T, the natural map $\big(\mathrm{colim}_i\, F(T_i)\big) \to F(T)$ is a bijection. The following is [4, Corollary 2.2].

Proposition 2.3.17 *Let* X *be a locally Noetherian scheme and* $x \in X$ *such that* $\mathcal{O}_{X,x}$ *is a* G-ring *in the sense of [10, Tag 07GH]. Let* $F\colon (Sch/X)^{\mathrm{opp}} \to Sets$ *be a limit-preserving contravariant functor and* $\varphi \in F(\mathrm{Spec}(\mathcal{O}^{\wedge}_{X,x}))$*. Then for any natural number* N*, there exists an elementary étale neighborhood* $(U,u) \to (X,x)$ *and an element* $\Phi \in F(U)$ *such that* φ *and* Φ *map to the same element in* $F\big(\mathrm{Spec}(\mathcal{O}^{\wedge}_{U,u}/\mathfrak{m}_u^N)\big)$*.*

Proof We may assume that $N \geq 2$. By [10, Tag 02Y2], F corresponds to a category fibered in sets $p\colon \mathcal{S}_F \to (Sch/X)$. Under this correspondence, the fact that F is limit-preserving translates to the fact that p is limit-preserving on objects in the sense of [10, Tag 06CT]. Now [10, Tag

07XB] (applied to $\mathcal{X} = \mathcal{S}_F$ and $R = \mathcal{O}^\wedge_{X,x}$) gives a ring A, a morphism $U := \mathrm{Spec}(A) \to X$ of finite type, a closed point $u \in U$ lying over x, and an object $\Phi \in \mathcal{S}_F(U)$ such that the map $U \to X$ induces an isomorphism $\mathcal{O}^\wedge_{X,x}/\mathfrak{m}_x^N \to \mathcal{O}_{U,u}/\mathfrak{m}_u^N$. Over this isomorphism, it guarantees the existence of an isomorphism $\varphi|_{\mathrm{Spec}(\mathcal{O}^\wedge_{X,x}/\mathfrak{m}_x^N)} \simeq \Phi|_{\mathrm{Spec}(\mathcal{O}_{U,u}/\mathfrak{m}_u^N)}$. Lastly, there exists an isomorphism of graded $\mathcal{O}_{X,x}/\mathfrak{m}_x$-algebras $\mathrm{Gr}_{\mathfrak{m}_x}(\mathcal{O}^\wedge_{X,x}) \to \mathrm{Gr}_{\mathfrak{m}_u}(\mathcal{O}_{U,u})$, where $\mathrm{Gr}_{\mathfrak{m}}(B) := \bigoplus_{\ell \geq 0} \mathfrak{m}^\ell/\mathfrak{m}^{\ell+1}$ denotes the graded ring attached to a local ring B with maximal ideal $\mathfrak{m}$; note that this graded algebra is generated in degree 1.

The natural map $\mathcal{O}_{X,x} \to \mathcal{O}_{U,u}$ induces isomorphisms $\mathrm{Gr}^\ell_{\mathfrak{m}_x}(\mathcal{O}^\wedge_{X,x}) \to \mathrm{Gr}^\ell_{\mathfrak{m}_u}(\mathcal{O}_{U,u})$ for $\ell = 0, 1$ because $N \geq 2$. Since the graded algebras are generated in degree 1 and their graded pieces have the same dimension (by the existence of the abstract isomorphism), $\mathcal{O}_{X,x} \to \mathcal{O}_{U,u}$ must in fact induce isomorphisms $\mathrm{Gr}^\ell_{\mathfrak{m}_x}(\mathcal{O}^\wedge_{X,x}) \to \mathrm{Gr}^\ell_{\mathfrak{m}_u}(\mathcal{O}_{U,u})$ for all ℓ. For $\ell = 1$, we see that $\kappa(x) \to \kappa(u)$ is an isomorphism. Via induction and repeated application of the five lemma to the diagram of natural morphisms

$$\begin{array}{ccccccccc} 0 & \longrightarrow & \mathrm{Gr}^\ell_{\mathfrak{m}}(\mathcal{O}_{X,x}) & \longrightarrow & \mathcal{O}_{X,x}/\mathfrak{m}_x^{\ell+1} & \longrightarrow & \mathcal{O}_{X,x}/\mathfrak{m}_x^\ell & \longrightarrow & 0 \\ \downarrow & & \downarrow & & \downarrow & & \downarrow & & \downarrow \\ 0 & \longrightarrow & \mathrm{Gr}^\ell_{\mathfrak{m}}(\mathcal{O}_{U,u}) & \longrightarrow & \mathcal{O}_{U,u}/\mathfrak{m}_u^{\ell+1} & \longrightarrow & \mathcal{O}_{U,u}/\mathfrak{m}_u^\ell & \longrightarrow & 0, \end{array}$$

the map $\mathcal{O}_{X,x}/\mathfrak{m}_x^\ell \to \mathcal{O}_{U,u}/\mathfrak{m}_u^\ell$ is an isomorphism for all ℓ. Hence, $\mathcal{O}^\wedge_{X,x} \to \mathcal{O}^\wedge_{U,u}$ is an isomorphism, and $U \to X$ is étale at u by [10, Tag 039N, Tag 039M]. After shrinking U if necessary, $(U, u) \to (X, x)$ is an elementary étale neighborhood. Finally, under the correspondence of $\mathcal{S}_F$ with F, the isomorphism $\varphi|_{\mathrm{Spec}(\mathcal{O}^\wedge_{U,u}/\mathfrak{m}_u^N)} \simeq \Phi|_{\mathrm{Spec}(\mathcal{O}^\wedge_{U,u}/\mathfrak{m}_u^N)}$ translates to the statement that φ and Φ map to the same element in $F(\mathrm{Spec}(\mathcal{O}^\wedge_{U,u}/\mathfrak{m}_u^N))$, so the proposition is proved. □

Now we will add group actions to many of the tags used in the proof of [10, Tag 0CBY]. The following is an equivariant analog of [10, Tag 0CAU] and [10, Tag 0CAV].

Lemma 2.3.18 *Let G be a finite group and let S be a locally Noetherian scheme. Let X and Y be G-schemes with G-invariant maps $X \to S$, $Y \to S$, locally of finite type. Choose $x \in X^G$ and $y \in Y^G$ lying over the same point $s \in S$ with $\mathcal{O}_{S,s}$ a G-ring in the sense of [10, Tag 07GH]. Suppose we are given a G-equivariant local $\mathcal{O}_{S,s}$-algebra map*

$$\varphi\colon \mathcal{O}_{Y,y} \to \mathcal{O}^\wedge_{X,x}.$$

Then, for each $N \geq 1$, there is a G-equivariant elementary étale neighborhood $\mu\colon (U,u) \to (X,x)$, with $u \in U^G$, such that U^g surjects onto $\mu(U) \cap X^g$ for all $g \in G$, and a G-equivariant S-morphism $f\colon U \to Y$ mapping u to y such that the following diagram commutes modulo $\mathfrak{m}_u^N$:

$$\begin{array}{ccc} \mathcal{O}^{\wedge}_{X,x} & \longrightarrow & \mathcal{O}^{\wedge}_{U,u} \\ {\scriptstyle\varphi}\uparrow & & \uparrow \\ \mathcal{O}_{Y,y} & \xrightarrow{f_u^{\#}} & \mathcal{O}_{U,u} \end{array} \tag{7}$$

Moreover, if φ induces an isomorphism $\mathcal{O}^{\wedge}_{Y,y} \simeq \mathcal{O}^{\wedge}_{X,x}$, then we can choose f such that $(U,u) \to (Y,y)$ is an elementary étale neighborhood.

Proof Define a contravariant functor F on X-schemes as follows. For $U \to X$ let

$$F(U) = \mathrm{Hom}_S^G(U^{\mathrm{e}}, Y)$$

be the set of G-equivariant morphisms of S-schemes from the equivariantization U^{e} (see Definition 2.2.21) to Y. If $U \to V$ is a morphism of X-schemes, the induced map of equivariantizations $U^{\mathrm{e}} \to V^{\mathrm{e}}$ defines a restriction map $F(V) \to F(U)$ by precomposition. We check that F is limit-preserving; that is, if $\{U_i\}$ is a directed inverse system of affine X-schemes, we show that the canonical map

$$\varinjlim \mathrm{Hom}_S^G(U_i^{\mathrm{e}}, Y) \longrightarrow \mathrm{Hom}_S^G((\varprojlim U_i)^{\mathrm{e}}, Y) \tag{8}$$

is a bijection. First observe that equivariantization is a type of limit and hence commutes with limits. Next observe that (8) is the G-invariant part of the canonical morphism

$$\varinjlim \mathrm{Hom}_S(U_i^{\mathrm{e}}, Y) \longrightarrow \mathrm{Hom}_S(\varprojlim U_i^{\mathrm{e}}, Y), \tag{9}$$

where to commute the G-invariants and the colimit on the left-hand side we use that finite limits commute with filtered colimits ([10, 002W]). But (9) is a bijection since the functor $U \mapsto \mathrm{Hom}_S(U,Y)$ is locally finitely presented by [10, Tag 01TX] and [10, Tag 01ZC].

The given morphism φ defines an element of $\mathrm{Hom}_S^G(\mathrm{Spec}(\mathcal{O}^{\wedge}_{X,x}), Y)$.

We compute

$$\begin{aligned} \operatorname{Spec}(\mathcal{O}_{X,x}^{\wedge})^{\mathrm{e}} &= \Big[\varinjlim \operatorname{Spec}(\mathcal{O}_{X,x}/\mathfrak{m}_x^n)\Big]^{\mathrm{e}} \\ &= \varinjlim \operatorname{Spec}(\mathcal{O}_{X,x}/\mathfrak{m}_x^n)^{\mathrm{e}} \\ &= \varinjlim \operatorname{Spec}(\mathcal{O}_{X,x}/\mathfrak{m}_x^n) \\ &= \operatorname{Spec}(\mathcal{O}_{X,x}^{\wedge}) \end{aligned} \tag{10}$$

where the first and last equalities are the definition of completion,[2] the second again uses [10, Tag 002W], and the third is Lemma 2.2.25 together with [10, Tag 01L6]. Hence φ produces an element of $F(\operatorname{Spec}(\mathcal{O}_{X,x}^{\wedge}))$. Now Proposition 2.3.17 produces for any $N \geq 1$ an elementary étale neighborhood $\mu\colon (U,u) \to (X,x)$ and an element $\Phi \in F(U)$ such that φ and Φ map to the same element of $F(\operatorname{Spec}(\mathcal{O}_{U,u}^{\wedge}/\mathfrak{m}_u^N))$.

Since $\mathcal{O}_{U,u}$ is Noetherian, by [10, Tag 031C] we have $\mathcal{O}_{U,u}^{\wedge}/\mathfrak{m}_u^N \simeq \mathcal{O}_{U,u}/\mathfrak{m}_u^N$. Unwinding the definition of F, we see that we have an equivariant map $U^{\mathrm{e}} \to Y$ fitting into the black part of the following G-equivariant commutative diagram:

$$\begin{array}{ccccc} \operatorname{Spec}(\mathcal{O}_{X,x}^{\wedge}) & \longleftarrow & \operatorname{Spec}(\mathcal{O}_{U^{\mathrm{e}},u^{\mathrm{e}}}^{\wedge}/\mathfrak{m}_{u^{\mathrm{e}}}^N) & \xrightarrow{\sim} & \operatorname{Spec}(\mathcal{O}_{U^{\mathrm{e}},u^{\mathrm{e}}}/\mathfrak{m}_{u^{\mathrm{e}}}^N) \\ \sim \Big| (10) & & & & \Big| \sim \\ \operatorname{Spec}(\mathcal{O}_{X,x}^{\wedge})^{\mathrm{e}} & \longleftarrow & \operatorname{Spec}(\mathcal{O}_{U,u}^{\wedge}/\mathfrak{m}^N)^{\mathrm{e}} & \xrightarrow{\sim} & \operatorname{Spec}(\mathcal{O}_{U,u}/\mathfrak{m}^N)^{\mathrm{e}} \\ \downarrow & & & & \downarrow \\ Y & \longleftarrow & & & U^{\mathrm{e}} \end{array} \tag{11}$$

We have $u^{\mathrm{e}} \in U^{\mathrm{e}}$ defined as in (2) and $\mu^{\mathrm{e}}\colon (U^{\mathrm{e}},u^{\mathrm{e}}) \to (X,x)$ is an elementary étale neighborhood. Note that the black part of (11) implies that u^{e} maps to y. For the elementary étale neighborhood of the lemma, we take $((\mu^{\mathrm{e}})^{-1}(V), u^{\mathrm{e}})$ as constructed in Lemma 2.2.24.

To show that (7) commutes, we claim that we can construct the entire diagram in (11) to be commutative, so that the composition of the vertical arrows on the right is the canonical map $\operatorname{Spec}(\mathcal{O}_{U^{\mathrm{e}},u^{\mathrm{e}}}/\mathfrak{m}_{u^{\mathrm{e}}}^N) \to U^{\mathrm{e}}$. Granting this, (7) is obtained from (11) by taking the induced maps on local rings.

[2] These direct limits are taken in the category of schemes (not locally ringed spaces). In fact, they can be computed in the category of affine schemes since a map from $\operatorname{Spec}(\mathcal{O}_{X,x}/\mathfrak{m}_x^n)$ to an arbitrary scheme Y factors through every affine open neighborhood of the image of $\mathfrak{m}$. See, e.g., [6].

We now derive (11). First, we have a commutative diagram

$$\begin{array}{ccccc} & & \mathrm{Spec}(\mathcal{O}^{\wedge}_{U,u}/\mathfrak{m}_u^N) & \xrightarrow{\sim} & \mathrm{Spec}(\mathcal{O}_{U,u}/\mathfrak{m}_u^N) \\ & \swarrow & \uparrow\wr & & \uparrow\wr \\ \mathrm{Spec}(\mathcal{O}^{\wedge}_{X,x}) & \longleftarrow & \mathrm{Spec}(\mathcal{O}^{\wedge}_{U^{\mathrm{e}},u^{\mathrm{e}}}/\mathfrak{m}_{u^{\mathrm{e}}}^N) & \xrightarrow{\sim} & \mathrm{Spec}(\mathcal{O}_{U^{\mathrm{e}},u^{\mathrm{e}}}/\mathfrak{m}_{u^{\mathrm{e}}}^N) \end{array} \tag{12}$$

induced by the map $\mathrm{pr}_1 : (U^{\mathrm{e}}, u^{\mathrm{e}}) \to (U, u)$ of elementary étale neighborhoods over (X, x), where pr_1 was defined in Lemma 2.2.25. In this diagram, the left vertical map is an isomorphism by [10, Tag 0AGX] and the right vertical map is an isomorphism because the other three are isomorphisms. Next, the equivariantization of (12) maps to (12) under pr_1, and we call the resulting three-dimensional commuting diagram D.

The map $\mathrm{pr}_1 : \mathrm{Spec}(\mathcal{O}_{U^{\mathrm{e}},u^{\mathrm{e}}}/\mathfrak{m}_{u^{\mathrm{e}}}^N)^{\mathrm{e}} \to \mathrm{Spec}(\mathcal{O}_{U^{\mathrm{e}},u^{\mathrm{e}}}/\mathfrak{m}_{u^{\mathrm{e}}}^N)$ in the diagram D is an isomorphism by Lemma 2.2.25 (using the factorization $\mathrm{Spec}(\mathcal{O}_{U^{\mathrm{e}},u^{\mathrm{e}}}/\mathfrak{m}_{u^{\mathrm{e}}}^N) \to \mathrm{Spec}(\mathcal{O}_{X,x}/\mathfrak{m}_x^N) \to X$, where the first map is an isomorphism by the same reasoning as we used for the right vertical map of (12), and the second is a monomorphism by [10, Tag 01L6]).

From D one may extract the gray part of (11). Specifically, the bottom line of (12) is the top (gray) line of (11), and the equivariantization of the top line of (12) is the middle line of (11). By examining D, one may also check that the composition of the vertical arrows on the right-hand side of (11) is equal to the composition

$$\mathrm{Spec}(\mathcal{O}_{U^{\mathrm{e}},u^{\mathrm{e}}}/\mathfrak{m}_{u^{\mathrm{e}}}^N) \xrightarrow{\tilde{\Delta}} \left[\mathrm{Spec}(\mathcal{O}_{U^{\mathrm{e}},u^{\mathrm{e}}}/\mathfrak{m}_{u^{\mathrm{e}}}^N \to U^{\mathrm{e}} \xrightarrow{\mathrm{pr}_1} U\right]^{\mathrm{e}}$$

where the notation $[-]^{\mathrm{e}}$ means that we have applied the equivariantization functor to the canonical maps in the square brackets, and $\tilde{\Delta}$ was defined in Lemma 2.2.25. After replacing X and U with affine schemes one can write down the corresponding ring map and see that it induces the canonical map $\mathrm{Spec}(\mathcal{O}_{U^{\mathrm{e}},u^{\mathrm{e}}}/\mathfrak{m}_{u^{\mathrm{e}}}^N) \to U^{\mathrm{e}}$.

For the "moreover" part of the theorem, we first note that by taking $N \geq 2$ we can choose f to induce an isomorphism on residue fields. Next, the proof of [10, Tag 0CAV] shows that when φ induces an isomorphism of completed local rings, the morphism f is étale at u. Hence there is an open subset U' of U containing u where f is étale, and we can replace U with $(U')^{\mathrm{e}}$. □

The next lemma is the analog of [10, Tag 0GDX].

Lemma 2.3.19 *Let G be a finite group acting on schemes X and Y.*

Moreover, let $S, T, x, y, s, t, \sigma, y_\sigma$, and φ be given as follows: we have G-invariant morphisms of schemes

$$\begin{array}{cc} X & Y \\ \downarrow & \downarrow \\ S & T \end{array} \quad \text{with } G\text{-fixed points} \quad \begin{array}{cc} x & y \\ \downarrow & \downarrow \\ s & t \end{array}$$

such that S is locally Noetherian, T is of finite type over $\mathbf{Z}$, the morphisms $X \to S$ and $Y \to T$ are locally of finite type, and $\mathcal{O}_{S,s}$ is a G-ring *in the sense of [10, Tag 07GH].*

The map

$$\sigma \colon \mathcal{O}_{T,t} \to \mathcal{O}^\wedge_{S,s}$$

is a local homomorphism. Set $Y_\sigma = Y \times_{T,\sigma} \operatorname{Spec}(\mathcal{O}^\wedge_{S,s})$. The point y_σ is a point of Y_σ mapping to y and the closed point of $\operatorname{Spec}(\mathcal{O}^\wedge_{S,s})$. Finally,

$$\varphi \colon \mathcal{O}^\wedge_{X,x} \xrightarrow{\sim} \mathcal{O}^\wedge_{Y_\sigma, y_\sigma}$$

is an isomorphism of $\mathcal{O}^\wedge_{S,s}$-algebras. In this situation there exists a commutative diagram

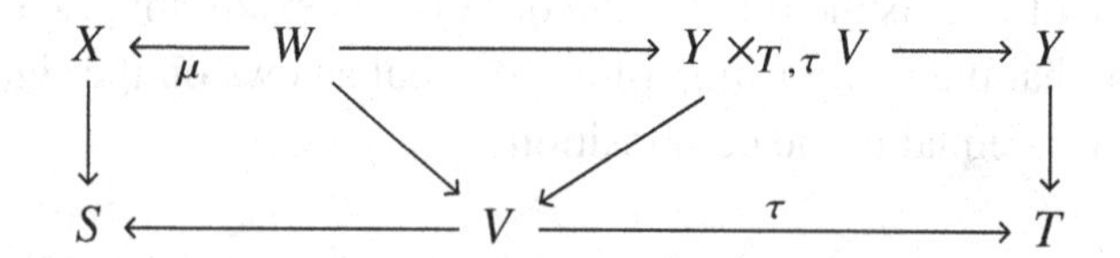

where W is a G-scheme, V is a scheme, all morphisms are G-equivariant, and points $w \in W^G$ and $v \in V$ are such that

(i) *$(V, v) \to (S, s)$ is an elementary étale neighborhood,*

(ii) *$\mu \colon (W, w) \to (X, x)$ is an elementary étale neighborhood and W^g surjects onto $\mu(W) \cap X^g$ for all $g \in G$,*

(iii) *$\tau(v) = t$.*

Let $y_\tau \in Y \times_{T,\tau} V$ correspond to y_σ via the identification $(Y_\sigma)_s = (Y \times_T V)_v$. Then

(iv) *$(W, w) \to (Y \times_{T,\tau} V, y_\tau)$ is an elementary étale neighborhood.*

Proof The proof of [10, Tag 0GDX] works equivariantly. We will outline that proof here (omitting critical details that were checked in the original!) and explain what changes occur in the equivariant setting.

One initially constructs the black part of the diagram

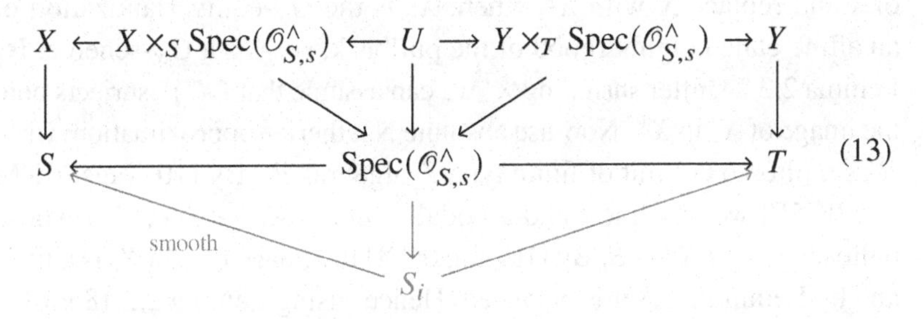

(13)

with $U \to X \times_S \operatorname{Spec}(\mathcal{O}^{\wedge}_{S,s})$ and $U \to U \times_T \operatorname{Spec}(\mathcal{O}^{\wedge}_{S,s})$ étale, using Lemma 2.3.18 for the existence of U. In particular, the top row of (13) consists of G-schemes and equivariant homomorphisms, $u \in U^G$ maps to $x \in X$ and $y \in Y$, and U^g surjects onto the image of U in $(X \times_S \operatorname{Spec}(\mathcal{O}^{\wedge}_{S,s}))^g$. Next one uses our hypothesis that $\mathcal{O}_{S,s}$ is a G-ring in order to write $\operatorname{Spec}(\mathcal{O}^{\wedge}_{S,s})$ as the limit of smooth affine S-schemes S_i; after increasing i, by [10, Tag 01ZC], we may assume we have the gray part of the diagram (13). By [10, Tags 01ZM and 07RP] we can, after increasing i, find a diagram of G-schemes over S_i

$$\begin{array}{ccccc} X \times_S S_i & \longleftarrow & U_i & \longrightarrow & Y \times_T S_i \\ & \searrow & \downarrow & \swarrow & \\ & & S_i & & \end{array} \tag{14}$$

with $U_i \to X \times_S S_i$, $U_i \to Y \times_T S_i$ étale, such that it pulls back to the relevant part of (13) and the map $U \to U_i$ is G-equivariant. (To see that we can choose U_i to have a G-action, observe that by [10, 01ZM], for each $g \in G$, we can increase i to find an automorphism g_i of U_i that restricts to g. After further increasing i we can assume that all necessary diagrams relating to group actions and equivariance of maps commute.) After increasing i again, using Lemma 2.2.18 and [10, Tag 07RR] we can ensure that U_i^g surjects onto the image of U in $(X \times_S S_i)^g$. Set $u_i \in U_i$ as the image of $u \in U$; since u was G-fixed so is u_i. Finally we use [10, Tag 057G] to find a closed subscheme $V \subset S_i$ containing the image of u_i such that $V \to S$ is étale. We replace (14) with its pullback to V to get the desired diagram of the lemma. In particular, we set $W := V \times_{S_i} U_i$, and $w := u_i \in U_i^G$ is contained in this closed subscheme. The restriction to V preserves surjections and g-fixed loci by Lemma 2.2.18, so for all $g \in G$ W^g surjects onto $\mu(W) \cap X^g$ in the final diagram. □

Proof of Proposition 2.3.15 We follow the proof of [10, Tag 0CBY], so we will be brief here. Replace S with an affine open neighborhood $\mathrm{Spec}(R)$ of s and replace X with X', where X' is the G_x-equivariantization of an affine étale neighborhood of the pullback, so X is a G_x-scheme. By Lemma 2.2.24, after shrinking X' we can assume that $(X')^g$ surjects onto the image of X' in X^g. Now use absolute Noetherian approximation: write R as a filtered colimit of finite type $\mathbf{Z}$-algebras R_i. By [10, Tags 01ZM and 0C5F] we can find i and a nodal G-curve $X_i \to \mathrm{Spec}(R_i)$ whose pullback to S is $X \to S$. By [10, Tag 0C3I] the image of x in X_i is a node and by Lemma 2.3.8 it is balanced. Hence (using Lemma 2.2.18 for the properties of X^g) it suffices to prove the lemma for $X_i \to \mathrm{Spec}(R_i)$.

We have reduced to the case where S is affine and of finite type over $\mathbf{Z}$. By Lemma 2.2.3 and [10, Tag 02LF] there is an affine étale neighborhood $(V, v) \to (S, s)$ such that, if X_v denotes the base change of X to v, there is a split node $x_v \in X_v$ with image x. By Lemma 2.3.8, x_v is balanced since x is balanced.

We have reduced to the situation where x is a balanced split node of its fiber and $G = G_x$.

By Lemma 2.3.14, we can after another étale base change we can assume that $X \to S$ satisfies all the assumptions of Lemma 2.3.13. We conclude the argument as in the proof of [10, Tag 0CBY], using Lemma 2.3.13 in place of [10, Tag 0CBX] and Lemma 2.3.19 in place of [10, Tag 0GDX]. □

We give a first application of Lemma 2.3.19 and Proposition 2.3.15 to the étale local structure of nodal G-curves.

Lemma 2.3.20 *Let X be a nodal G-curve over an algebraically closed field k and let $x \in X(k)$.*

(i) *The point x is smooth if and only if there is a G_x-equivariant isomorphism $\mathcal{O}^{\mathrm{sh}}_{X,x} \simeq k[u]^{\mathrm{sh}}_{(u)}$ with the G_x-action on the target given by $g \cdot u = \xi(g)u$ for some faithful $\xi\colon G_x \to k^\times$.*

(ii) *The point x is a balanced node if and only if there is a G_x-equivariant isomorphism $\mathcal{O}^{\mathrm{sh}}_{X,x} \simeq \left(k[u,v]/(uv)\right)^{\mathrm{sh}}_{(u,v)}$, with the G_x-action on the target given by $g \cdot u = \xi(g)u$ and $g \cdot v = \xi^{-1}(g)v$ for some faithful $\xi : G_x \to k^\times$.*

Proof (i). If x is a smooth point, Lemma 2.3.4 gives a faithful character $\xi\colon G_x \to k^\times$ and a G_x-equivariant isomorphism $\mathcal{O}^{\wedge}_{X,x} \simeq k[\![u]\!]$. Let $Y = \mathrm{Spec}\, k[u]$ with k-linear G_x-action given by ξ on u. Lemma 2.3.19 applied to X and Y over k (together with the fact that $\mathrm{Spec}\, k$ has no nontrivial connected étale covers) then guarantees the existence of a pointed G_x-scheme (W, w) together with G_x-equivariant étale morphisms $(W, w) \longrightarrow (X, x)$ and $(W, w) \longrightarrow (Y, (u))$. Since étale morphisms induce isomorphisms on strict henselizations, we get the desired description of $\mathscr{O}^{\mathrm{sh}}_{X,x}$. Conversely, assume $\mathscr{O}^{\mathrm{sh}}_{X,x}$ has the form given in (i) above. Since $\mathbf{A}^1_k$ is smooth and $\mathscr{O}^{\mathrm{sh}}_{X,x} \simeq k[u]^{\mathrm{sh}}_{(u)} \simeq \mathscr{O}^{\mathrm{sh}}_{\mathbf{A}^1,0}$, [10, Tag 00TV] and [10, Tag 06LN] show that x is a smooth point of X.

(ii). If x is a node, then an argument similar to that used in (i) above shows that $\mathscr{O}^{\mathrm{sh}}_{X,x}$ has the desired form (use Proposition 2.3.15 in place of Lemmas 2.3.4 and 2.3.19). Conversely, assume $\mathscr{O}^{\mathrm{sh}}_{X,x}$ has the form given in (ii) above. Completing the isomorphism $\mathscr{O}^{\mathrm{sh}}_{X,x} \simeq \left(k[u,v]/(uv)\right)^{\mathrm{sh}}_{(u,v)}$ yields a G_x-equivariant isomorphism $\mathscr{O}^{\wedge}_{X,x} \simeq k[\![u,v]\!]/(uv)$, by [10, 06LJ(b)] and the fact that henselizations and strict henselizations coincide for local rings whose residue fields are algebraically closed (as follows from [10, Tag 0BSL]). This shows that x is a balanced node in the sense of Definition 2.3.6. □

2.3.3 The stack of prestable balanced curves

Let G be a finite group. Recall that if $X \longrightarrow S$ is a prestable G-curve then S is a $\mathrm{Spec}(\mathbf{Z}[1/|G|])$-scheme. In this section we construct a locally closed substack of $\mathfrak{M}_{\mathrm{Spec}(\mathbf{Z}[1/|G|])}$ parametrizing balanced prestable G-curves. We will repeatedly use Proposition 2.3.15 to work étale-locally (see Remark 2.3.16 for a discussion of why a flat local picture is not enough).

Remark 2.3.21 As a warmup for the next lemma, recall from [10, Tag 034H] the definition of the inertia stack $\mathscr{I}$ of an algebraic stack $\mathscr{X}$. An element of $\mathscr{I}(S)$ is a pair (x, α) with $x \in \mathscr{X}(S)$ and $\alpha \in \mathrm{Aut}_S(x)$ (by definition, α is an arrow $x \longrightarrow x$ lying over $\mathrm{id}\colon S \longrightarrow S$). A morphism $(x_1, \alpha_1) \longrightarrow (x_2, \alpha_2)$ between objects of $\mathscr{I}(S)$ is an arrow $f\colon x_1 \longrightarrow x_2$ such that $f \circ \alpha_1 = \alpha_2 \circ f$.

Lemma 2.3.22 *Let $\mathscr{X}$ be an algebraic stack with separated diagonal. Let $\mathscr{I}$ be the inertia stack of $\mathscr{X}$ and let $\prod_G \mathscr{I}$ be the fiber product $\mathscr{I} \times_{\mathscr{X}} \cdots \times_{\mathscr{X}} \mathscr{I}$ of $|G|$ copies of $\mathscr{I}$. Then there is a closed substack $\mathscr{X}(G) \subset \prod_G \mathscr{I}$ given on S-valued points by*

$$\mathscr{X}(G)(S) := \{(x, \theta) \mid x \in \mathscr{X}(S),\ \theta \in \mathrm{Hom}^{\mathrm{gp}}_S(G_S, \mathrm{Aut}_S(x))\} \qquad (15)$$

where $\mathrm{Hom}^{\mathrm{gp}}_S(G_S, \mathrm{Aut}_S(x))$ is the set of homomorphisms of group algebraic spaces over S.

Remark 2.3.23 An object of $\prod_G \mathscr{I}(S)$ may be written as $(x, (\alpha_g)_{g \in G})$

where $x \in \mathcal{X}(S)$ and $\alpha_g \in \mathrm{Aut}_S(x)$ for each $g \in G$. Thus, in Lemma 2.3.22, the groupoid fiber $\mathcal{X}(G)(S)$ is viewed as a full subcategory of $\prod_G \mathcal{I}(S)$ by identifying an object $(x,\theta) \in \mathcal{X}(G)(S)$ with the object $(x,(\theta(g))_{g\in G}) \in \prod_G \mathcal{I}(S)$. In particular, an arrow $(x_1,\theta_1) \to (x_2,\theta_2)$ between objects of $\mathcal{X}(G)(S)$ is given by a "G-equivariant arrow" $x_1 \to x_2$; i.e., for each $g \in G$, the arrow $x_1 \to x_2$ commutes with the automorphisms specified by $\theta_1(g)$ and $\theta_2(g)$.

Proof of Lemma 2.3.22 For an object x of $\mathcal{X}(S)$, a morphism $\theta \in \mathrm{Hom}_S(G_S, \mathrm{Aut}_S(x))$ is equivalent to sections $\theta_g \colon S \to \mathrm{Aut}_S(x)$ for every $g \in G$. Let $1 \in G$ be the identity and let $1_S \colon S \to \mathrm{Aut}_S(x)$ be the identity section, $\iota\colon \mathrm{Aut}_S(x) \to \mathrm{Aut}_S(x)$ the involution, and $c\colon \mathrm{Aut}_S(x)\times\mathrm{Aut}_S(x) \to \mathrm{Aut}_S(x)$ the composition. That the map θ is a homomorphism translates to the identities (finite in number)

$$\begin{aligned} \theta_1 = 1_S, \qquad\qquad & \theta_{g^{-1}} = \iota\circ\theta_g \text{ for each } g\in G, \\ \theta_{gh} = c(\theta_g,\theta_h) & \text{ for each } (g,h)\in G\times G. \end{aligned} \tag{16}$$

Since the diagonal of $\mathcal{X}$ is separated, the algebraic space $\mathrm{Aut}_T(x) \to T$ is separated, and its diagonal is a closed embedding. So the proof of [10, Tag 01KM] shows that the locus where any one of the equalities in (16) holds is represented by a closed subscheme of S. Intersecting these, we get a closed subscheme of S representing the property that θ restricts to a group homomorphism. □

Setting $\mathcal{X} = \mathfrak{N}_{\mathrm{Spec}(\mathbf{Z}[1/|G|])}$ in Lemma 2.3.22 we obtain the following corollary.

Corollary 2.3.24 *There is an algebraic stack $\mathfrak{N}_{\mathrm{Spec}(\mathbf{Z}[1/|G|])}(G)$ whose objects over a scheme S are prestable G-curves over S, and whose arrows are given by G-equivariant morphisms of G-curves over S.*

Define $\mathfrak{N}^{\mathrm{bal}} \subset \mathfrak{N}_{\mathrm{Spec}(\mathbf{Z}[1/|G|])}(G)$ to be the full subcategory whose objects are balanced prestable G-curves.

Lemma 2.3.25 *The subcategory $\mathfrak{N}^{\mathrm{bal}} \subset \mathfrak{N}_{\mathrm{Spec}(\mathbf{Z}[1/|G|])}(G)$ is an open substack. In particular, $\mathfrak{N}^{\mathrm{bal}}$ is algebraic.*

Proof We show that $\mathfrak{N}^{\mathrm{bal}} \subset \mathfrak{N}_{\mathrm{Spec}(\mathbf{Z}[1/|G|])}(G)$ is an open substack by using the discussion in [10, Tag 0E0E]. That is, we show that if $\pi\colon X \to S$ is a G-space over S and a prestable curve, there is a largest open subscheme $S^{\mathrm{bal}} \subset S$ such that the pullback $X \times_S S^{\mathrm{bal}}$ is a balanced G-curve, and the formation of S^{bal} commutes with arbitrary base change. Define $|S^{\mathrm{bal}}| \subset |S|$ to be the locus of points $s \in S$ such that X_s is a balanced prestable G-curve. By Remark 2.3.3 and Lemma 2.3.8 the locus $|S^{\mathrm{bal}}|$ is preserved by arbitrary base change, so it remains to show that $|S^{\mathrm{bal}}|$ is open.

Note that each composition $X^g \to X \to S$ is proper. By [10, Tag 0D4Q], there is an open subscheme $S' \subset S$ whose underlying set consists of points $s \in S$ such that X^g_s has dimension ≤ 0 for each $1 \neq g \in G$; i.e., X_s is a prestable G-curve. If $s \in S'$ is contained in $|S^{\mathrm{bal}}|$, Lemma 2.3.8(iii) and Proposition 2.3.15 give for each node $x \in X_s$ an etale map $f_x \colon U_x \to X$ such that $f(U_x) \to \pi \circ f(U_x)$ is a balanced nodal G-curve. Let $X'' = X^{\mathrm{sm}} \cup \left(\bigcup_x f_x(U_x)\right)$ where the union is over nodes x of X_s, and let $Z := X \setminus X''$. Now $S'' := S' \setminus \pi(Z) \subseteq S'$ is open (because π is proper) and it is the maximal subscheme over which the restriction of X to $\pi^{-1}(S'')$ is balanced prestable. □

Lemma 2.3.26 *Let $X \to S$ be a balanced prestable G-curve with coarse quotient $f \colon X \to X/G$ and let $x \in |X|$. If x is a smooth point (resp. node) of its fiber, then $f(x)$ is also a smooth point (resp. node) of its fiber.*

Proof Since the statement may be checked on geometric fibers of $X \to X/G \to S$ (using Remark 2.2.8) and since the map $X \to X/G$ commutes with arbitrary base change (Lemma 2.2.31), we reduce to the case where $S = \mathrm{Spec}(k)$ is the spectrum of an algebraically closed field $k = \bar{k}$. Let $V = \mathrm{Spec}(C)$ be an affine étale neighborhood of $f(x) \in X/G$. Let $U = V \times_{X/G} X$ and let $u \in U$ be a closed point mapping to x. Let $v \in V$ be the image of u.

Note that U is also affine, so $U = \mathrm{Spec}(B)$ for some ring B and $C = B^G$ by Lemma 2.2.30. By Lemma 2.2.31 we have

$$\mathcal{O}^{\wedge}_{V,v} = (\mathcal{O}^{\wedge}_{V,v} \otimes_C B)^G.$$

Let $u_1, \ldots, u_n$ be the points in the fiber of v, numbered so that $u = u_1$, and note that G acts transitively on these points. By [10, Tag 07N9] we have

$$\mathcal{O}^{\wedge}_{V,v} \otimes_C B = \mathcal{O}^{\wedge}_{U,u_1} \oplus \cdots \oplus \mathcal{O}^{\wedge}_{U,u_n}.$$

Moreover, the projection

$$(\mathcal{O}^{\wedge}_{U,u_1} \oplus \cdots \oplus \mathcal{O}^{\wedge}_{U,u_n})^G \to (\mathcal{O}^{\wedge}_{U,u_1})^{G_{u_1}}$$

is an isomorphism. Since Lemmas 2.2.18 and 2.2.19 show that $G_{u_1} \simeq G_x$, we conclude that

$$\mathcal{O}^{\wedge}_{V,v} \simeq (\mathcal{O}^{\wedge}_{U,u})^{G_x}. \tag{17}$$

Set $n := |G_x|$. Assume x is a (necessarily balanced) node of its fiber. By Lemma 2.3.8(iii), u is also a balanced node and, by [10, Tag 0C57], $f(x)$ is a node if and only if v is. By (17) we have $\mathcal{O}^{\wedge}_{V,v} \xrightarrow{\simeq} (\mathcal{O}^{\wedge}_{U,u})^{G_x} \to \mathcal{O}^{\wedge}_{U,u}$, and hence by Lemma 2.3.8 we have $\mathcal{O}^{\wedge}_{V,v} \simeq k[\![t^n, s^n]\!]/(t^n s^n)$. In particular, $\mathcal{O}^{\wedge}_{V,v}$ is of the form in Definition 2.2.1(i) and hence v is a node.

Now assume x is a smooth point of its fiber. Since $U \to X$ is étale, u is a smooth point of its fiber. Using Lemma 2.3.4 and arguing as in the nodal case, we get $\mathcal{O}^{\wedge}_{V,v} = (\mathcal{O}^{\wedge}_{U,u})^{G_x} \simeq k[\![t^n]\!]$. This implies that v is smooth ([10, Tag 00TV, Tag 07NY]) and so by [10, Tag 05AX] $f(x)$ is smooth. □

Lemma 2.3.27 *If $X \to S$ is a balanced prestable G-curve, then the coarse quotient $X/G \to S$ is a prestable curve.*

Proof The map $X/G \to S$ is flat, proper, and of finite presentation by Lemma 2.2.31(ii) and Lemma 2.2.32. The remaining conditions may be checked on geometric fibers, so by Lemma 2.2.31(i) we may assume $S = \mathrm{Spec}(k)$ for some algebraically closed field $k = \bar{k}$. Now X/G is a nodal curve by Lemma 2.3.26 and it has finitely many connected components because X does and $X \to X/G$ is surjective (Lemma 2.2.29). Finally, X/G has pure dimension 1 because X does and $X \to X/G$ is finite and surjective; cf. [10, Tag 0ECG]. □

The following technical lemma will be used to analyze the geometry of X^g.

Lemma 2.3.28 *Let $X \to S$ be an algebraic space over a scheme S with $X_1, X_2 \subset X$ closed subspaces that are finite étale over S. Assume $|X_1| \subset |X_2|$. For each $x \in |X_1|$ there exists an open neighborhood $V \subset X$ of x such that $V \cap X_1 = V \cap X_2$.*

Proof Since $X_i \to S$ is finite, the subspace X_i is representable by a scheme (by definition [10, Tag 03ZP]). Applying [10, Tag 04HN] twice, we find an étale map $S' \to S$ such that the fibers $(X_1)_{S'}$ and $(X_2)_{S'}$ are both isomorphic to disjoint unions of schemes that are isomorphic to S' by the projection maps. The point x is in one of these. Removing the others from $X_{S'}$, we get an open neighborhood U of x in $X_{S'}$ such that $U \cap (X_1)_{S'} = U \cap (X_2)_{S'}$. If two closed subspaces of an algebraic space agree on an étale cover, then (by descent) they agree. Hence the desired neighborhood is the image of U in X. □

Lemma 2.3.29 *Let $X \to S$ be a balanced prestable G-curve.*

(i) *For every nonidentity $g \in G$, the map $(X^g \cap X^{\mathrm{sm}}) \to S$ is finite étale.*
(ii) *For each $x \in |X^{\mathrm{sm}}|$ there is an open neighborhood of x where*

$$\bigcup_{g \in (G \setminus \{1\})} (X^g \cap X^{\mathrm{sm}}) = X^h \cap X^{\mathrm{sm}}$$

for a generator h of G_x. In particular $\bigcup_{g\in(G\setminus\{1\})}(X^g \cap X^{\mathrm{sm}}) \to S$ is finite étale.

Proof (i) We claim there is an open $X' \subset X$ containing the points of X that are nodes in their fibers such that the stabilizer of any smooth point in X' is trivial. Granting this, the complement of X' is a closed set containing $X^g \cap X^{\mathrm{sm}}$ and contained in X^{sm}. This shows that for $g \neq 1$ the map $(X^g \cap X^{\mathrm{sm}}) \to S$ is proper; moreover it is finite because its fibers are closed subsets of X^{sm} that do not contain any irreducible component; cf. [10, Tag 0A4X].

We now construct X'. For a point $x \in |X|$ that is a node in its fiber, we have a diagram as in Proposition 2.3.15, where in particular U^g surjects onto $\mu(U) \cap X^g$ for all $g \in G_x$. On the right-hand side of the diagram we may directly compute that for any $g \in (G_x \setminus \{1\})$ the fixed locus contains no smooth point, so by Lemma 2.2.18 and Remark 2.2.8, the locus $(W^{\mathrm{sm}})^g$ is empty. This forces $(U^{\mathrm{sm}})^g$ to be empty as well (since the smooth locus is preserved by étale covers), and hence the image of U contains no g-fixed point of X^{sm}.

Intersecting these images for all nonidentity $g \in G_x$ gives a neighborhood of x that contains no smooth g-fixed point for each $g \in (G_x \setminus \{1\})$, and furthermore, by removing X^h for $h \notin G_x$, we get an open neighborhood of x whose intersection with $X^g \cap X^{\mathrm{sm}}$ is empty for each $g \in (G \setminus \{1\})$. Let $X' \subset X$ be the union of these open sets (one for each node of X).

Finally, we show that $(X^g \cap X^{\mathrm{sm}}) \to S$ is étale. We may assume $S = \mathrm{Spec}(R)$. As in the proof of Proposition 2.3.15 we can find a subring $R_i \subset R$ finitely generated over $\mathbf{Z}$ and a prestable G-curve $X_i \to \mathrm{Spec}(R_i)$ whose base change to S is $X \to S$. Since base change preserves $X^g \cap X^{\mathrm{sm}}$ by Remark 2.2.8 and Lemma 2.2.18 and smoothness can be checked locally on the target [10, Tag 02VL], we have reduced to the case where S is Noetherian.

By [10, Tag 056U], it suffices to prove that the morphism is étale at every closed point $x \in X$ with image $s \in S$ such that the induced field extension $\kappa(s) \subseteq \kappa(x)$ is finite separable. Let $x \in X$ be of that form. By [10, Tag 02LF, Tag 01JT], we can find an étale neighborhood

$(S', s') \to (S, s)$ and $x' \in X'_S$ closed such that $\kappa(s') = \kappa(x') = \kappa(x)$. We thus reduce to the case where $\kappa(x) = \kappa(s)$. Similarly, by Lemma 2.3.14, we can assume that the hypotheses of Lemma 2.3.13 are satisfied; taking G_x-invariants in (4) and using [10, Tag 00MA(2)], this implies that the natural map $\mathcal{O}^\wedge_{S,s} \to \mathcal{O}^\wedge_{X^g,x}$ is an isomorphism.

One may now apply [10, Tag 039N, Tag 039M] to see that $X \to S$ is étale at x.

(ii) It suffices to consider the case $X = X^{\mathrm{sm}}$. Fix $x \in |X|$ and let $(G \setminus G_x) \subset G$ be the subset equal to the complement of G_x. Then $U = X \setminus \left(\bigcup_{g \in G \setminus G_x} X^g\right)$ is an open subset containing x. Now let h be a generator of G_x. For every $n = 1, \dots, (|G_x| - 1)$ we have an inclusion $X^h \subset X^{h^n}$ of finite étale spaces over S with $x \in |X^h|$. By Lemma 2.3.28 we have an open set $V_n \subset X$ where $X^h \cap V_n = X^{h^n} \cap V_n$. The desired open neighborhood is $U \cap \left(\bigcap_{n=1}^{|G_x|-1} V_n\right)$. □

2.3.4 Construction of the moduli stack of admissible covers

For the rest of this chapter we fix a finite group G and work over $\mathrm{Spec}(\mathbf{Z}[1/|G|])$. Recall the moduli of marked connected prestable curves $\mathfrak{M}_\star$ from Definition 2.2.12. Also, we will write $\mathfrak{M}_\star$ for $(\mathfrak{M}_\star)_{\mathrm{Spec}(\mathbf{Z}[1/|G|])}$. We observe that $\mathfrak{N}^{\mathrm{bal}}$ is already defined over $\mathrm{Spec}(\mathbf{Z}[1/|G|])$. Set

$$\mathfrak{H}_0 = \underline{\mathrm{Hom}}_{\mathfrak{N}^{\mathrm{bal}} \times \mathfrak{M}_\star}(\mathfrak{C}_{\mathfrak{N}^{\mathrm{bal}}}, \mathfrak{C}_{\mathfrak{M}_\star}),$$

where $\underline{\mathrm{Hom}}$ denotes the Hom stack from Lemma 2.2.14. Let S be a scheme over $\mathrm{Spec}(\mathbf{Z}[1/|G|])$. An object in $\mathfrak{N}^{\mathrm{bal}} \times \mathfrak{M}_\star$ over S is a morphism $S \to \mathfrak{N}^{\mathrm{bal}} \times \mathfrak{M}_\star$, or equivalently a pair of morphisms $S \to \mathfrak{N}^{\mathrm{bal}}$ and $S \to \mathfrak{M}_\star$. Set

$$X := (\mathfrak{C}_{\mathfrak{N}^{\mathrm{bal}}})_S := \mathfrak{C}_{\mathfrak{N}^{\mathrm{bal}}} \times_{\mathfrak{N}^{\mathrm{bal}}} S \quad \text{and} \quad Y := (\mathfrak{C}_{\mathfrak{M}_\star})_S := \mathfrak{C}_{\mathfrak{M}_\star} \times_{\mathfrak{M}_\star} S.$$

An object of $\mathfrak{H}_0$ over this pair is a morphism $X \to Y$ of S-algebraic spaces. The stack $\mathfrak{H}_0$ is algebraic by Lemmas 2.2.13, 2.3.25, 2.2.11, and 2.2.14. Note that if $f\colon X \to Y$ is an admissible G-cover as defined in Definition 2.1.1, then X is a balanced nodal G-curve by Lemma 2.3.20(ii). Hence we may define the stack of admissible covers as a full subcategory of $\mathfrak{H}_0$.

Definition 2.3.30 The *stack of admissible covers* $\mathcal{A}dm(G)$ is the full subcategory of $\mathfrak{H}_0$ whose objects over a scheme S are given by

$$\mathcal{A}dm(G)(S) := \{(f\colon X \to Y) \in \mathfrak{H}_0(S) \mid f \text{ is an admissible } G\text{-cover}\}.$$

We now define a sequence of full subcategories $\mathfrak{H}_i \subset \mathfrak{H}_0$ and show that each is an algebraic stack, in fact a locally closed substack, of $\mathfrak{H}_0$. Briefly, if $f\colon X \to Y$ is an object of $\mathfrak{H}_0$, then

- it is in $\mathfrak{H}_1$ if f is G-invariant,
- it is in $\mathfrak{H}_2$ if Y is the coarse quotient X/G, and
- it is in $\mathfrak{H}_3$ if f sends smooth points of X with nontrivial stabilizer to marked points of Y.

In Theorem 2.3.40 we will show that $\mathfrak{H}_3$ is equal to $\mathcal{A}dm(G)$, thus proving Theorem A, that $\mathcal{A}dm(G)$ is algebraic.

Let $\mathfrak{H}_1 \subset \mathfrak{H}_0$ be the full subcategory of G-invariant morphisms. The S-valued points of $\mathfrak{H}_1$ are given by

$$\mathfrak{H}_1(S) = \{(f\colon X \to Y) \in \mathfrak{H}_0 \mid f \circ g = f \text{ for all } g \in G\}.$$

Lemma 2.3.31 *The full subcategory $\mathfrak{H}_1$ is an algebraic stack and the inclusion $\mathfrak{H}_1 \to \mathfrak{H}_0$ is a closed immersion.*

Proof Let S be a scheme and let $f\colon X \to Y$ be an object of $\mathfrak{H}_0$. We can find a closed subscheme $S' \subset S$ with the following universal property: a morphism $T \to S$ factors through S' if and only if the restriction of f to $X \times_S T$ is G-invariant. Granted this, the statement is a consequence of Lemma 2.2.10 and [10, Tag 04YL].

Define maps $\Gamma, \Gamma'\colon S \to \prod_G \mathrm{Hom}_S(X,Y)$ given by $\Gamma = (f \circ g)_{g \in G}$ and $\Gamma' = (f)_{g \in G}$. The desired locus S' is the equalizer of Γ and Γ'. This is closed as in [10, Tag 01KM] since $\mathrm{Hom}_S(X,Y)$ is separated ([10, Tag 0DPN]). □

If $f\colon X \to Y$ is an object of $\mathfrak{H}_1(S)$, then f is G-invariant and therefore factors through the coarse moduli space X/G of Definition 2.2.26. Define $\mathfrak{H}_2 \subset \mathfrak{H}_1$ to be the full subcategory whose objects over a scheme S are given by

$$\mathfrak{H}_2(S) = \left\{ (f\colon X \to Y) \in \mathfrak{H}_1(S) \,\middle|\, \begin{array}{c} \text{the induced map } \bar{f}\colon X/G \to Y \\ \text{is an isomorphism} \end{array} \right\}.$$

Lemma 2.3.32 *The full subcategory $\mathfrak{H}_2$ is an algebraic stack and the inclusion $\mathfrak{H}_2 \to \mathfrak{H}_1$ is an open immersion.*

Proof Let S be an arbitrary scheme and $f\colon X \to Y$ be an object of $\mathfrak{H}_1(S)$. We need to show that there exists an open subscheme $S' \subset S$ with the following universal property: a morphism $\alpha\colon T \to S$ factors through S' if and only if the morphism $X_T/G \to Y_T$ induced by $\alpha_T\colon X_T \to Y_T$ is an isomorphism.

This morphism factors through $X_T/G \to X/G \times_S T$, which is always an isomorphism by Lemma 2.2.31. Thus, it suffices to find an open $S' \subset S$ such that $\alpha\colon T \to S$ factors through S' if and only if the natural morphism $X/G \times_S T \to Y \times_S T$ is an isomorphism. Since $X/G \to S$ is a family of prestable curves by Lemma 2.3.27 and $Y \to S$ is a family of prestable curves by assumption, the existence of S' follows from [10, Tag 05XD].

Now the statement is a consequence of Lemmas 2.3.31 and 2.2.10 and [10, Tag 04YL]. □

If $Y \to S$ is a marked connected prestable curve, we define $Y_\star \subset Y$ to be the locus of marked points (the union of the images of the sections). Likewise if $X \to S$ is a balanced prestable G-curve, let $X_\star := \bigcup_{g \in (G \setminus \{1\})} X^g \cap X^{\mathrm{sm}}$. This is a closed subspace of X by the proof of Lemma 2.3.29(i). Define $\mathfrak{H}_3 \subset \mathfrak{H}_2$ to be the full subcategory whose objects over a scheme S are

$$\mathfrak{H}_3(S) = \{(f\colon X \to Y) \in \mathfrak{H}_2(S) \mid f(|X_\star|) \subseteq |Y_\star|\}.$$

Lemma 2.3.33 *The full subcategory $\mathfrak{H}_3$ is an algebraic stack and the inclusion $\mathfrak{H}_3 \to \mathfrak{H}_2$ is a closed immersion.*

Proof Let S be a scheme and $f\colon X \to Y$ be an object of $\mathfrak{H}_2(S)$. We need to show that there exists a closed subscheme $S' \subseteq S$ with the following universal property: a morphism $T \to S$ factors through S' if and only if $f_T(|X_{T,\star}|) \subseteq |Y_{T,\star}|$. This is equivalent to the condition that $f_T(X_{T,\star}) \subseteq Y_{T,\star}$ by Lemma 2.3.29(ii) and Lemma 2.3.28 (note that $Y_{T,\star}$ is finite étale over T since it is the union of the images of some disjoint sections of $Y_T \to T$). Since $(X_\star)_T = X_{T,\star}$ by Remark 2.2.8 and Lemma 2.2.18, this is in turn equivalent to the condition that the closed immersion $j\colon (X_\star)_T \hookrightarrow X$ factors through $f^{-1}(Y_\star)$.

Let $\mathcal{I}$ be the ideal sheaf of $f^{-1}(Y_\star)$ inside X. By [10, Tag 03MB] the condition is equivalent to the composition $j^*\mathcal{I} \to \mathcal{O}_{(X_\star)_T} = j^*\mathcal{O}_{X_\star}$ being zero. By Lemma 2.3.29, the $\mathcal{O}_X$-module $\mathcal{O}_{X_\star}$ is flat and proper over S. Therefore [10, Tag 083M] shows that this locus is represented by a closed subscheme $S' \subseteq S$. Now the statement is a consequence of Lemmas 2.3.32 and 2.2.10 and [10, Tag 04YL]. □

All that remains is to show that the algebraic stack $\mathfrak{H}_3$ is isomorphic to the category $\mathcal{A}dm(G)$ of admissible G-covers. Observe that both $\mathfrak{H}_3$ and $\mathcal{A}dm(G)$ are full subcategories of $\mathfrak{H}_0$, so it suffices to prove that the objects of $\mathfrak{H}_3(S)$ and $\mathscr{A}dm(G)(S)$ give the same subset of $\mathfrak{H}_0(S)$. To that end, we first clarify what was meant by a G-torsor in Definition 2.1.1.

Definition 2.3.34 Let X be an algebraic space with an action by G and let $\mathcal{Y}$ be a Deligne–Mumford stack. A G-invariant morphism $X \to \mathcal{Y}$ is a G-torsor if

(i) the map $(\mathrm{act}, \mathrm{pr}_2)\colon G \times_{\mathcal{Y}} X \to X \times_{\mathcal{Y}} X$ is an isomorphism, and
(ii) there is an étale cover $Y \to \mathcal{Y}$ such that the pullback $X \times_{\mathcal{Y}} Y \to Y$ admits a section.

Remark 2.3.35 A G-torsor in our sense defines a G-torsor on the small étale site of $\mathcal{Y}$ in the sense of [10, Tag 03AH].

Remark 2.3.36 By [10, Tag 03AI], a G-torsor $X \to \mathcal{Y}$ is isomorphic to $G \times \mathcal{Y} \to \mathcal{Y}$ with the natural left G-action if and only if it admits a section. In particular, Definition 2.3.34(ii) guarantees that a G-torsor $X \to \mathcal{Y}$ is an étale morphism.

Example 2.3.37 If X is an algebraic space with an action by G then $X \to [X/G]$ is a G-torsor. Condition (i) from Definition 2.3.34 is [10, Tag 04M9]. Condition (ii) also follows from [10, Tag 04M9] using Remark 2.3.36 and the fact that $X \to [X/G]$ is étale. This last fact may be deduced from [10, Tag 04M9] using [10, Tag 0CIQ(3)] and the fact that $X \to [X/G]$ is smooth by [10, Tag 04X0].

The next couple of lemmas show that an admissible cover $X \to Y$ over S belongs to $\mathfrak{H}_3(S)$.

Lemma 2.3.38 *Let S be a scheme. Let $X \to S$ and $Y \to S$ be two prestable curves and $f\colon X \to Y$ a morphism over S. Assume that for every $x \in |X|$, if x is smooth (resp. nodal) in its fiber, then $f(x)$ is smooth (resp. nodal) in its fiber. Finally, suppose there is an open subset $U \subset Y$ such that, for every geometric point $\bar{s}$ of S, the fiber $U_{\bar{s}}$ is dense in every irreducible component of the fiber $Y_{\bar{s}}$, and $f^{-1}(U_{\bar{s}})$ is dense in every component of $X_{\bar{s}}$. If $f\colon f^{-1}(U) \to U$ is an isomorphism, then f is an isomorphism.*

Proof By [10, Tag 05XD], there is an open subscheme $S' \subseteq S$ such that a map $T \to S$ factors through S' if and only if $f_T : X_T \to Y_T$ is an isomorphism. We want to prove that S' contains all points of S. Thus, it suffices to show that f is an isomorphism on every geometric fiber, so we may assume that $S = \operatorname{Spec} k$ for an algebraically closed field k.

Let $\nu_X : X^\nu \to X$ and $\nu_Y : Y^\nu \to Y$ denote the normalization maps.

Then f induces a map $f^\nu\colon X^\nu \to Y^\nu$ between disjoint unions of integral, normal curves using [10, Tag 035Q(4)] and the fact that every component of X dominates a component of Y.

We first show that f^ν is an isomorphism. Since f gives a bijection between irreducible components of $f^{-1}(U)$ and U (or equivalently X and Y by density), f^ν induces a bijection between connected components. Moreover, our assumptions imply that f^ν defines birational maps on the various connected components, hence isomorphisms by [10, Tag 0BY1], and we see that f^ν is an isomorphism as desired. Since ν_X and ν_Y become isomorphisms after restriction to X^{sm} and Y^{sm}, respectively, we see that f restricts to an isomorphism $X^{\mathrm{sm}} \to Y^{\mathrm{sm}}$.

Next we show that f induces a bijection on points. Surjectivity follows since f^ν and ν_Y are both surjective. For injectivity, since f preserves nodes and smooth points and f^{sm} is an isomorphism, we only need to show that a node $y \in Y$ has a unique node in X mapping to it. By [10, Tag 0CBW] applied to Y, and the fact that f^ν is an isomorphism, there are exactly two points in X^ν mapping to y. By [10, Tag 0CBW] applied to X, there can only be one node in X mapping to y. So f is a bijection on points, as desired.

Now let $x_1, \ldots, x_n$ be the points of X that are nodes in their fibers and let $y_1, \ldots, y_n$ be their images in Y. Since f is surjective, and using [10, Tag 00MB], we have that $Y^{\mathrm{sm}} \sqcup \bigsqcup_{i=1}^n \mathrm{Spec}(\mathcal{O}^\wedge_{Y,y_i}) \to Y$ is an fpqc covering of Y (see [10, Tag 03NW]). By [10, Tags 02L4 and 07N9], the fact that f^{sm} is an isomorphism, and the fact that f is injective on points, to complete the proof of the lemma we only have to show that the map $\mathcal{O}^\wedge_{Y,y_i} \to \mathcal{O}^\wedge_{X,x_i}$ induced by f is an isomorphism.

For this last step, let $A = \mathcal{O}^\wedge_{X,x_i}$ and $B = \mathcal{O}^\wedge_{Y,y_i}$, and let A' (resp. B') be the integral closure of A (resp. B) in its total ring of fractions. By [10, Tag 0C3V] and the fact that f^ν is an isomorphism, we have that f induces an isomorphism $f^\sharp\colon B' \xrightarrow{\sim} A'$. The proof of Lemma 2.3.5 describes A (resp. B) as the explicit subring of A' (resp. B') equal to the wedge $(A')_{\mathfrak{m}_1} \wedge (A')_{\mathfrak{m}_2}$ (resp. $(B')_{\mathfrak{n}_1} \wedge (B')_{\mathfrak{n}_2}$), where $\mathfrak{m}_1, \mathfrak{m}_2$ (resp. $\mathfrak{n}_1$, $\mathfrak{n}_2$) are the two maximal ideals lying over the maximal ideal $\mathfrak{m}_{x_i} \subset A$ (resp. $\mathfrak{n}_{y_i} \subset B$). Since $f^\sharp$ is an isomorphism, it sends the maximal ideals $\mathfrak{m}_1, \mathfrak{m}_2$ of A' to the maximal ideals $\mathfrak{n}_1, \mathfrak{n}_2$ of B', and hence induces an isomorphism of subrings $B \xrightarrow{\sim} A$.

□

Lemma 2.3.39 *If $f\colon X \to Y$ is an admissible G-cover in the sense* *of Definition 2.1.1, then the induced morphism $q\colon X/G \to Y$ is an isomorphism.*

Proof Let $Y^{\mathrm{gen}} \subset Y$ be the complement of the nodes and markings in Y. It is an open subset. Lemma 2.2.31 gives the following pullback diagram of open immersions:

$$\begin{array}{ccccc} f^{-1}(Y^{\mathrm{gen}}) & \longrightarrow & f^{-1}(Y^{\mathrm{gen}})/G & \xrightarrow{q^{\mathrm{gen}}} & Y^{\mathrm{gen}} \\ \downarrow & & \downarrow & & \downarrow \\ X & \longrightarrow & X/G & \xrightarrow{q} & Y \end{array}$$

We claim q^{gen} is an isomorphism. Indeed, by assumption $f^{-1}(Y^{\mathrm{gen}}) \to Y^{\mathrm{gen}}$ is a G-torsor. By Definition 2.3.34(i) and Remark 2.3.36, after passing to an étale cover of Y^{gen} we may assume that $f^{-1}(Y^{\mathrm{gen}}) \simeq G \times Y^{\mathrm{gen}}$, so $f^{-1}(Y^{\mathrm{gen}})/G \simeq Y^{\mathrm{gen}}$ in this situation. Now q^{gen} is an isomorphism by [10, Tags 04ZP and 04XD].

The locus $f^{-1}(Y^{\mathrm{gen}})$ is dense in every irreducible component of every geometric fiber $X_{\bar{s}}$; if not, some irreducible component of $X_{\bar{s}}$ would map to a node or marking of $Y_{\bar{s}}$, contradicting the finiteness of f [10, Tag 0397]. Now we apply Lemmas 2.3.27 and 2.3.38. □

Theorem 2.3.40 *As full subcategories of $\mathfrak{H}_0$, we have $\mathfrak{H}_3 = \mathcal{A}dm(G)$. In particular, $\mathcal{A}dm(G)$ is an algebraic stack.*

Proof Let S be a test scheme. We show that $\mathfrak{H}_3(S) = \mathcal{A}dm(G)(S)$ as full subcategories of $\mathfrak{H}_0(S)$. Since an admissible G-cover $f\colon X \to Y$ is G-invariant it belongs to $\mathfrak{H}_1(S)$. By Lemma 2.3.39, it is in $\mathfrak{H}_2(S)$. Lastly, let $x \in X_\star$. As G acts freely on a G-torsor, x cannot be contained in an open subset of X that is a G-torsor over its image. Thus, $f(x)$ is a node or marked point of Y by Definition 2.1.1(i). Lemma 2.3.26 rules out that $f(x)$ is a node, hence $f(x) \in Y_\star$. In other words, f is in $\mathfrak{H}_3(S)$, showing "$\supseteq$".

Conversely, assume that $f\colon X \to Y$ is an object of $\mathfrak{H}_3(S)$. Then f is finite by Lemma 2.2.29. If $x \in X$ maps to $y \in Y$, then by Lemma 2.3.26, x is a node (resp. smooth) if and only if y is a node (resp. smooth). In particular, part (ii) of Definition 2.1.1 holds. By Lemma 2.2.31 the map $X_{\bar{k}} \to Y_{\bar{k}}$ is still a coarse quotient, so by Lemma 2.2.30 and [10, Tag 05WR] we have

$$\mathcal{O}^{\mathrm{sh}}_{Y_{\bar{k}},y} = (\mathcal{O}^{\mathrm{sh}}_{X_{\bar{k}},x_1} \times \cdots \times \mathcal{O}^{\mathrm{sh}}_{X_{\bar{k}},x_n})^G$$

where $x_1, \ldots x_n$ are the points in the fiber of y. Since the right-hand side is canonically isomorphic to $(\mathscr{O}^{\mathrm{sh}}_{X_{\bar{k},x_i}})^{G_{x_i}}$ for any i, by the proof of Lemma 2.3.26, parts (iii) and (iv) of Definition 2.1.1 follow from Lemma 2.3.20.

Finally, let $U := f^{-1}(Y^{\mathrm{sm}} \smallsetminus Y_\star)$ be the locus of X away from the markings and nodes of Y. By Lemma 2.2.31 and the condition defining $\mathfrak{H}_2$, the restriction $f|_U : U \to Y^{\mathrm{sm}} \smallsetminus Y_\star$ can be identified with $U \to U/G$. Moreover, the condition defining $\mathfrak{H}_3$ guarantees that G acts freely on the locus U. Therefore, the map from the inertia stack $\mathscr{I}_{[U/G]} \to [U/G]$ is an isomorphism [10, Tag 06PB] so that $[U/G]$ is an algebraic space [10, Tag 04SZ] and $[U/G] \simeq U/G$ by the universal property of coarse moduli spaces. In particular, by Example 2.3.37 the map $U \to U/G$ is a G-torsor, giving part (i) of Definition 2.1.1 and "$\subseteq$". □

Acknowledgements We thank Pieter Belmans, Aise Johan de Jong, and Wei Ho for organizing the Stacks Project Workshop 2020, during which we started this project. Special thanks go to our group mentor, Ravi Vakil, for suggesting that we work on this topic and for all his help and guidance during the preparation of this chapter. We also benefited from helpful conversations with Martin Olsson. A careful referee suggested many corrections and improvements to the chapter.

During the preparation of this material, N.D. was supported by the INSPIRE Fellowship (IF160348) of the Department of Science and Technology, Government of India, E.R. was supported by the National Science Foundation under Grant No. DMS-1926686, and R.W. was supported by an NSF Postdoctoral Research Fellowship, award number 200213.

References

[1] Dan Abramovich, Alessio Corti, and Angelo Vistoli. Twisted bundles and admissible covers. *Comm. Algebra*, 31(8):3547–3618, 2003. Special issue in honor of Steven L. Kleiman. doi:10.1081/AGB-120022434.

[2] Dan Abramovich, Martin Olsson, and Angelo Vistoli. Tame stacks in positive characteristic. *Ann. Inst. Fourier (Grenoble)*, 58(4):1057–1091, 2008.

[3] Allen B. Altman and Steven L. Kleiman. Compactifying the Picard scheme. *Adv. Math.*, 35(1):50–112, 1980.

[4] Michael Artin. Algebraic approximation of structures over complete local rings. *Inst. Hautes Études Sci. Publ. Math.*, (36):23–58, 1969.

[5] José Bertin and Matthieu Romagny. Champs de Hurwitz. *Mém. Soc. Math. Fr. (N.S.)*, 125–126:219 p., 2011. doi:10.24033/msmf.437.

[6] Matt Emerton. Colimits of schemes. Answer in https://mathoverflow.net/q/10000 (version: 2010-12-10).

[7] Joe Harris and David Mumford. On the Kodaira dimension of the moduli space of curves. *Invent. Math.*, 67(1):23–88, 1982. With an appendix by William Fulton. doi:10.1007/BF01393371.

[8] Shinichi Mochizuki. The geometry of the compactification of the Hurwitz scheme. *Publ. Res. Inst. Math. Sci.*, 31(3):355–441, 1995. doi:10.2977/prims/1195164048.

[9] David Ishii Smyth. Towards a classification of modular compactifications of $\mathscr{M}_{g,n}$. *Invent. Math.*, 192(2):459–503, 2013. doi:10.1007/s00222-012-0416-1.

[10] The Stacks Project Authors. *The Stacks project*. https://stacks.math.columbia.edu.

[11] Stefan Wewers. *Construction of Hurwitz spaces*. University of Essen, 1998.

3

Projectivity of the moduli space of vector bundles on a curve

Jarod Alper
University of Washington

Pieter Belmans
University of Luxembourg

Daniel Bragg
University of California, Berkeley

Jason Liang
University of Michigan

Tuomas Tajakka
Stockholm University

Dedicated to the memory of
Conjeevaram Srirangachari Seshadri

Abstract We discuss the projectivity of the moduli space of semistable vector bundles on a curve of genus $g \geq 2$. This is a classical result from the 1960s, obtained using geometric invariant theory. We outline a modern approach that combines the recent machinery of good moduli spaces with determinantal line bundle techniques. The crucial step producing an ample line bundle follows an argument by Faltings with improvements by Esteves and Popa. We hope to promote this approach as a blueprint for other projectivity arguments.

Introduction

In [17], Kollár took a modern approach to the construction of the moduli space $\overline{\mathrm{M}}_g$ of stable curves of genus $g \geq 2$ as a projective variety, avoiding methods from geometric invariant theory. The approach can be summarized in three steps.

(i) Prove that the stack of stable curves is a proper Deligne–Mumford stack $\overline{\mathcal{M}}_g$ [7].

(ii) Use the Keel–Mori theorem to show that $\overline{\mathcal{M}}_g$ has a coarse moduli space $\overline{\mathcal{M}}_g \to \overline{\mathrm{M}}_g$, where $\overline{\mathrm{M}}_g$ is a proper algebraic space [15].

(iii) Prove that some line bundle on $\overline{\mathcal{M}}_g$ descends to an *ample* line bundle on $\overline{\mathrm{M}}_g$ [17, 6].

In [5], Chapter 1 of this volume, the projectivity of $\overline{\mathrm{M}}_g$ was established by giving a stack-theoretic treatment of Kollár's paper.

We take a similar approach to constructing the moduli space $\mathrm{M}^{\mathrm{ss}}_X(r, d)$ of semistable vector bundles of rank r and degree d on a smooth, proper curve X as a projective variety. There are again three steps in the proof.

(i) Prove that $\mathcal{M}^{\mathrm{ss}}_X(r, d)$ is a universally closed algebraic stack of finite type [19].

(ii) Use an analogue of the Keel–Mori theorem to show that there exists a *good moduli space* $\mathcal{M}^{\mathrm{ss}}_X(r, d) \to \mathrm{M}^{\mathrm{ss}}_X(r, d)$ [2], where $\mathrm{M}^{\mathrm{ss}}_X(r, d)$ is a proper algebraic space.

Unlike the case of $\overline{\mathcal{M}}_g$, the stack $\mathcal{M}^{\mathrm{ss}}_X(r, d)$ is never Deligne–Mumford and the morphism to $\mathrm{M}^{\mathrm{ss}}_X(r, d)$ will identify nonisomorphic vector bundles, introducing the notion of S-equivalence.

(iii) Prove that a suitable determinantal line bundle on $\mathcal{M}^{\mathrm{ss}}_X(r, d)$ descends to an ample line bundle on $\mathrm{M}^{\mathrm{ss}}_X(r, d)$.

In Sections 3.1–3.3 we define and study the moduli stack $\mathcal{M}^{\mathrm{ss}}_X(r, d)$ and establish steps (i) and (ii). The main points are summarized in Theorems 3.2.17 and 3.3.12.

In Section 3.5 we discuss step (iii) and give a detailed exposition of the construction of an ample line bundle on $\mathrm{M}^{\mathrm{ss}}_X(r, d)$ using techniques from Esteves and Popa [9]. This method is more recent than the first GIT-free construction, due to Faltings [10], which received expository accounts by Seshadri (and Nori) in [22] and for a special case by Hein in [13]. All these constructions build upon the notion of determinantal line bundles, which we introduce in Section 3.4.

Because the existence results for good moduli spaces from [2] are currently restricted to characteristic 0, we make that restriction as well.

3.1 The moduli stack of all vector bundles

Let k be an algebraically closed field of characteristic 0, and let X be a smooth, projective, connected curve over k.

We denote by $\underline{\mathrm{Coh}}_X$ the stack of all coherent sheaf on X. For integers

$r > 0$ and d, we let $\mathcal{M}_X(r,d)$ denote the stack parameterizing vector bundles on X of rank r and degree d. An object of $\mathcal{M}_X(r,d)$ over a scheme S is a vector bundle $\mathcal{E}$ on $X \times S$ such that for every geometric point $\operatorname{Spec} K \to S$, the restriction $\mathcal{E}|_{X \times \operatorname{Spec} K}$ has rank r and degree d.

Proposition 3.1.1 *The stack $\mathcal{M}_X(r,d)$ is an algebraic stack, locally of finite type over k with affine diagonal.*

Proof The stack $\underline{\mathrm{Coh}}_X$ is algebraic by [23, Tag 09DS]. Given a map $T \to \underline{\mathrm{Coh}}_X$ corresponding to a sheaf $\mathcal{E}$ on $T \times X$ flat over T, the locus $Z \subseteq T \times X$ where $\mathcal{E}$ is not locally free is closed, so $U = T \setminus \mathrm{pr}_1(Z) \subseteq T$ is open by properness, showing that the substack $\mathcal{M} \subset \underline{\mathrm{Coh}}_X$ parameterizing vector bundles on X is an open substack. The Hilbert polynomial P_E of a coherent sheaf E on X is determined by its rank and degree, and by the Riemann–Roch theorem it is given by

$$\mathrm{P}_E(n) = \mathrm{rk}(E)n + \deg(E) + \mathrm{rk}(E)(1-g). \tag{1}$$

It is locally constant in flat families, so by the discussion in [23, Tag 0DNF] there is an open and closed substack $\mathcal{M}_X(r,d) \subset \mathcal{M}$ parameterizing vector bundles of rank r and degree d.

Since $\underline{\mathrm{Coh}}_X$ is locally of finite type over k by [23, Tag 0DLZ], and has affine diagonal by [23, Tag 0DLY], so does $\mathcal{M}_X(r,d)$. □

There is a universal vector bundle $\mathcal{E}_{\mathrm{univ}}$ on $X \times \mathcal{M}_X(r,d)$. Specializing to the rank-1 case, we obtain the Picard stack $\mathcal{P}\mathrm{ic}_X^d$ parameterizing line bundles of degree d on X. In this case the universal bundle is called the Poincaré bundle and denoted $\mathcal{P}$. The line bundle $\det(\mathcal{E}_{\mathrm{univ}})$ on $X \times \mathcal{M}_X(r,d)$ induces a determinant morphism

$$\det \colon \mathcal{M}_X(r,d) \to \mathcal{P}\mathrm{ic}_X^d,$$

with the property that $(\mathrm{id}_X \times \det)^* \mathcal{P} \cong \det(\mathcal{E}_{\mathrm{univ}})$.

Let L be a line bundle on X of degree d corresponding to a closed point of $\mathcal{P}\mathrm{ic}_X^d$, and let $[L] \colon \operatorname{Spec} k \to \mathcal{P}\mathrm{ic}_X^d$ be the corresponding morphism. The algebraic stack parameterizing vector bundles with fixed determinant L is defined as the fiber product

$$\begin{array}{ccc} \mathcal{M}_X(r,L) & \longrightarrow & \mathcal{M}_X(r,d) \\ \downarrow & \square & \downarrow{\scriptstyle \det} \\ \operatorname{Spec} k & \xrightarrow{[L]} & \mathcal{P}\mathrm{ic}_X^d . \end{array} \tag{2}$$

Explicitly, an object of $\mathcal{M}_X(r,L)$ over a k-scheme S is a pair $(\mathcal{E},\varphi)$, where $\mathcal{E}$ is a vector bundle on $X \times S$ with rank r and degree d on the fibers, and $\varphi\colon \det \mathcal{E} \xrightarrow{\sim} L|_{X\times S}$ is an isomorphism.

An alternative approach to fixing the determinant uses the residual gerbe $\mathrm{B}\mathbb{G}_\mathrm{m} \hookrightarrow \mathcal{P}ic_X^d$ instead. We compare these in Example 3.3.2.

Proposition 3.1.2 *The stack $\mathcal{M}_X(r, L)$ is an algebraic stack, locally of finite type over k with affine diagonal.*

Proof This follows by applying [23, Tag 04TF] for the algebraicity and [23, Tag 06FU] for being locally of finite type. To see that $\mathcal{M}_X(r, L)$ has affine diagonal, we use the fact that $\det\colon \mathcal{M}_X(r, d) \to \mathcal{P}ic_X^d$ has affine diagonal (because both stacks do) and that having affine diagonal is stable under base change. □

Let $\mathcal{X}$ be an algebraic stack. Let $x_0 \in \mathcal{X}(k)$ be a k-point. By [23, Tag 07WZ], the tangent space $\mathrm{T}_{\mathcal{X},x_0}$ to $\mathcal{X}$ at x_0 is the set of isomorphism classes of objects in the groupoid $\mathcal{X}_{x_0}(k[\varepsilon]/(\varepsilon^2))$, where $\mathcal{X}_{x_0}$ denotes the slice category. Using deformation theory one can prove the following result about the moduli stacks.

Proposition 3.1.3

(i) *The algebraic stacks $\mathcal{M}_X(r, d)$ and $\mathcal{M}_X(r, L)$ are smooth over k.*

(ii) *If E is a vector bundle of rank r and $\varphi\colon L \cong \det E$ is an isomorphism, then there are identifications of the Zariski tangent space*

$$\mathrm{T}_{\mathcal{M}_X(r,d),[E]} \cong \mathrm{Ext}^1_X(E, E), \tag{3}$$

$$\mathrm{T}_{\mathcal{M}_X(r,L),[(E,\varphi)]} \cong \ker\left(\mathrm{Ext}^1_X(E, E) \xrightarrow{\mathrm{tr}} \mathrm{H}^1(X, \mathcal{O}_X)\right). \tag{4}$$

Remark 3.1.4 Let E be a *simple* vector bundle of rank r and degree d, that is, $\dim \mathrm{Hom}_X(E, E) = 1$. We will see in Lemma 3.2.5 that stable bundles are simple. By the Riemann–Roch theorem, we have

$$\begin{aligned}\dim \mathrm{Ext}^1_X(E, E) &= \dim \mathrm{Hom}_X(E, E) - \chi(X, E^\vee \otimes E)\\ &= 1 - \deg(E^\vee \otimes E) - \mathrm{rk}(E^\vee \otimes E)(1 - g)\\ &= r^2(g - 1) + 1.\end{aligned}$$

Next, note that since we are working over a field of characteristic 0, the trace map $E^\vee \otimes E \to \mathcal{O}_X$ is split by the section $\frac{1}{d}\,\mathrm{id}_E\colon \mathcal{O}_X \to E^\vee \otimes E \cong \mathcal{H}om_X(E, E)$. Thus, the surjection $\mathrm{Ext}^1_X(E, E) \to \mathrm{H}^1(X, \mathcal{O}_X)$ in (4) is split, and so its kernel has dimension

$$\begin{aligned}\dim \mathrm{Ext}^1_X(E, E) - \dim \mathrm{H}^1(X, \mathcal{O}_X) &= r^2(g - 1) + 1 - g\\ &= (r^2 - 1)(g - 1).\end{aligned}$$

3.2 Semistability

In order to obtain a more well-behaved geometric object, we will restrict ourselves to an open substack $\mathcal{M}_X^{\mathrm{ss}}(r,d)$ of the stack $\mathcal{M}_X(r,d)$. This substack parameterizes the *semistable* vector bundles. The Jordan–Hölder and Harder–Narasimhan filtrations explain how it is sufficient to consider these vector bundles in order to describe all vector bundles on X:

- the Harder–Narasimhan filtration allows one to uniquely break up any vector bundles into semistable constituents;
- the Jordan–Hölder filtration allows one to break up a semistable vector bundle into *stable* bundles.

The essential properties of semistability, which will be formalized in Theorem 3.2.17, are

- openness: $\mathcal{M}_X^{\mathrm{ss}}(r,d)$ is an open substack of $\mathcal{M}_X(r,d)$;
- boundedness: $\mathcal{M}_X^{\mathrm{ss}}(r,d)$ is a quasicompact;
- irreducibility of the semistable locus.

For more information, in particular on the boundedness of semistable sheaves in higher dimensions and in arbitrary characteristic, see [11], Chapter 4 in this volume.

3.2.1 Definition and basic properties

Definition 3.2.1 Given a nonzero vector bundle E on X, we define the *slope* of E to be the rational number

$$\mu(E) := \frac{\deg(E)}{\mathrm{rk}(E)}.$$

Recall that on a smooth curve, any subsheaf of a locally free sheaf is locally free.

Definition 3.2.2 A vector bundle E on X is

- *semistable* if, for every nonzero subsheaf $E' \subset E$, we have $\mu(E') \leq \mu(E)$,
- *stable* if it is semistable and the only subsheaf $E' \subset E$ with $\mu(E') = \mu(E)$ is $E' = E$,
- *polystable* if it is isomorphic to a finite direct sum $\bigoplus_i E_i$, where each E_i is stable with $\mu(E_i) = \mu(E)$.

Below are some basic properties of semistable vector bundles.

Lemma 3.2.3 *For a line bundle L, a vector bundle E is (semi)stable if and only if $E \otimes L$ is (semi)stable.*

Proof Note that, for any vector bundle E, we have

$$\deg(E \otimes L) = \deg(E) + \mathrm{rk}(E) \cdot \deg(L).$$

Therefore, by definition of the slope,

$$\mu(E \otimes L) = \mu(E) + \deg(L).$$

Hence, for any subsheaf $E' \subset E$, we have

$$\mu(E) - \mu(E') = \mu(E \otimes L) - \mu(E' \otimes L).$$

By definition, E is (semi)stable if and only if for all $E' \subset E$, the above quantity is positive (resp. non-negative), which holds if and only if $E \otimes L$ is (semi)stable. □

Lemma 3.2.4 *If E and F are semistable vector bundles with $\mu(E) > \mu(F)$, then* $\mathrm{Hom}_X(E, F) = 0$.

Proof Suppose that on the contrary $E \to F$ is a nonzero map, and let K and I be its kernel and image respectively. Since E is semistable, $\mu(K) \leq \mu(E)$. Hence,

$$\begin{aligned}\mu(I)\,\mathrm{rk}(I) &= \mu(E)\,\mathrm{rk}(E) - \mu(K)\,\mathrm{rk}(K)\\ &\geq \mu(E)\,\mathrm{rk}(E) - \mu(E)\,\mathrm{rk}(K)\\ &= \mu(E)\,\mathrm{rk}(I)\end{aligned}$$

so we see that $\mu(I) \geq \mu(E) > \mu(F)$. But now I cannot be a subsheaf of F because F is semistable. □

The following is proved similarly.

Lemma 3.2.5 *Any nonzero endomorphism of a stable bundle E is invertible. In particular, as k is algebraically closed we have* $\mathrm{End}_X(E) \cong k$.

We will use the following lemma in the proof of Proposition 3.2.15 to establish boundedness.

Lemma 3.2.6 *If E is a semistable vector bundle on X and $\mu(E) > 2g-1$, then E is globally generated and* $\mathrm{H}^1(X, E) = 0$.

Proof Let $x \in X(k)$ be a closed point. Let $\mathcal{O}_x$ denote the skyscraper sheaf at x. By taking the sheaf cohomology of the short exact sequence

$$0 \to E \otimes \mathcal{O}_X(-x) \to E \to \mathcal{O}_x^{\oplus \mathrm{rk}(E)} \to 0,$$

we see that it suffices to show that $\mathrm{H}^1(X, E \otimes \mathcal{O}_X(-x)) = 0$. By Serre duality, this is equivalent to $\mathrm{Hom}_X(E \otimes \mathcal{O}_X(-x), \omega_X) = 0$. Now, $\mu(E \otimes \mathcal{O}_X(-x)) = \mu(E) - 1 > 2g - 2 = \mu(\omega_X)$ so the result follows by Lemma 3.2.4 since ω_X is semistable as a line bundle. □

3.2.2 Jordan–Hölder and Harder–Narasimhan filtrations

Every semistable vector bundle E on X admits a *Jordan–Hölder filtration*

$$0 = E_0 \subset E_1 \subset E_2 \subset \cdots \subset E_k = E$$

where each subquotient $G_i = E_i/E_{i-1}$ is a *stable* vector bundle with $\mu(G_i) = \mu(E)$. This filtration is not unique, but the list of subquotients is unique by the following result. See [20] for a proof.

Proposition 3.2.7 (Jordan–Hölder filtration) *Let E be a semistable vector bundle on X. Suppose that*

$$0 = E_0 \subset E_1 \subset E_2 \subset \cdots \subset E_k = E$$

and

$$0 = E'_0 \subset E'_1 \subset E'_2 \subset \cdots \subset E'_\ell = E$$

are two Jordan–Hölder filtrations with associated graded bundles $G_i = F_i/F_{i-1}$ and $G'_i = F'_i/F'_{i-1}$. We have $\ell = k$, and there exists a permutation $\sigma \in \mathrm{Sym}_k$ such that $G'_i \cong G_{\sigma(i)}$.

Hence the *associated graded* $\mathrm{gr}_E = \bigoplus_{i=1}^k G_i$ of a semistable bundle E is well defined up to isomorphism. This allows us to make the following definition, where the S refers to Seshadri.

Definition 3.2.8 Semistable vector bundles E and E' of the same rank and degree are called *S-equivalent* if their associated graded vector bundles are isomorphic.

We can think of the Jordan–Hölder filtration as stating that semistable vector bundles are built out of stable vector bundles of the same slope. Meanwhile, the result below tells us that *all* vector bundles are built out of semistable bundles in a *unique* way. See again [20] for a proof.

Proposition 3.2.9 (Harder–Narasimhan filtration) *Let F be a vector bundle on X. There exists a unique filtration of F by subbundles*

$$0 = F_0 \subset F_1 \subset F_2 \subset \cdots \subset F_k = F$$

such that $G_i = F_i/F_{i-1}$ *is semistable for each* i *and*

$$\mu(G_1) > \mu(G_2) > \cdots > \mu(G_k).$$

This filtration is called the Harder–Narasimhan filtration.

Example 3.2.10 ($X = \mathbb{P}^1$) By a theorem of Birkhoff and Grothendieck, vector bundles on $X = \mathbb{P}^1$ split as direct sums of line bundles. Writing $E \cong \bigoplus_{i=1}^{k} O_{\mathbb{P}^1}(d_i)^{\oplus m_i}$ with $d_1 > d_2 > \cdots > d_k$, the associated graded bundles of the Harder–Narasimhan filtration are $G_i = O_{\mathbb{P}^1}(d_i)^{\oplus m_i}$ and their slopes are $\mu(G_i) = d_i$. In particular E is semistable if and only if $k = 1$, and stable if and only if it is a line bundle.

3.2.3 Openness of semistability

We can now show that semistability is an open property in families. We will use the results from [23, Tag 0DM1] on Quot, where we will denote by $\mathrm{Quot}^{P(t)}_{\mathcal{E}/X/B}$ the object from [23, Tag 0DP6] for a numerical polynomial $P(t)$, a morphism $f\colon X \to B$ of schemes, and a quasicoherent sheaf $\mathcal{E}$ on X.

Proposition 3.2.11 (Openness of semistability) *Let* S *be a* k*-scheme and* $\mathcal{E}$ *be a vector bundle of rank* r *and relative degree* d *on* $X \times S \to S$. *The points* $p \in S$ *for which* $\mathcal{E}_p$ *is semistable form an open subscheme.*

Proof By standard approximation techniques we can assume that S is of finite type over k. We may also assume that S is connected. We will show that the locus of points $s \in S$ for which $\mathcal{E}_s$ is *not semistable* is closed by expressing it as a finite union of images of proper morphisms. Because our family is flat, the Hilbert polynomial $\mathrm{P}_{\mathcal{E}_s}(t)$ is independent of $p \in S$.

By definition, $\mathcal{E}_s$ is not semistable if and only if there exists some subsheaf $E' \hookrightarrow \mathcal{E}_s$ with $\mu(E') > d/r$. Considering the cokernel of $E' \hookrightarrow \mathcal{E}_s$, we see that $\mathcal{E}_s$ is not semistable if and only if $\mathcal{E}_s$ has a quotient with Hilbert polynomial $\mathrm{P}_{\mathcal{E}_s}(t) - \mathrm{P}_{E'}(t)$ for some E' with $\mu(E') > d/r$. In other words, the set of $s \in S$ such that $\mathcal{E}_s$ is not semistable is the union of images of relative Quot schemes $\mathrm{Quot}^{Q(t)}_{\mathcal{E}/X\times S/S}$ for the projection $X \times S \to S$, where $Q(t)$ ranges over the set

$$\left\{Q(t) = \mathrm{P}_{\mathcal{E}_s}(t) - \mathrm{P}_{E'}(t) \mid \exists s \in S,\, E' \subset \mathcal{E}_s \text{ with } \mu(E') > d/r\right\}.$$

Relative Quot schemes are proper by [23, Tag 0DPC], so to show that the nonsemistable locus is closed, it suffices to show that the above set of polynomials is finite.

The Hilbert polynomial $\mathrm{P}_{E'}(t)$ is determined by the rank and degree of E'. Being a subsheaf of E there are only finitely many possibilities for the rank of E'. The requirement $\mu(E') > d/r$ puts a lower bound on the degree of E'. It remains to show that the degrees of all subsheaves of $\mathcal{E}_s$ as s varies over S are bounded above.

To see this, choose a finite morphism $X \to \mathbb{P}^1$ and consider the induced morphism $f\colon X \times S \to \mathbb{P}^1 \times S$. If $E' \subset \mathcal{E}_p$ is a subsheaf, then $f_* E' \subset f_* \mathcal{E}_p$ is a subsheaf.

According to Example 3.2.10, vector bundles on $\mathbb{P}^1$ split as direct sums of line bundles. Only finitely many splitting types occur among the $f_* \mathcal{E}_s$ for $s \in S$, since otherwise $\mathrm{h}^0(\mathbb{P}^1, f_* \mathcal{E}_s)$ would achieve infinitely many values, which is impossible by semicontinuity. Moreover, for each vector bundle on $\mathbb{P}^1$, the set of Euler characteristics of all its subsheaves is bounded above.

Because f is finite, $\chi(X, f_* E') = \chi(X, E')$. It follows that the Euler characteristics of all subsheaves $E' \subset \mathcal{E}_s$ are bounded above and, hence, the degrees of all subsheaves $E' \subset \mathcal{E}_s$ are bounded above. □

3.2.4 Boundedness and irreducibility

To obtain a finite-type moduli space, the objects parameterized by the moduli problem must be bounded in the following sense.

Definition 3.2.12 (Boundedness) A collection of isomorphism classes of vector bundles on X is called *bounded* if there exists a scheme S of finite type and a vector bundle $\mathcal{F}$ on $X \times S \to S$ such that every isomorphism class in the collection is represented by $\mathcal{F}_s$ for some k-point $s \in S$.

When ranging over a bounded collection of isomorphism classes, the set of values of any discrete algebraic invariant must be finite. However, as soon as the rank is at least 2, fixing the rank and degree is not sufficient to obtain a bounded family.

Example 3.2.13 (An unbounded family) Let $x \in X$ be a point, and consider the collection $\{F_n\}$ of rank $r \geq 2$, degree d vector bundles

$$F_n = \mathcal{O}_X(-nx) \oplus \mathcal{O}_X((n+d)x) \oplus \mathcal{O}_X^{\oplus(r-2)}.$$

The range of values of $\mathrm{h}^1(X, F_n)$ is unbounded in this family. Hence, the collection of all isomorphism classes of vector bundles of fixed rank $r \geq 2$ and degree d cannot be bounded.

One important source of bounded families comes from quotients of a fixed vector bundle.

Example 3.2.14 (Boundedness of quotients) Fix a vector bundle F on X and a polynomial $P(t) \in \mathbb{Z}[t]$. The Quot scheme $\mathrm{Quot}^{P(t)}_{F/X/\operatorname{Spec} k}$ parameterizing quotients of F with Hilbert polynomial $P(t)$ is of finite type by [23, Tag 0DP9]. Therefore, the collection of isomorphism classes of vector bundles that appear as a quotient of F with Hilbert polynomial $P(t)$ is bounded: restricting the universal quotient sheaf on $X \times \mathrm{Quot}^{P(t)}_{F/X/\operatorname{Spec} k}$ to the open locus where the quotient is locally free provides a family of vector bundles realizing each isomorphism class.

Example 3.2.14 helps us see that semistable bundles of fixed rank and degree are bounded.

Proposition 3.2.15 (Boundedness of semistability) *The collection of isomorphism classes of semistable vector bundles of rank r and degree d on X is bounded.*

Proof Let $x \in X$ be a k-point and let n be an integer such that $d/r + n > 2g-1$. For every semistable vector bundle E, the vector bundle $E \otimes \mathcal{O}_X(nx)$ is semistable by Lemma 3.2.3 and has slope greater than $2g - 1$. Hence, by Lemma 3.2.6, the vector bundle $E \otimes \mathcal{O}_X(nx)$ is globally generated and $\mathrm{h}^1(X, E \otimes \mathcal{O}_X(nx)) = 0$. In other words, $E \otimes \mathcal{O}_X(nx)$ is a quotient of $V \otimes \mathcal{O}_X$ where V is a vector space of dimension $\mathrm{h}^0(X, E \otimes \mathcal{O}_X(nx))$, and this dimension is independent of E. It follows that every semistable bundle E of rank r and degree d arises as a quotient

$$V \otimes \mathcal{O}_X(-nx) \to E.$$

Boundedness of this family now follows from Example 3.2.14 with $F = V \otimes \mathcal{O}_X(-nx)$ and $P(t) = d + r(t + 1 - g)$. □

Proposition 3.2.16 (Irreducibility of the moduli stack) *The moduli stack $\mathcal{M}^{\mathrm{ss}}_X(r, d)$ is irreducible. The same is true for $\mathcal{M}^{\mathrm{ss}}_X(r, L)$.*

Proof For a closed point $x \in X(k)$, Lemma 3.2.3 says that E is semistable if and only if $E \otimes \mathcal{O}_X(x)$ is semistable. This gives rise to an isomorphism $\mathcal{M}^{\mathrm{ss}}_X(r, d) \to \mathcal{M}^{\mathrm{ss}}_X(r, d + r)$. Therefore, it suffices to prove the claim when $d > r(2g - 1)$. In other words, we may assume that the slope is at least $2g - 1$, and so all semistable bundles of our fixed rank and degree are globally generated by Lemma 3.2.6.

We will construct an irreducible variety V^{ss} that maps to $\mathcal{M}^{\mathrm{ss}}_X(r, d)$ surjectively. Because V^{ss} is a scheme the morphism is automatically representable, and so we get an induced surjection on the level of topological spaces by [23, Tag 04XI]. But then the topological space is irreducible, and hence so is the stack. Our space V^{ss} will parameterize semistable extensions of a degree-d line bundle by a rank-$(r-1)$ trivial bundle. For a fixed line bundle L, the extensions

$$0 \to \mathcal{O}_X^{r-1} \to E \to L \to 0 \tag{1}$$

are parameterized by $\mathrm{Ext}^1_X(L, \mathcal{O}_X^{r-1}) = \mathrm{H}^1(X, L^\vee)^{r-1}$. To parameterize such extensions of all degree-d line bundles, let $\mathcal{L}$ be a Poincaré line bundle on $X \times \mathrm{Pic}^d_X$ and let $\pi\colon X \times \mathrm{Pic}^d_X \to \mathrm{Pic}^d_X$ be the projection. By the theorem on cohomology and base change, the fiber of the vector bundle $V = (\mathrm{R}^1\pi_*\mathcal{L}^\vee)^{r-1}$ over $L \in \mathrm{Pic}^d_X$ is the space of extensions $\mathrm{H}^1(X, L^\vee)^{r-1}$, and there is a universal extension on $X \times V$. By Proposition 3.2.11, the locus V^{ss} where the universal extension is semistable is an open subvariety, so we obtain a map $V^{ss} \to \mathcal{M}^{ss}_X(r, d)$. Clearly V is irreducible, hence so is the open subvariety V^{ss}.

To show that this map is *surjective*, we must show that every semistable bundle E arises as an extension as in (1). By the first paragraph of the proof, E is globally generated. For such an E, we claim there exists a subspace $\Gamma \subset \mathrm{H}^0(X, E)$ of dimension $r-1$, so that the evaluation map

$$\Gamma \otimes \mathcal{O}_X \to E \tag{2}$$

has rank $r-1$ at all points of X. To see this, we use a dimension count. At each point $x \in X$, we consider the map of vector spaces $\mathrm{H}^0(X, E) \to E(x)$, where $E(x)$ denotes the fiber of E at x. We wish to select a Γ in the Grassmannian $\mathrm{Gr}(r-1, \mathrm{H}^0(X, E))$ which does not meet the kernel nontrivially. Since the kernel has codimension r, the linear spaces meeting it form a codimension-2 Schubert variety in $\mathrm{Gr}(r-1, \mathrm{H}^0(X, E))$. Varying x over X, we see that the collection of Γ's where (2) has rank less than $r-1$ has codimension at least 1. Hence, there exists some Γ where (2) has rank $r-1$. For such Γ, we obtain an exact sequence

$$0 \to \Gamma \otimes \mathcal{O}_X \to E \to \det(E) \to 0.$$

This shows that E occurs as an extension of a degree d line bundle by a rank-$(r-1)$ trivial bundle. Hence, E is in the image of $V^{ss} \to \mathcal{M}^{ss}_X(r, d)$.

The proof for $\mathcal{M}^{ss}_X(r, L)$ is similar, using an open subscheme of $\mathrm{Ext}^1_X(L, \mathcal{O}_X^{r-1})$ surjecting onto the stack. □

We summarize the results of this section as follows.

Theorem 3.2.17 *The moduli stacks $\mathcal{M}_X^{ss}(r,d)$ and $\mathcal{M}_X^{ss}(r,L)$ are irreducible algebraic stacks, of finite type over k with affine diagonal. They are moreover smooth, and their Zariski tangent spaces are given in Proposition 3.1.3.*

Remark 3.2.18 One can show that if the genus of X is at least 2, then for any integer $r > 0$ and line bundle $L \in \mathrm{Pic}_X$ there exist stable vector bundles of rank r and determinant L; see, e.g., [21, Lemma 4.3]. Hence the open substacks $\mathcal{M}_X^{s}(r,d) \subseteq \mathcal{M}_X^{ss}(r,d)$ and $\mathcal{M}_X^{s}(r,L) \subseteq \mathcal{M}_X^{ss}(r,L)$ of *stable* bundles are nonempty and dense.

3.3 Good moduli spaces

To study the properties of an algebraic stack, we often want to find the best possible approximation as a scheme or algebraic space. For a Deligne–Mumford stack such a procedure is given by the Keel–Mori theorem, producing a coarse moduli space. For an algebraic stack the first author introduced the notion of a good moduli space in [1], and in [2] criteria to verify its existence were established. The goal of this section is to briefly discuss how $\mathcal{M}_X^{ss}(r,d)$ fits into this general framework.

The following example explains why the stack $\mathcal{M}_X^{ss}(r,d)$ is not Deligne–Mumford and the Keel–Mori theorem does not apply.

Example 3.3.1 The automorphism group of any semistable vector bundle contains a canonical copy of $\mathbb{G}_m$ acting by scalars, and moreover $\mathrm{Aut}(F) = \mathbb{G}_m$ when F is a stable bundle by Lemma 3.2.5. In particular, the stack $\mathcal{M}_X^{ss}(r,d)$ is not Deligne–Mumford, as the automorphism groups of all points are infinite.

A prototypical example of a strictly semistable vector bundle of rank 2 and degree 0 is the polystable bundle $F = \mathcal{O}_X \oplus \mathcal{O}_X$. Now $\mathrm{Aut}(F) = \mathrm{GL}_2$, so the dimension of the automorphism group can jump up when going from the stable locus to the strictly semistable locus.

The automorphism groups do not have to be reductive either. If F fits in a non-split exact sequence

$$0 \to \mathcal{O}_X \xrightarrow{i} F \xrightarrow{q} \mathcal{O}_X \to 0,$$

then the map $(a,b) \mapsto a \cdot \mathrm{id}_F + ab \cdot i \circ q$ gives an an isomorphism $\mathbb{G}_m \times \mathbb{G}_a \cong \mathrm{Aut}(F)$. Such extensions are parameterized by the projective space $\mathbb{P}\mathrm{Ext}^1_X(\mathcal{O}_X, \mathcal{O}_X) \cong \mathbb{P}^{g-1}$.

Example 3.3.2 For $\mathcal{M}_X^{ss}(r, L)$ as defined in (2) one has generically a finite stabilizer, because for a *stable* vector bundle with fixed determinant the automorphisms need to preserve the isomorphism, and hence are isomorphic to μ_r.

If instead we took the fiber product along the closed substack $\mathrm{B}\mathbb{G}_m \hookrightarrow \mathcal{P}\mathrm{ic}_X^d$ then the resulting closed substack would have the same stabilizers as $\mathcal{M}_X^{ss}(r, d)$, and in particular they would never be finite.

Observe that, with the first definition of $\mathcal{M}_X^{ss}(r, L)$, if $\gcd(r, \deg L) = 1$ then there are no strictly semistable vector bundles, and $\mathcal{M}_X^{ss}(r, L)$ is Deligne–Mumford. In this case $\mathcal{M}_X^{ss}(r, L)$ is a μ_r-gerbe over its coarse moduli space. With the second definition the stack is, rather, a $\mathbb{G}_m$-gerbe over this coarse moduli space.

Example 3.3.3 The moduli stack $\mathcal{M}_X^{ss}(r, d)$ is *not* separated, despite its many other nice properties. Indeed, if it were separated, its diagonal would be proper. But the diagonal is affine, hence it would be finite, and so the stabilizers would be finite as they are the fibers of the diagonal. This contradicts the previous example. Likewise, the moduli stack $\mathcal{M}_X^{ss}(r, L)$ is *not* separated if the strictly semistable locus is nonempty.

Definition 3.3.4 Let $f : \mathcal{X} \to X$ be a quasi-compact and quasi-separated morphism over k, where $\mathcal{X}$ is an algebraic stack and X is an algebraic space. We say that f is a *good moduli space* if

(i) f is cohomologically affine, i.e., $f_* : \mathrm{Qcoh}\,\mathcal{X} \to \mathrm{Qcoh}\,X$ is exact;
(ii) the morphism $\mathcal{O}_X \to f_*\mathcal{O}_{\mathcal{X}}$ is an isomorphism.

We now summarize some of the fundamental properties of good moduli spaces that will be relevant for us. See [1] for a complete statement.

Theorem 3.3.5 *Let $\mathcal{X}$ be an algebraic stack of finite type over k, and let $f : \mathcal{X} \to X$ be a good moduli space.*

(i) *The map f is surjective, universally closed, and initial for maps to algebraic spaces.*
(ii) *The space X is of finite type over k.*
(iii) *The map f identifies two k-points $x, y : \mathrm{Spec}\,k \to \mathcal{X}$ if and only if the closures of $\{x\}$ and $\{y\}$ in $|\mathcal{X}|$ intersect. In particular, f induces a bijection on closed points.*
(iv) *The pullback f^* induces an equivalence of categories between vector bundles on X and those vector bundles $\mathcal{F}$ on $\mathcal{X}$ such that the stabilizer of any closed point $x : \mathrm{Spec}\,k \to \mathcal{X}$ acts trivially on $x^*\mathcal{F}$.*

The following existence theorem for good moduli spaces is a specialized version of [2, Theorem A], in the form in which we will apply it. The terms “S-complete” and “Θ-reductive” in the statement refer to certain valuative criteria that we will discuss below.

Theorem 3.3.6 *Let $\mathcal{X}$ be an algebraic stack of finite type over k with affine diagonal. There exists a good moduli space $\mathcal{X} \to X$ where X is a separated algebraic space over k if and only if*

(i) *$\mathcal{X}$ is S-complete, and*
(ii) *$\mathcal{X}$ is Θ-reductive.*

Moreover, X is proper if and only if $\mathcal{X}$ is universally closed, or equivalently satisfies the existence part of the valuative criterion for properness.

3.3.1 Langton's semistable reduction theorem

We first discuss the last part of Theorem 3.3.6. Since the stacks $\mathcal{M}_X^{\mathrm{ss}}(r,d)$ and $\mathcal{M}_X^{\mathrm{ss}}(r,L)$ are quasicompact and have affine diagonal, hence are quasiseparated, it follows from [23, Tags 0CLW and 0CLX] that universal closedness is equivalent to the existence part of the valuative criterion for properness.

The valuative criterion states the following. If R is a discrete valuation ring R with residue field κ and fraction field K, then for every 2-commutative diagram

$$\begin{array}{ccc} \operatorname{Spec} K & \longrightarrow & \mathcal{X} \\ \downarrow & & \downarrow \\ \operatorname{Spec} R & \longrightarrow & \operatorname{Spec} k \end{array}$$

there exists a field extension K'/K and a discrete valuation ring $R' \subseteq K'$ dominating R, such that there exists a morphism $\operatorname{Spec} R' \to \mathcal{X}$ making the diagram

$$\begin{array}{ccccc} \operatorname{Spec} K' & \longrightarrow & \operatorname{Spec} K & \longrightarrow & \mathcal{X} \\ \downarrow & & \downarrow & \nearrow & \downarrow \\ \operatorname{Spec} R' & \longrightarrow & \operatorname{Spec} R & \longrightarrow & \operatorname{Spec} k \end{array}$$

2-commutative.

It turns out that for $\mathcal{M}_X^{\mathrm{ss}}(r,d)$ and $\mathcal{M}_X^{\mathrm{ss}}(r,L)$ it is not necessary to consider field extensions of K, by the following classical theorem of Langton [19].

Theorem 3.3.7 *Let E be a semistable vector bundle of rank r and degree d on X_K. There exists a vector bundle $\mathcal{E}$ on X_R such that $\mathcal{E}_K \cong E$ and $\mathcal{E}_\kappa$ is semistable. Thus, the stacks $\mathcal{M}_X^{\mathrm{ss}}(r,d)$ and $\mathcal{M}_X^{\mathrm{ss}}(r,L)$ are universally closed.*

According to Example 3.3.3, the stack $\mathcal{M}_X^{\mathrm{ss}}(r,d)$ is not separated, and there will in general be many extensions of E to X_R. However, for any two extensions, the central fibers will be *S-equivalent*, as we will see in Theorem 3.3.12, justifying the notion of S-equivalence. In particular, over the stable locus $\mathcal{M}_X^{\mathrm{s}}(r,d)$ any two extensions are isomorphic, although not uniquely.

3.3.2 Θ-reductivity and S-completeness

We now turn to the more contemporary ingredients of Theorem 3.3.6. Let R denote a discrete valuation ring with field of fractions K and residue field k, and let π be a uniformizer. We define the quotient stacks

- $\Theta_R = [\operatorname{Spec} R[t]/\mathbb{G}_{\mathrm{m}}]$ where the action has weight -1,
- $\overline{\mathrm{ST}}_R := [(\operatorname{Spec} R[s,t]/(st-\pi))/\mathbb{G}_{\mathrm{m}}]$ where the action has weight 1 on s and -1 on t.

Both Θ-reductivity and S-completeness are lifting criteria for *codimension-2* punctures in Θ_R and $\overline{\mathrm{ST}}_R$. The notion of Θ-reductivity is encoded by the lifting criterion in the left diagram, whereas S-completeness is the lifting criterion in the right diagram:

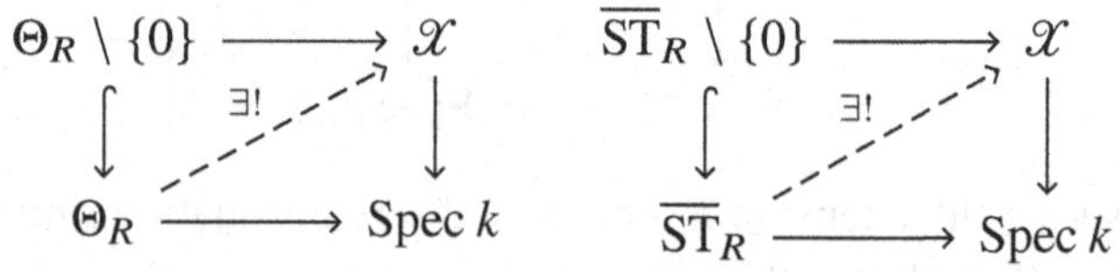

$$
\begin{array}{ccc}
\Theta_R \setminus \{0\} & \longrightarrow & \mathcal{X} \\
\downarrow & \exists! \nearrow & \downarrow \\
\Theta_R & \longrightarrow & \operatorname{Spec} k
\end{array}
\qquad
\begin{array}{ccc}
\overline{\mathrm{ST}}_R \setminus \{0\} & \longrightarrow & \mathcal{X} \\
\downarrow & \exists! \nearrow & \downarrow \\
\overline{\mathrm{ST}}_R & \longrightarrow & \operatorname{Spec} k
\end{array}
\tag{1}
$$

The notion of Θ-reductivity was introduced in [12] and that of S-completeness in [2, Section 3.5]. If $\mathcal{X}$ is an algebraic stack with affine diagonal, then in either criterion, a lift is automatically unique by [2, Proposition 3.17].

Each of the two lifting criteria encodes the extension of a certain type of filtration over the puncture, which we will now describe.

Θ-reductivity

We let $0 \in \Theta_R$ be the unique closed point defined by the vanishing of t and the uniformizer $\pi \in R$. Observe that $\Theta_R \setminus 0 = \Theta_K \cup_{\operatorname{Spec} K} \operatorname{Spec} R$. We have the following two cartesian diagrams which describe the geometry of Θ_R:

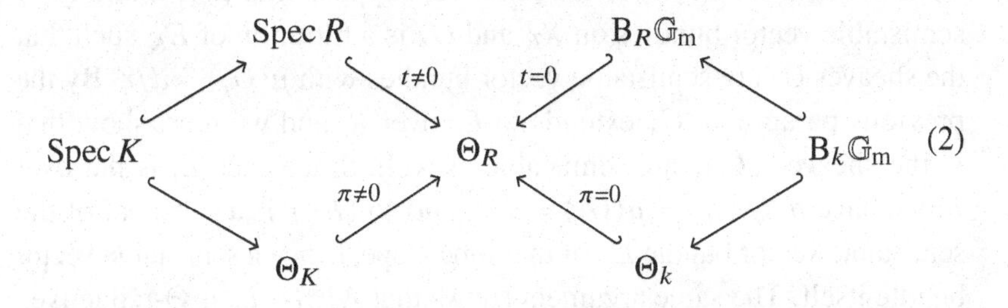

Here the maps on the left are open embeddings and those on the right are closed embeddings.

A morphism $\rho\colon \Theta \to \underline{\mathrm{Coh}}_X$ corresponds to the data of a coherent sheaf E on X and a filtration $0 = E_0 \subset E_1 \subset \cdots \subset E_n = E$. Under this correspondence, $\rho(1) = E$ and $\rho(0) = \bigoplus_i E_{i+1}/E_i$ is the associated graded. The map $\rho\colon \Theta \to \underline{\mathrm{Coh}}_X$ factors through $\mathcal{M}_X(r,d)$ (resp. $\mathcal{M}_X^{\mathrm{ss}}(r,d)$) if and only if E is a vector bundle (resp. semistable vector bundle) of rank r and degree d and each factor E_{i+1}/E_i is a vector bundle (resp. semistable vector bundle of slope $\mu = d/r$). There are analogous descriptions for morphisms from Θ_A for a k-algebra A.

Under this correspondence, the Θ-reductivity of these stacks translates into a valuative criterion for filtrations. For $\mathcal{M}_X^{\mathrm{ss}}(r,d)$, Θ-reductivity translates into the following: for every discrete valuation ring R with fraction field K and semistable vector bundle E on X_R of rank r and degree d, any filtration $0 = G_0 \subset G_1 \subset \cdots \subset G_n = E_K$ of the generic fiber where each factor G_{i+1}/G_i is semistable of slope $\mu = d/r$ extends to a filtration $0 = E_0 \subset E_1 \subset \cdots \subset E_n = E$ where each factor E_{i+1}/E_i is a family of semistable vector bundles on X over R. Note that the factors G_{i+1}/G_i are semistable of slope d/r for all i if and only if G_i is semistable of slope d/r for all i if and only if $\mu(G_i) = d/r$ for all i.

Proposition 3.3.8 *The stacks $\underline{\mathrm{Coh}}_X$, $\mathcal{M}_X^{\mathrm{ss}}(r,d)$, and $\mathcal{M}_X^{\mathrm{ss}}(r,L)$ are Θ-reductive.*

Proof First consider $\underline{\mathrm{Coh}}_X$. Let E be an R-flat coherent sheaf on X_R and let $0 = G_0 \subset G_1 \subset \cdots \subset G_n = E_K$ be a filtration of the generic fiber. By descending induction starting with $j = n$, suppose that we have constructed a filtration $E_j \subset E_{j+1} \subset \cdots \subset E_n = E$ extending $G_\bullet$. Since the relative Quot scheme of quotients of E_j with the same Hilbert polynomial as G_j/G_{j-1} is proper over R, there is a unique subsheaf

$E_{j-1} \subset E_j$ with $(E_{j-1})_K = G_{j-1}$ and E_j/E_{j-1} flat over R. This gives the next step in the filtration $E_\bullet$.

To see that $\mathcal{M}_X^{\mathrm{ss}}(r,d)$ is Θ-reductive, suppose that E is a family of semistable vector bundles on X_R and $G_\bullet$ is a filtration of E_K such that the sheaves G_i are semistable vector bundles with $\mu(G_i) = d/r$. By the previous paragraph, $G_\bullet$ extends to $E_\bullet$ over R, and we must show that in the sheaves $(E_i)_k$ are semistable as well. Since each E_i is flat over R, we have $\mu((E_i)_k) = \mu(G_i) = d/r$, and so $(E_i)_k$ is a subsheaf of the semistable vector bundle E_k of the same slope, hence a semistable vector bundle itself. The same argument shows that $\mathcal{M}_X^{\mathrm{ss}}(r,L)$ is Θ-reductive. □

S-completeness

The stack $\overline{\mathrm{ST}}_R$ can be viewed as a local model of the quotient $[\mathbb{A}^2/\mathbb{G}_\mathrm{m}]$ where $\mathbb{A}^2$ has coordinates s and t with weights 1 and -1; indeed, $\overline{\mathrm{ST}}_R$ is the base change of the good moduli space $[\mathbb{A}^2/\mathbb{G}_m] \to \operatorname{Spec} k[st]$ along $\operatorname{Spec} R \to \operatorname{Spec} k[st]$ defined by $st \mapsto \pi$.

We let $0 \in \overline{\mathrm{ST}}_R$ be the unique closed point defined by the vanishing of s and t. Observe that $\overline{\mathrm{ST}}_R \setminus 0$ is the non-separated union $\operatorname{Spec} R \cup_{\operatorname{Spec} K} \operatorname{Spec} R$. We have the following two cartesian diagrams which describe the geometry of $\overline{\mathrm{ST}}_R$:

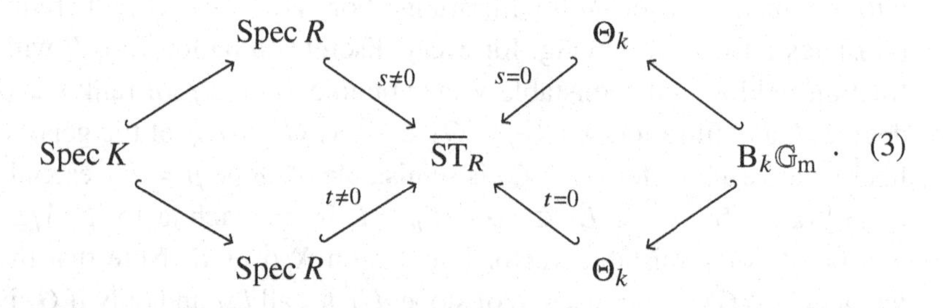

(3)

Here the maps on the left are open embeddings and those on the right are closed embeddings.

Remark 3.3.9 If G is a linear algebraic group over k, then BG is S-complete if and only if G is reductive (see [2, Proposition 3.45 and Remark 3.46]). Moreover, as S-completeness is preserved under closed substacks, it follows that every closed point (corresponding to a polystable object) in an S-complete algebraic stack with affine diagonal has a *reductive stabilizer*.

We will see in Lemma 3.3.11 below that the closed points of $\mathcal{M}_X^{\mathrm{ss}}(r,d)$ correspond to polystable vector bundles, so although some k-points of

Property	$\mathcal{M}_X^{\text{ss}}(r,d)$	$\mathcal{M}_X(r,d)$	$\underline{\text{Coh}}_X$
affine diagonal	yes	yes, Proposition 3.1.1	yes, [23, Tag 0DLY]
locally of finite type	yes	yes, Proposition 3.1.1	yes, [23, Tag 0DLZ]
quasicompact	yes, Proposition 3.2.15	no, Example 3.2.13	no
Θ-reductive	yes, Proposition 3.3.8	no	yes, Proposition 3.3.8
S-complete	yes, Proposition 3.3.10	no	yes, Proposition 3.3.10
existence part of valuative criterion	yes	yes	yes
separated	no, Example 3.3.3	no	no
existence of good moduli space	yes, Theorem 3.3.12	no	no

Table 3.1 *Properties of the stacks* $\mathcal{M}_X^{\text{ss}}(r,d) \subset \mathcal{M}_X(r,d) \subset \underline{\text{Coh}}_X$

$\mathcal{M}_X^{\text{ss}}(r,d)$ may have nonreductive stabilizers as in Example 3.3.1, such points necessarily correspond to non-polystable bundles.

Proposition 3.3.10 *The stacks* $\underline{\text{Coh}}_X$, $\mathcal{M}_X^{\text{ss}}(r,d)$, *and* $\mathcal{M}_X^{\text{ss}}(r,L)$ *are S-complete.*

We omit the proof of this proposition, which can be proven along the same lines as Proposition 3.3.8 using the interpretation of morphisms from $\overline{\text{ST}}_R$ to $\mathcal{M}_X^{\text{ss}}(r,d)$ from [2, Remark 3.36].

We have summarized in Table 3.1 the properties of $\mathcal{M}_X^{\text{ss}}(r,d)$, $\mathcal{M}_X(r,d)$, and $\underline{\text{Coh}}_X$ discussed in the first three sections.

3.3.3 Existence of good moduli spaces

In this subsection we will establish the existence and basic properties of good moduli spaces for the stacks $\mathcal{M}_X^{\text{ss}}(r,d)$ and $\mathcal{M}_X^{\text{ss}}(r,L)$, but before doing so we make an observation regarding their topology.

Lemma 3.3.11

(i) *If E is a semistable vector bundle, then the corresponding k-point*

$[E] \in \mathcal{M}_X^{\mathrm{ss}}(r,d)(k)$ contains the point $[\mathrm{gr}_E]$ in its closure, where gr_E is the graded bundle associated to a Jordan–Hölder filtration.

(ii) *A point $[E] \in \mathcal{M}_X^{\mathrm{ss}}(r,d)(k)$ is closed if and only if E is polystable.*

The same holds for the stack $\mathcal{M}_X^{\mathrm{ss}}(r,L)$.

Proof (i) If E is semistable but not stable, there exists a non-split extension $0 \to E' \to E \to E'' \to 0$ of semistable vector bundles of the same slope. Let $\mathcal{E}$ be the universal family over the affine line in $\mathrm{Ext}_X^1(E'',E')$ spanned by this extension, so that $\mathcal{E}$ is a family of semistable vector bundles on X parameterized by $\mathbb{A}^1$ such that $\mathcal{E}_t \cong E$ if $t \neq 0$ and $\mathcal{E}_0 \cong E' \oplus E''$. The resulting map

$$[\mathcal{E}] \colon \mathbb{A}^1 \to \mathcal{M}_X^{\mathrm{ss}}(r,d)$$

satisfies $t \mapsto [E]$ if $t \neq 0$ and $0 \mapsto [E' \oplus E'']$. It follows that $[E' \oplus E'']$ is contained in the closure of $[E]$. Iterating this construction for E' and E'' shows that $[\mathrm{gr}_E]$ is in the closure of $[E]$.

(ii) For a contradiction, suppose E is a polystable vector bundle such that $[E]$ is not closed, and let $[F]$ be in its closure. By (i), $[\mathrm{gr}_F]$ is in the closure of $[F]$ and, since no two points can be in the closure of each other, we must have $E \not\cong \mathrm{gr}_F$.

On the other hand, if E_i is stable with the same slope as E, then E_i appears as a direct summand of E with multiplicity $\hom_X(E_i,E) = \dim_k \mathrm{Hom}_X(E_i,E)$ and similarly for gr_F. For any E_i, the function $\hom_X(E_i,-)$ is upper semicontinuous in the second variable, since $[\mathrm{gr}_F]$ is in the closure of $[E]$, so we have $\hom_X(E_i,E) \leq \hom_X(E_i,\mathrm{gr}_F)$. This means that any stable summand of E appears in gr_F with at least the same multiplicity. But E and F have the same rank, so we must have $E \cong \mathrm{gr}_F$, a contradiction. Thus, $[E]$ is closed. □

Theorem 3.3.12

(i) *There exist good moduli spaces*

$$\mathcal{M}_X^{\mathrm{ss}}(r,d) \to \mathrm{M}_X^{\mathrm{ss}}(r,d),$$
$$\mathcal{M}_X^{\mathrm{ss}}(r,L) \to \mathrm{M}_X^{\mathrm{ss}}(r,L).$$

(ii) *These good moduli space maps induce bijections between the k-points of the good moduli space and S-equivalence classes of semistable sheaves.*

(iii) *The good moduli spaces* $\mathrm{M}_X^{\mathrm{ss}}(r,d)$ *and* $\mathrm{M}_X^{\mathrm{ss}}(r,L)$ *are irreducible, proper algebraic spaces of dimensions*

$$\dim \mathrm{M}_X^{\mathrm{ss}}(r,d) = r^2(g-1)+1,$$
$$\dim \mathrm{M}_X^{\mathrm{ss}}(r,L) = (r^2-1)(g-1).$$

Proof

(i) This follows from Propositions 3.3.8 and 3.3.10 and the first part of Theorem 3.3.6.

(ii) By Theorem 3.3.5(iii), two k-points $[E],[E'] \in \mathcal{M}_X^{\mathrm{ss}}(r,d)$ map to the same point in $\mathrm{M}_X^{\mathrm{ss}}(r,d)$ if and only if the closures of $\{[E]\}$ and $\{[E']\}$ in $\mathcal{M}_X^{\mathrm{ss}}(r,d)$ intersect. On the one hand, if E is any semistable vector bundle, then by Lemma 3.3.11(i) $[E]$ contains $[\mathrm{gr}_E]$ in its closure, so both points map to the same point in $\mathrm{M}_X^{\mathrm{ss}}(r,d)$. On the other hand, if E and E' are polystable and nonisomorphic, then by Lemma 3.3.11(ii), the corresponding points in $\mathcal{M}_X^{\mathrm{ss}}(r,d)$ are closed and distinct, hence map to distinct points in $\mathrm{M}_X^{\mathrm{ss}}(r,d)$.

(iii) The stacks are irreducible by Proposition 3.2.16 and the good moduli space maps are surjective by Theorem 3.3.5(i), so the good moduli spaces are irreducible as well. They are proper by Proposition 3.3.7 and the second part of Theorem 3.3.6.

To compute the dimensions, one can show using (ii) and Lemma 3.2.5 that over the stable locus $\mathcal{M}_X^{\mathrm{s}}(r,d) \subseteq \mathcal{M}_X^{\mathrm{ss}}(r,d)$ the good moduli space map is a $\mathbb{G}_{\mathrm{m}}$-gerbe over its image $\mathrm{M}_X^{\mathrm{s}}(r,d)$, and similarly for $\mathcal{M}_X^{\mathrm{ss}}(r,L)$. Since the stable loci are moreover nonempty, by Remark 3.2.18, the dimensions now follow from Proposition 3.1.3 and Remark 3.1.4. □

3.4 Determinantal line bundles

Let X be a smooth, projective and connected curve over k. For an algebraic stack $\mathcal{S}$ over k we consider the diagram

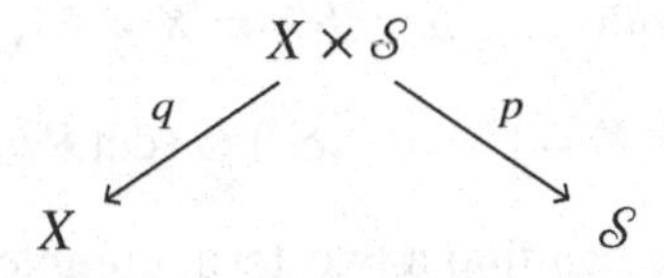

In this setting we can freely apply cohomology and base change arguments, because p is representable by schemes. If $\mathcal{E}$ is a vector bundle on $X \times \mathcal{S}$, then the derived direct image $\mathbf{R}p_*\mathcal{E}$ is a perfect complex on $\mathcal{S}$ with

amplitude in $[0, 1]$. In fact, something stronger is true. We will use the construction of [14, Proposition 2.1.10]. Let $\mathcal{E}$ be a vector bundle on $X \times \mathcal{M}_X(r, d)$. Then there exists a short exact sequence

$$0 \to \mathcal{E}^{-1} \to \mathcal{E}^0 \to \mathcal{E} \to 0$$

of vector bundles such that $\mathrm{R}^0 p_* \mathcal{E}^j = 0$ and $K^j := \mathrm{R}^1 \mathcal{E}^{j+1}$ is locally free for $j = 0, 1$. In particular, we have a quasi-isomorphism $K^\bullet := [K^0 \to K^1] \to \mathbf{R}p_*\mathcal{E}$. We make the following definition.

Definition 3.4.1 (Determinants and sections) If $\mathcal{E}$ is a vector bundle on $X \times \mathcal{S}$ and $[K^0 \to K^1]$ is a two-term complex of locally free sheaves such that $\mathbf{R}p_*\mathcal{E} \sim [K^0 \to K^1]$, we define the line bundle

$$\det \mathbf{R}p_*\mathcal{E} := \det(K^0) \otimes \det(K^1)^\vee.$$

If $\mathrm{rk}(\mathbf{R}p_*\mathcal{E}) = 0$, then $\mathrm{rk}\, K^0 = \mathrm{rk}\, K^1$ and the dual $(\det \mathbf{R}p_*\mathcal{E})^\vee$ is equipped with a section, locally given by the determinant of the map $K^0 \to K^1$.

As described in [23, Tag 0FJI], the definition of $\det \mathbf{R}p_*\mathcal{E}$ is independent of the choice of quasi-isomorphism with a two-term complex.

We will apply this construction in the case $\mathcal{S} = \mathcal{M}_X(r, d)$ and $\mathcal{E} = \mathcal{E}_{\mathrm{univ}} \otimes q^*V$, where $\mathcal{E}_{\mathrm{univ}}$ is the universal vector bundle on $X \times \mathcal{M}_X(r, d)$ and V is a vector bundle on X. The universal vector bundle exists tautologically because we are working with the moduli stack. Consider the diagram

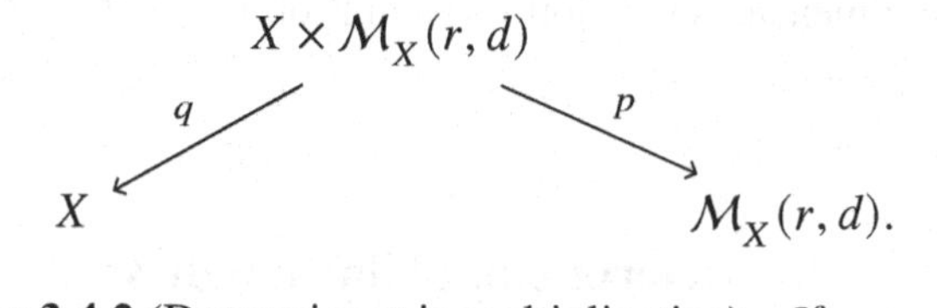

Proposition 3.4.2 (Determinant is multiplicative) *If*

$$0 \to \mathcal{E}' \to \mathcal{E} \to \mathcal{E}'' \to 0,$$

is an exact sequence of vector bundles on $X \times \mathcal{M}_X(r, d)$, then

$$\det \mathbf{R}p_*\mathcal{E} = (\det \mathbf{R}p_*\mathcal{E}') \otimes (\det \mathbf{R}p_*\mathcal{E}'').$$

Proof As before, we can find a two-term complex $\mathcal{E}^\bullet = [\mathcal{E}^{-1} \to \mathcal{E}^0]$ supported in degrees $[-1, 0]$ and a quasi-isomorphism $\mathcal{E}^\bullet \to \mathcal{E}$ with the properties described above. We choose $\mathcal{E}'^\bullet$ and $\mathcal{E}''^\bullet$ similarly. In fact, it follows from the construction in the proof of [14, Proposition 2.1.10] that we may choose these resolutions compatibly, so that we have a short exact sequence

$$0 \to \mathcal{E}'^{\bullet} \to \mathcal{E}^{\bullet} \to \mathcal{E}''^{\bullet} \to 0$$

of complexes, compatible with the quasi-isomorphisms to the members in our original exact sequence. Taking cohomology, we find a short exact sequence

$$0 \to K'^{\bullet} \to K^{\bullet} \to K''^{\bullet} \to 0$$

of complexes of locally free sheaves on $\mathcal{M}_X(r,d)$. The result follows from the multiplicativity of the determinant in short exact sequences of locally free sheaves. □

Definition 3.4.3 (Determinantal line bundles and sections) For a vector bundle V on X, we define the *determinantal line bundle*

$$\mathcal{L}_V := (\det \mathbf{R}p_*(q^*V \otimes \mathcal{E}_{\mathrm{univ}}))^\vee$$

on $\mathcal{M}_X(r,d)$ associated to V. If $\chi(X, V \otimes E) = 0$ for all $[E] \in \mathcal{M}_X(r,d)$, or equivalently by the Riemann–Roch theorem

$$d\,\mathrm{rk}(V) + r\deg(V) + r\,\mathrm{rk}(V)(1-g) = 0, \tag{1}$$

then the rank of $\mathbf{R}p_*(q^*V \otimes \mathcal{E}_{\mathrm{univ}})$ is zero and we define the section $s_V \in \Gamma(\mathcal{M}_X(r,d), \mathcal{L}_V)$ as in Definition 3.4.1.

Remark 3.4.4 Since $\mathbf{R}p_*(q^*V \otimes \mathcal{E}_{\mathrm{univ}})$ is perfect, its construction commutes with base change. In particular, its restriction to a k-point $[E] \in \mathcal{M}_X(r,d)$ is identified with the two-term complex $\mathbf{R}\Gamma(X, E \otimes V)$. If moreover $\chi(X, V \otimes E) = \mathrm{h}^0(X, V \otimes E) - \mathrm{h}^1(X, V \otimes E) = 0$, we see that indeed $\mathrm{rk}(\mathbf{R}p_*(q^*V \otimes \mathcal{E}_{\mathrm{univ}})) = 0$.

In order to state some basic properties of the construction in Definition 3.4.3, we recall that, by the universal property of the Picard stack $\mathcal{P}\mathrm{ic}_X^d$, the determinant $\det(\mathcal{E}_{\mathrm{univ}})$ of the universal vector bundle on $X \times \mathcal{M}_X(r,d)$ induces a morphism

$$\det \colon \mathcal{M}_X(r,d) \to \mathcal{P}\mathrm{ic}_X^d$$

such that $(\det \times \mathrm{id}_X)^*\mathscr{P} \cong \det(\mathcal{E}_{\mathrm{univ}})$, where $\mathscr{P}$ is the Poincaré bundle on $\mathcal{P}\mathrm{ic}_X^d \times X$.

Proposition 3.4.5 (Properties of the determinantal line bundle) *The following hold:*

(i) *The assignment* $V \mapsto \mathcal{L}_V$ *induces a group homomorphism*

$$\mathrm{K}_0(X) \to \mathrm{Pic}(\mathcal{M}_X(r,d)). \tag{2}$$

Consequently, the isomorphism class of the line bundle $\mathcal{L}_V$ *depends only on* $\mathrm{rk}(V)$ *and* $\det(V)$.

(ii) *If* V *and* W *are vector bundles with equal rank and degree, then there exists a line bundle* $\mathcal{N}$ *on* $\mathcal{P}\mathrm{ic}^d_X$ *such that*

$$\mathcal{L}_W \cong \mathcal{L}_V \otimes \det^* \mathcal{N}.$$

Proof An exact sequence $0 \to V_1 \to V_2 \to V_3 \to 0$ of vector bundles on X induces an exact sequence of vector bundles

$$0 \to q^*V_1 \otimes \mathcal{E}_{\mathrm{univ}} \to q^*V_2 \otimes \mathcal{E}_{\mathrm{univ}} \to q^*V_3 \otimes \mathcal{E}_{\mathrm{univ}} \to 0,$$

which by Proposition 3.4.2 induces an isomorphism

$$\mathcal{L}_{V_2} \cong \mathcal{L}_{V_1} \otimes \mathcal{L}_{V_3}.$$

Thus, we have a group homomorphism $\mathrm{K}_0(X) \to \mathrm{Pic}(\mathcal{M}_X(r,d))$. The second statement follows from the isomorphism $\mathrm{K}_0(X) \cong \mathbb{Z} \otimes \mathrm{Pic}(X)$ given by $V \mapsto (\mathrm{rk}(V), \det(V))$. This proves (i).

By (i), the isomorphism type of $\mathcal{L}_V$ depends only on the rank r_V and determinant of V, so we may assume that $V = \mathcal{O}_X^{\oplus r_V - 1} \oplus \mathcal{O}_X(D)$ for some divisor D on X, and similarly for W. Moreover, writing $D = D_1 - D_2$ as a difference of effective divisors and using standard exact sequences, we see that $[\mathcal{O}_X(D)] = [\mathcal{O}_X] + [\mathcal{O}_{D_1}] - [\mathcal{O}_{D_2}]$ in $\mathrm{K}_0(X)$. This implies that the classes of V and W in $\mathrm{K}_0(X)$ differ only by the class of a divisor of degree 0. Thus, by the additivity of the determinantal construction, it suffices to prove that

$$\det \mathbf{R}p_*(\mathcal{E}_{\mathrm{univ}} \otimes q^*\mathcal{O}_x) \cong \det^* \mathcal{N}'$$

for some line bundle $\mathcal{N}'$ on $\mathcal{P}\mathrm{ic}^d_X$, where $x \in X$ is a closed point. If we view $\mathcal{E}_x = \mathcal{E}_{\mathrm{univ}}|_{\mathcal{M}^{\mathrm{ss}}_X(r,d)\times\{x\}}$ and $\mathcal{P}_x = \mathcal{P}|_{\mathcal{P}\mathrm{ic}^d_X\times\{x\}}$ as sheaves on $\mathcal{M}^{\mathrm{ss}}_X(r,d)$ and $\mathcal{P}\mathrm{ic}^d_X$ respectively, then we have

$$\det \mathbf{R}p_*(\mathcal{E}_{\mathrm{univ}} \otimes q^*\mathcal{O}_x) = \det \mathcal{E}_x = \det^* \mathcal{P}_x.$$

This proves (ii). □

Definition 3.4.6 Let E be a vector bundle on X. We say that E is *cohomology-free* if $\mathrm{h}^0(X,E) = \mathrm{h}^1(X,E) = 0$.

The following proposition relates cohomology-freeness to the properties of sections of determinantal line bundles. It will be an essential tool in the ampleness proof.

Proposition 3.4.7 *If $\chi(X, E \otimes V) = 0$ for all $[E] \in \mathcal{M}_X(r, d)$, then the following are equivalent:*

- *the section $s_V \in \Gamma(\mathcal{M}_X(r, d), \mathcal{L}_V)$ is nonzero at a vector bundle $[E] \in \mathcal{M}_X(r, d)$;*
- *$E \otimes V$ is cohomology-free.*

Proof In the setup of Definition 3.4.3 the morphism of line bundles $\det j\colon \det(K_0) \to \det(K_1)$ is nonzero at the point $[E] \in \mathcal{M}_X(r, d)$ if and only if the morphism of vector bundles $j\colon K_0 \to K_1$ is an isomorphism at $[E]$. Since the derived direct image sheaf is the cohomology of this complex, this occurs if and only if $\mathrm{h}^0(X, E \otimes V) = \mathrm{h}^1(X, E \otimes V) = 0$. □

Remark 3.4.8 We emphasize that while $\mathcal{L}_V$ only depends on $\mathrm{rk}(V)$ and $\det(V)$, the section s_V *does* depend on V itself. We will leverage this fact to produce enough sections of $\mathcal{L}_V$ to establish ampleness. Notice also that, under the assumption $\chi(X, E \otimes V) = 0$, the vanishing of $\mathrm{H}^0(X, E \otimes V)$ is equivalent to the vanishing of $\mathrm{H}^1(X, E \otimes V)$.

We now specialize the construction to the open substack $\mathcal{M}_X^{\mathrm{ss}}(r, d) \subseteq \mathcal{M}_X(r, d)$. As explained in the introduction, our goal is to prove that the proper algebraic space $\mathrm{M}_X^{\mathrm{ss}}(r, d)$ obtained in Theorem 3.3.12 is actually a projective scheme. For this we need to produce an ample line bundle on it, and the determinantal line bundle we have constructed a priori lives on $\mathcal{M}_X^{\mathrm{ss}}(r, d)$. To remedy this with the following.

Proposition 3.4.9 *The determinantal line bundle $\mathcal{L}_V$ of Proposition 3.4.7 associated to a vector bundle V descends to $\mathrm{M}_X^{\mathrm{ss}}(r, d)$, that is, there exists a unique line bundle $\mathrm{L}_V \in \mathrm{Pic}(\mathrm{M}_X^{\mathrm{ss}}(r, d))$ such that $\mathcal{L}_V \cong \phi^*\mathrm{L}_V$.*

Proof By Theorem 3.3.5(iv), we must show that stabilizers of $\mathcal{M}_X^{\mathrm{ss}}(r, d)$ act trivially on the fibers of $\mathcal{L}_V$. By Theorem 3.3.12(ii), the closed points of $\mathcal{M}_X^{\mathrm{ss}}(r, d)$ correspond to polystable bundles, and for a polystable bundle $E = \bigoplus_{j=1}^n E_j^{\oplus m_j}$, where the E_i are pairwise nonisomorphic, we have $\mathrm{Aut}(E) \cong \mathrm{GL}_{m_1} \times \cdots \times \mathrm{GL}_{m_n}$. The fiber of $\mathcal{L}_V|_{[E]}$ is identified with

$$\det \mathbf{R}\Gamma(X, E \otimes V) = \prod_{i=1}^{n} (\det \mathrm{H}^i(X, E \otimes V))^{\otimes(-1)^i}.$$

An element $(g_1, \ldots, g_n) \in \mathrm{Aut}(E)$ acts on

$$\det \mathrm{H}^i(X, E \otimes V) \cong \bigotimes_{j=1}^{n} \left(\det \mathrm{H}^i(X, E_j \otimes V)\right)^{\otimes m_j}$$

by multiplication with $\prod_{j=1}^{n} \det(g_1)^{\dim \mathrm{H}^i(X, E_j \otimes V)}$, and thus on $\mathcal{L}_V|_{[E]}$ by multiplication with $\prod_{j=1}^{n} \det(g_j)^{\chi(X, E_j \otimes V)}$. But each E_j has slope equal to d/r, so by the assumption on V we have $\chi(X, E_j \otimes V) = 0$, and so the action is trivial. □

Corollary 3.4.10 *For a line bundle L of degree d on X, the restriction of the determinantal line bundle L_V to $\mathrm{M}_X^{\mathrm{ss}}(r, L)$ only depends on the rank and degree of V.*

Proof We have commuting diagrams

$$\begin{array}{ccc} \mathcal{M}_X^{\mathrm{ss}}(r, L) & \longrightarrow & \mathcal{M}_X^{\mathrm{ss}}(r, d) \\ \downarrow & \square & \downarrow \mathrm{det} \\ \mathrm{Spec}\, k & \xrightarrow{[L]} & \mathcal{P}ic_X^d \end{array} \quad \text{and} \quad \begin{array}{ccc} \mathcal{M}_X^{\mathrm{ss}}(r, L) & \longrightarrow & \mathcal{M}_X^{\mathrm{ss}}(r, d) \\ \downarrow & & \downarrow \\ \mathrm{M}_X^{\mathrm{ss}}(r, L) & \hookrightarrow & \mathrm{M}_X^{\mathrm{ss}}(r, d). \end{array}$$

where in the second square, both vertical arrows are good moduli space maps. If V and W are vector bundles of equal rank and degree on X, both satisfying condition (1), then by Proposition 3.4.7(ii), there exists a line bundle $\mathcal{N}$ on $\mathcal{P}ic_X^d$ such that $\mathcal{L}_W \cong \mathcal{L}_V \otimes \det^* \mathcal{N}$. The left diagram shows that, restricting to $\mathcal{M}_X^{\mathrm{ss}}(r, L)$, this isomorphism becomes $\mathcal{L}_W \cong \mathcal{L}_V$. The right diagram now shows that the restrictions of L_V and L_W to $\mathrm{M}_X^{\mathrm{ss}}(r, L)$ become isomorphic after pulling back to $\mathcal{M}_X^{\mathrm{ss}}(r, L)$, so the restrictions must be isomorphic to begin with, by Theorem 3.3.5(iv). □

Using geometric invariant theory, Drézet and Narasimhan gave in [8, Théorèmes B, C] the following description of the Picard groups of the good moduli spaces.

Theorem 3.4.11 (Drézet–Narasimhan) *There exist isomorphisms*

$$\begin{aligned} \mathrm{Pic}(\mathrm{M}_X^{\mathrm{ss}}(r, d)) &\cong \mathrm{Pic}(\mathrm{Pic}_X^d) \oplus \mathbb{Z}, \\ \mathrm{Pic}(\mathrm{M}_X^{\mathrm{ss}}(r, L)) &\cong \mathbb{Z}. \end{aligned} \tag{3}$$

In the second line $\mathbb{Z}$ is generated by the determinantal line bundle L_V, where V is chosen to be of minimal rank.

This explains how the isomorphism type of the line bundle L_V on the good moduli space depends on the invariants of the vector bundle V satisfying $\chi(X, E \otimes V) = 0$, building upon Proposition 3.4.5.

3.5 Projectivity

In this section we will prove the following result.

Theorem 3.5.1 *Let X be a smooth projective curve of genus $g \geq 2$. The good moduli spaces $\mathrm{M}_X^{\mathrm{ss}}(r,d)$ and $\mathrm{M}_X^{\mathrm{ss}}(r,L)$ are projective varieties of dimension $r^2(g-1)+1$ and $(r^2-1)(g-1)$ respectively.*

We begin by reducing the projectivity of $\mathrm{M}_X^{\mathrm{ss}}(r,d)$ to that of $\mathrm{M}_X^{\mathrm{ss}}(r,L)$. The determinant morphism

$$\det\colon \mathcal{M}_X^{\mathrm{ss}}(r,d) \to \mathcal{P}ic_X^d$$

discussed above descends to a map

$$\det\colon \mathrm{M}_X^{\mathrm{ss}}(r,d) \to \mathrm{Pic}_X^d$$

of good moduli spaces whose fiber over $[L] \in \mathrm{Pic}_X^d$ is the moduli space $\mathrm{M}_X^{\mathrm{ss}}(r,L)$. It is a classical fact that Pic_X^d is a projective variety isomorphic to the Jacobian of X. Moreover, since $\mathrm{M}_X^{\mathrm{ss}}(r,d)$ is a proper algebraic space, to conclude that it is a projective scheme, it suffices by [23, Tag 0D36] to produce a line bundle on it whose restriction to each fiber $\mathrm{M}_X^{\mathrm{ss}}(r,L)$ is ample. We will show that the determinantal line bundle L_V will work for an arbitrary vector bundle V satisfying condition (1) below.

From now on, we will fix a line bundle L of degree d on X and focus on the moduli space $\mathrm{M}_X^{\mathrm{ss}}(r,L)$.

Setup Denote $h = \gcd(r,d)$ and set $r_1 = r/h$ and $d_1 = d/h$. Let V be a vector bundle on X, with invariants

$$\mathrm{rk}(V) = mr_1, \quad \deg(V) = mr_1(g-1) - md_1 \tag{1}$$

for some $m \geq 1$. By the Riemann–Roch theorem this ensures that $\chi(X, V \otimes E) = 0$ for a vector bundle E of rank r and degree d. Hence by Section 3.4 the associated determinantal line bundle L_V on $\mathrm{M}_X^{\mathrm{ss}}(r,L)$ depends only on $\mathrm{rk}(V)$ and $\deg(V)$, and comes with a section s_V, depending on V, such that $s_V([E]) \neq 0$ if and only if $V \otimes E$ is cohomology-free.

We will prove the theorem using the following two propositions.

Proposition 3.5.2 *Let E be a stable vector bundle of rank r and degree d. For large enough m, a general vector bundle V with*

$$\mathrm{rk}(V) = mr_1, \quad \deg(V) = mr_1(g-1) - md_1$$

satisfies

$$\mathrm{H}^0(X, V \otimes E) = \mathrm{H}^1(X, V \otimes E) = 0. \tag{2}$$

In particular, the determinantal line bundle L_V *on* $\mathrm{M}_X^{\mathrm{ss}}(r, L)$ *is* semi-ample.

Remark 3.5.3 In the approach of Faltings this statement (albeit for $\mathrm{M}_X^{\mathrm{ss}}(r, d)$) corresponds to [10, Theorem I.2] (note that op. cit. is concerned about the more complicated notion of stable Higgs bundles) and [22, Lemma 3.1], but the method of proof will be different.

Proposition 3.5.4 *Let* E *and* E' *be polystable vector bundles of rank* r *and degree* d *such that* E *has a stable summand that is not a summand of* E'*. There exists an integer* $m > 0$ *and a stable vector bundle* V *of rank* mr_1 *and degree* $mr_1(g-1) - md_1$ *such that*

$$\mathrm{H}^0(X, V \otimes E) \neq 0, \quad \mathrm{H}^0(X, V \otimes E') = 0. \tag{3}$$

In particular, the sections s_V *of the determinantal line bundle* L_V *on* $\mathrm{M}_X^{\mathrm{ss}}(r, L)$ *separate points determined by polystable bundles for which at least one stable summand is different.*

Remark 3.5.5 It is not true that for any two polystable bundles E and E' we can find a vector bundle V of the appropriate rank and degree for which $\mathrm{H}^0(X, V \otimes E) \neq 0$ and $\mathrm{H}^0(X, V \otimes E') = 0$. Namely, consider two polystable vector bundles

$$E = E_1 \oplus E_1 \oplus E_2 \qquad \text{and} \qquad E' = E_1 \oplus E_2 \oplus E_2,$$

where E_1 and E_2 are nonisomorphic stable vector bundles of the same rank and degree. However, it can be shown that sections of the form s_V span the space of global sections of L_V for $m \gg 0$ and induce a closed embedding $\mathrm{M}_X^{\mathrm{ss}}(r, d) \hookrightarrow \mathbb{P}^N$; see [3, Corollary 7.15].

Remark 3.5.6 The method of Faltings (see [10, Theorem I.4] and [22, Lemma 4.2]) takes a different approach at this point. He shows that, given a map $C \to \mathcal{M}_X^{\mathrm{ss}}(r, d)$ from a smooth projective curve C corresponding to a family $\mathcal{E}$ of semistable vector bundles on X, the degree of the determinantal line bundle $\mathcal{L}_V$ is nonnegative and is zero if and only if all the fibers $\{\mathcal{E}_t\}_{t \in C}$ are S-equivalent. This allows one to conclude that the morphism $\mathrm{M}_X^{\mathrm{ss}}(r, d) \to \mathbb{P}^N$ has finite fibers, and then the argument can proceed as in the proof of Theorem 3.5.1. Interestingly, Faltings's argument uses the stability of vector bundles on the curve C.

Proof of Theorem 3.5.1 By Proposition 3.5.2 the determinantal line bundle $\mathcal{L}_V$ associated to the appropriate choice of V is semiample, so choose some power for which it is basepoint-free and obtain a morphism $\phi\colon \mathrm{M}_X^{\mathrm{ss}}(r, d) \to \mathbb{P}^N$ for some N.

By Proposition 3.5.4 the morphism is actually quasi-finite. Indeed, for a given polystable bundle $E = E_1^{\oplus r_1} \oplus \cdots \oplus E_s^{\oplus r_s}$ where the $E_1, \ldots, E_s$ are nonisomorphic stable bundles, there are only finitely many other polystable bundles whose stable summands are precisely $E_1, \ldots, E_s$, and for any other polystable bundle E', there exists a section s_V separating E and E'.

Since $\mathrm{M}_X^{\mathrm{ss}}(r,d)$ is a proper algebraic space, the morphism $\mathrm{M}_X^{\mathrm{ss}}(r,d) \to \mathbb{P}^N$ is proper by [23, Tag 04NX] and hence finite by [23, Tag 0A4X]. But then it is also affine, hence representable, so $\mathrm{M}_X^{\mathrm{ss}}(r,d)$ is a scheme. Finally, pulling back an ample line bundle on $\mathbb{P}^N$ along the affine morphism produces an ample line bundle on $\mathrm{M}_X^{\mathrm{ss}}(r,d)$, by [23, Tag 0892]. □

We are left with proving Propositions 3.5.2 and 3.5.4. Their proofs will be based on a dimension-counting argument inspired by [9] (see also [18, Section 5] for similar methods), where the authors actually prove a more refined result that yields an effective bound on the rank of V in the theorem. To get started, we will need the following lemma, discussed in [4, Remark 4.2].

Lemma 3.5.7 *If $\{F_\alpha\}_{\alpha\in\Lambda}$ is a bounded family of vector bundles of rank r on X, there exists a scheme S of finite type over k of dimension at most $r^2(g-1)+1$ and a vector bundle $\mathcal{F}$ on $S \times X$ such that each F_α appears as the fiber of $\mathcal{F}$ over some point $s \in S$.*

In fact, the stable bundles of rank r and degree d can be parameterized by a smooth variety of dimension $r^2(g-1)+1$, and any bounded family of non-stable bundles by a variety of dimension at most $r^2(g-1)$.

The following lemma will be the technical heart for the proof of both propositions. We will apply it for $\epsilon = -1, 0, 1$, and think of it as a "fudge factor" perturbing the equality $\chi(X, E \otimes V) = 0$ for $\epsilon = 0$.

Lemma 3.5.8 *Let E be a vector bundle of rank r and degree d on X, and let ϵ be an integer. Let m be a sufficiently large integer and V a general stable vector bundle with*

$$\mathrm{rk}(V) = mr_1, \quad \deg(V) = mr_1(g-1) - md_1 + \epsilon.$$

(i) *There exist no maps $V^\vee \to E$ whose image F has slope $\mu(F) < \mu(E)$.*
(ii) *If moreover E is stable and $\epsilon \leq 0$, there exist no nonzero maps $V^\vee \to E$.*

Proof We can assume that $\mathrm{rk}(V) > r$. Note that $V^\vee$ satisfies

$$\mathrm{rk}(V^\vee) = mr_1, \quad \deg(V^\vee) = mr_1(1-g) + md_1 - \epsilon.$$

If $V^\vee \to E$ is a map with image F, then, since $V^\vee$ is stable, we must have

$$\mu(F) > \mu(V^\vee) = \mu(E) - g + 1 - \frac{\epsilon}{mr_1} \geq \mu(E) - g + 1 - \epsilon.$$

Thus, there are only finitely many options for the slope of the image F such that $\mu(F) \leq \mu(E)$. Since moreover we must have $1 \leq \mathrm{rk}(F) \leq r$, there are only finitely many options for the rank and degree of F.

To prove (i), fix integers $1 \leq k \leq r$ and l with $l/k < \mu(E)$. We will find an upper bound for the dimension of a variety that parameterizes all stable bundles V such that there exists a map $V^\vee \to E$ whose image F has rank k and degree l. Any such $V^\vee$ fits into an exact sequence

$$\begin{array}{ccccccccc} 0 & \longrightarrow & G & \longrightarrow & V^\vee & \longrightarrow & F & \longrightarrow & 0 \\ & & & & & \searrow & \downarrow & & \\ & & & & & & E & & \end{array} \tag{4}$$

where G satisfies

$$\mathrm{rk}(G) = mr_1 - k, \quad \deg(G) = mr_1(1-g) + md_1 - \epsilon - l.$$

The possible sheaves F that can appear in (4) form a bounded family since they are all subsheaves of E of fixed rank and degree. By Lemma 3.5.7 they are parameterized by a scheme $\mathcal{U}_1$ of finite type over k with

$$\dim \mathcal{U}_1 \leq k^2(g-1) + 1.$$

Similarly, since the family of stable bundles of fixed rank and degree is bounded, the sheaves G that can appear in (4) form a bounded family and so by Lemma 3.5.7 are parameterized by a scheme $\mathcal{U}_2$ with

$$\dim \mathcal{U}_2 \leq (mr_1 - k)^2(g-1) + 1.$$

Note that since $V^\vee$ in (4) is stable, we must have $\mathrm{Hom}_X(F, G) = 0$, because for any nonzero map $F \to G$ the composition

$$V^\vee \twoheadrightarrow F \to G \hookrightarrow V^\vee$$

would give a nonzero endomorphism of $V^\vee$ that is not an isomorphism. Now, for a fixed F and G, the possible $V^\vee$ appearing in (4) are parameterized by an open subset of $\mathbb{P}\mathrm{Ext}^1_X(F, G)$. By the Riemann–Roch theorem we have

$$\begin{aligned}
&\dim \operatorname{Ext}^1_X(F,G)\\
&\quad = -\chi(X, F^\vee \otimes G) = -\deg(F^\vee \otimes G) - \operatorname{rk}(F^\vee \otimes G)(1-g)\\
&\quad = \deg(F)\operatorname{rk}(G) - \operatorname{rk}(F)\deg(G) + \operatorname{rk}(F)\operatorname{rk}(G)(g-1)\\
&\quad = l(mr_1 - k) - k(mr_1(1-g) + md_1 - \epsilon - l) + k(mr_1 - k)(g-1)\\
&\quad = (lr_1 - kd_1)m + 2kr_1(g-1)m - k^2(g-1) + k\epsilon.
\end{aligned}$$

Note that by our assumption on k and l, we have

$$lr_1 - kd_1 = kr_1\left(\frac{l}{k} - \frac{d_1}{r_1}\right) = kr_1(\mu(F) - \mu(E)) < 0.$$

Thus, the possible $V^\vee$ that could appear are parameterized by an open subset of a projective bundle $\mathcal{P}$ over an open subset $\mathcal{U} \subseteq \mathcal{U}_1 \times \mathcal{U}_2$ of dimension

$$\begin{aligned}
\dim \mathcal{P} &\leq \dim \mathcal{U}_1 + \dim \mathcal{U}_2 + \dim \operatorname{Ext}^1_X(F,G) - 1\\
&\leq k^2(g-1) + 1 + (mr_1 - k)^2(g-1) + 1\\
&\quad + kr_1(\mu(F) - \mu(E))m + 2kr_1(1-g)m - k^2(g-1) - k\epsilon - 1\\
&= (mr_1)^2(g-1) + kr_1(\mu(F) - \mu(E))m + k^2(g-1) + k\epsilon + 1.
\end{aligned}$$

Since the coefficient of m in the last expression is negative, we see that for large enough m, we have $\dim \mathcal{P} < (mr_1)^2(g-1) + 1$, where by Theorem 3.3.12 the right-hand side is the dimension of the moduli space of stable bundles of rank mr_1 and fixed degree. Thus, a general stable bundle V cannot appear in an extension of the form (4). This proves (i).

To prove (ii), assume that E is stable and $\epsilon \leq 0$. By (i), if V is a general stable vector bundle of the given rank and degree and m is sufficiently large, there are no maps $V^\vee \to E$ whose image has slope less than $\mu(E)$. Thus, since E is stable, any nonzero map must be surjective, and so we are led to estimate the dimension of a variety parameterizing all stable bundles V such that there exists a short exact sequence

$$0 \to G \to V^\vee \to E \to 0.$$

Such $V^\vee$ are parameterized by a projective bundle $\mathcal{P}$ over a scheme $\mathcal{U}_2$, where $\mathcal{U}_2$ parameterizes bundles G with

$$\operatorname{rk}(G) = mr_1 - r, \quad \deg(G) = mr_1(1-g) + md_1 - \epsilon - d$$

and again $\operatorname{Hom}_X(E,G) = 0$ since $V^\vee$ is stable. The fiber of $\mathcal{P}$ over

$[G] \in \mathcal{U}_2$ is $\mathbb{P}\mathrm{Ext}^1_X(E,G)$, and the dimension calculations from above apply, so we have

$$\begin{aligned}\dim \mathcal{P} &\leq \dim \mathcal{U}_2 + \dim \mathrm{Ext}^1_X(E,G) - 1 \\ &= (mr_1 - r)^2(g-1) + 1 + 2rr_1(g-1)m - r^2(g-1) + r\epsilon - 1 \\ &= (mr_1)^2(g-1) + r\epsilon.\end{aligned}$$

Since $\epsilon \leq 0$, we have $\dim \mathcal{P} < (mr_1)^2(g-1)+1$, where the left-hand side is the dimension of the moduli space of stable bundles of rank mr_1 by Theorem 3.3.12. □

Proof of Proposition 3.5.2 Applying Lemma 3.5.8 with $\epsilon = 0$ to a stable vector bundle E produces an open subset in the relevant moduli space such that any V in the open subset satisfies the first claim of the proposition.

Now let E be a polystable vector bundle corresponding to a point in $\mathrm{M}^{\mathrm{ss}}_X(r,L)$. Taking the intersection of the open subsets for all the stable summands of E produces a nonempty open subset where one can find a vector bundle V such that $\mathrm{H}^i(X, E\otimes V) = 0$ for $i = 0, 1$. Hence the characterization from Proposition 3.4.7 shows that the associated section s_V does not vanish at $[E] \in \mathrm{M}^{\mathrm{ss}}_X(r,L)$.

□

Lemma 3.5.9 *Let $E_0, E_1, \ldots, E_n$ be stable vector bundles of slope $\mu = r/d$. Denote the rank of E_i by k_i. For m sufficiently large, a general stable vector bundle V with*

$$\mathrm{rk}(V) = mr_1 \quad \textit{and} \quad \deg(V) = mr_1(g-1) - md_1 + 1$$

satisfies the following:

(i) $\dim \mathrm{Hom}_X(V^\vee, E_i) = k_i$*, or equivalently* $\mathrm{Ext}^1_X(V^\vee, E_i) = 0$*;*
(ii) *for $0 \leq i \leq n$, every nonzero map $V^\vee \to E_i$ is surjective;*
(iii) *for $1 \leq i \leq n$, for all nonzero maps $V^\vee \to E_0$ and $V^\vee \to E_i$, the induced map $V^\vee \to E_0 \oplus E_i$ is surjective.*

Proof Note first that $\mu = d_1/r_1 = \deg(E_i)/k_i$. We have

$$\dim \mathrm{Hom}_X(V^\vee, E_i) = \mathrm{h}^0(V \otimes E_i) = \chi(X, V\otimes E_i) + \mathrm{h}^1(X, V \otimes E_i),$$

and by the Riemann–Roch theorem,

$$\begin{aligned}\chi(X, V\otimes E_i) &= \deg(V\otimes E_i) + \mathrm{rk}(V\otimes E_i)(1-g)\\ &= \deg(V)\,\mathrm{rk}(E_i) + \mathrm{rk}(V)\deg(E_i) + \mathrm{rk}(V)\,\mathrm{rk}(E_i)(1-g)\\ &= (mr_1(g-1) - md_1 + 1)k_i + mr_1k_i\mu + mr_1k_i(1-g)\\ &= k_i.\end{aligned}$$

To obtain (i) we must show that for a general stable V we have $\mathrm{H}^1(X, V\otimes E_i) = 0$ for $0 \le i \le n$. By Serre duality, we have

$$\mathrm{h}^1(X, V\otimes E_i) = \mathrm{h}^0(X, V^\vee \otimes \omega_X \otimes E_i^\vee) = \dim \mathrm{Hom}_X(V\otimes \omega_X^\vee, E_i^\vee).$$

Now $V^\vee \otimes \omega_X$ is a general stable bundle of rank mr_1 and degree

$$\begin{aligned}\deg(V^\vee\otimes\omega_X) &= -\deg(V) + \mathrm{rk}(V)\deg(\omega_X)\\ &= mr_1(1-g) + md_1 - 1 + mr_1(2g-2)\\ &= mr_1(g-1) + md_1 - 1,\end{aligned}$$

and $E_i^\vee$ is a stable bundle of slope $-d_1/r_1$, so we may apply Lemma 3.5.8(ii) with $\epsilon = -1$ to each $E_i^\vee$ to conclude that, for sufficiently large m and general stable V, we have $\mathrm{Hom}_X(V\otimes\omega_X^\vee, E_i^\vee) = 0$ for each i. This gives (i).

By applying Lemma 3.5.8(i) with $\epsilon = 1$ to each E_i and by increasing m if necessary, we may assume that for general stable V the image of every nonzero map $V^\vee \to E_i$ has slope equal to μ, and, since E_i is stable, such a map must be surjective. This gives (ii).

Finally, by applying Lemma 3.5.8(i) to each $E_0 \oplus E_i$ for $1 \le i \le n$ and increasing m if necessary, we may assume that for general stable V, every nonzero map $V^\vee \to E_0 \oplus E_i$ has image with slope equal to μ. For such a V and two nonzero maps $V^\vee \to E_0$ and $V^\vee \to E_i$, the image $F \subseteq E_0 \oplus E_i$ of the induced map $V^\vee \to E_0 \oplus E_i$ must thus have slope $\mu(F) = \mu$, and moreover F surjects onto both summands. But the only nonzero subsheaves of $E_0 \oplus E_i$ of slope μ are E_0, E_i, and $E_0 \oplus E_i$ and, since E_0 and E_i are by assumption nonisomorphic, we must have $F = E_0 \oplus E_i$. This gives (iii). □

Before we give the proof of Proposition 3.5.4 we introduce a method for constructing new vector bundles out of old ones.

Definition 3.5.10 Let E be a vector bundle on X and let $x \in X(k)$ be a closed point. Denote by $\mathcal{O}_x$ the skyscraper sheaf at x and $E(x)$ the fiber of E at x. Let $\alpha \in \mathrm{Hom}_X(E, \mathcal{O}_x) = E(x)^\vee$ be a nonzero morphism. The *elementary transformation* of E at x associated to α is the vector bundle E' defined as the kernel of α, so that we have the short exact sequence

$$0 \to E' \to E \xrightarrow{\alpha} \mathcal{O}_x \to 0. \tag{5}$$

The kernel of α does not change under rescaling, so elementary transformations of E at x correspond to closed points of $\mathbb{P}E(x)^\vee$.

If $\phi\colon F \to E$ is a morphism, then $\operatorname{im}\phi(x)$ is a subspace of $E(x)$, while $\ker\alpha(x)$ is a hyperplane in $E(x)$, which we will denote by H. If $\operatorname{im}\phi(x) \subseteq H$ then the composition $F \xrightarrow{\phi} E \xrightarrow{\alpha} \mathcal{O}_x$ vanishes and so $F \to E$ factors through E'. Conversely, if $\operatorname{im}\phi(x)$ is not contained in H then $F \to E$ does not factor through E'.

Proof of Proposition 3.5.4 Let $E_1, \ldots, E_n$ be the stable summands of E', and let E_0 be a stable summand of E not isomorphic to any of the E_i. As before, let k_i denote the rank of E_i. It suffices to find a vector bundle F such that

$$\mathrm{H}^0(X, F \otimes E_0) \neq 0 \quad \text{but} \quad \mathrm{H}^0(X, F \otimes E_i) = 0 \text{ for } i = 1, \ldots, n.$$

Let V be a stable vector bundle satisfying the conditions of Lemma 3.5.9, and fix a nonzero map $\phi\colon E_0^\vee \to V$. The goal is to find a subsheaf $F \subset V$ of degree $\deg(F) = \deg(V) - 1$ such that

- ϕ factors through F, but
- for $i = 1, \ldots, n$, no nonzero map $E_i^\vee \to V$ factors through F.

Let Q denote the cokernel of ϕ. It is locally free since by Lemma 3.5.9(ii) ϕ is the dual of a surjective map $V^\vee \to E_0$. Note that, for any nonzero map $\psi_i\colon E_i^\vee \to V$, the induced map $E_0^\vee \oplus E_i^\vee \to V$ is dual to a surjection by Lemma 3.5.9(iii) and hence injective. The inclusions

$$E_0^\vee \hookrightarrow E_0^\vee \oplus E_i^\vee \overset{(\phi,\psi_i)}{\hookrightarrow} V$$

induce a short exact sequence

$$0 \to E_i^\vee \xrightarrow{\overline{\psi}_i} Q \to V/E_0^\vee \oplus E_i^\vee \to 0,$$

and so the composition $\overline{\psi}_i\colon E_i^\vee \to V \to Q$ is injective with locally free cokernel, from which it follows that the restriction of $\overline{\psi}_i$ to a fiber is also injective.

Let $x \in X$ be a closed point, let $Q(x)$ and $V(x)$ denote the fibers of Q and V at x respectively, and let $\operatorname{Grass}(k_i, Q(x))$ denote the Grassmannian of k_i-dimensional subspaces of the vector space $Q(x)$. Let

$M_i \subseteq \operatorname{Grass}(k_i, Q(x))$ denote the image of the morphism

$$\mathbb{P}\mathrm{Hom}_X(E_i^\vee, V) \to \operatorname{Grass}(k_i, Q(x))$$

that sends a nonzero map ψ_i to its image under the composition

$$E_i^\vee \xrightarrow{\overline{\psi}_i} Q \to Q(x).$$

This morphism is well defined, because in the previous paragraph we established that $\overline{\psi}_i(x)$ is injective. We have $\dim M_i \leq k_i - 1$ by Lemma 3.5.9(i).

For a hyperplane $H \subseteq Q(x)$, let $Z_{H,i} \subseteq \operatorname{Grass}(k_i, Q(x))$ denote the Schubert variety parameterizing subspaces contained in H. The codimension of $Z_{H,i}$ in $\operatorname{Grass}(k_i, Q(x))$ is k_i, and so the expected dimension of the intersection $M_i \cap Z_{H,i}$ is -1. Thus, by the Bertini–Kleiman theorem [16, Corollary 4(i)] we obtain for the general hyperplane H that $M_i \cap Z_{H,i}$ is empty for $i = 1, \ldots, n$. Let us fix one such hyperplane H.

Let F denote the elementary transformation of V at x defined by the subspace H, i.e., it is the kernel of the morphism $V \to \mathcal{O}_x$ determined by the composition

$$V \to V(x) \to Q(x) \to Q(x)/H.$$

By construction $\phi\colon E_0^\vee \to V$ factors through F, and so

$$\mathrm{H}^0(X, E_0 \otimes F) = \mathrm{Hom}_X(E_0^\vee, F) \neq 0.$$

On the other hand, no nonzero map $E_i^\vee \to V$ factors through F, as the composition

$$E_i^\vee(x) \xrightarrow{\psi_i(x)} V(x) \to Q(x)/H$$

is nonzero by the choice of H. Thus,

$$\mathrm{H}^0(X, E_i \otimes F) = \mathrm{Hom}_X(E_i^\vee, F) = 0 \quad \text{for} \quad i = 1, \ldots, n.$$

Since $\operatorname{rk}(F) = \operatorname{rk}(V)$ and $\deg(F) = \deg(V) - 1$, this completes the proof. □

Acknowledgements We thank Hannah Larson and Shizhang Li for being part of our group during SPONGE, and the many interesting conversations during and after the workshop. We are grateful for the referee's comments. PB and TT would also like to thank Chiara Damiolini, Hans Franzen, Vicky Hoskins, and Svetlana Makarova for interesting discussions on a

GIT-free proof of the (quasi-)projectivity of moduli of semistable quiver representations. JA was partially supported by NSF grants DMS-1801976 and DMS-2100088. DB was supported by NSF Postdoctoral Research Fellowship DMS-1902875.

References

[1] Jarod Alper. Good moduli spaces for Artin stacks. *Ann. Inst. Fourier (Grenoble)*, 63(6):2349–2402, 2013.

[2] Jarod Alper, Daniel Halpern-Leistner, and Jochen Heinloth. Existence of moduli spaces for algebraic stacks, 2019. `arXiv:1812.01128v3`.

[3] Luis Álvarez Cónsul and Alastair King. A functorial construction of moduli of sheaves. *Invent. Math.*, 168(3):613–666, 2007.

[4] Leticia Brambila-Paz, Iwona Grzegorczyk, and Peter Newstead. Geography of Brill–Noether loci for small slopes. *J. Algebraic Geom.*, 6(4):645–669, 1997.

[5] Raymond Cheng, Carl Lian, and Takumi Murayama. Projectivity of the moduli of curves. In *Stacks project expository collection*, number 480 in London Mathematical Society Lecture Note Series, Chapter 1. Cambridge University Press, 2022.

[6] Maurizio Cornalba. On the projectivity of the moduli spaces of curves. *J. Reine Angew. Math.*, 443:11–20, 1993.

[7] Pierre Deligne and David Mumford. The irreducibility of the space of curves of given genus. *Inst. Hautes Études Sci. Publ. Math.*, (36):75–109, 1969.

[8] Jean-Marc Drézet and M. S. Narasimhan. Groupe de Picard des variétés de modules de fibrés semi-stables sur les courbes algébriques. *Invent. Math.*, 97(1):53–94, 1989.

[9] Eduardo Esteves and Mihnea Popa. Effective very ampleness for generalized theta divisors. *Duke Math. J.*, 123(3):429–444, 2004.

[10] Gerd Faltings. Stable G-bundles and projective connections. *J. Algebraic Geom.*, 2(3):507–568, 1993.

[11] Haoyang Guo, Sanal Shivaprasad, Dylan Spence, and Yueqiao Wu. Boundedness of semistable sheaves. In *Stacks project expository collection*, number 480 in London Mathematical Society Lecture Note Series, Chapter 4. Cambridge University Press, 2022.

[12] Daniel Halpern-Leistner. On the structure of instability in moduli theory, 2018. `arXiv:1411.0627v4`.

[13] Georg Hein. Faltings' construction of the moduli space of vector bundles on a smooth projective curve. In *Affine flag manifolds and principal bundles*, Trends Math., pages 91–122. Birkhäuser/Springer Basel AG, Basel, 2010.

[14] Daniel Huybrechts and Manfred Lehn. *The geometry of moduli spaces of*

sheaves. Cambridge Mathematical Library. Cambridge University Press, second edition, 2010.

[15] Seán Keel and Shigefumi Mori. Quotients by groupoids. *Ann. Math.*, 145(1):193–213, 1997.

[16] Steven L. Kleiman. The transversality of a general translate. *Compositio Math.*, 28:287–297, 1974.

[17] János Kollár. Projectivity of complete moduli. *J. Differential Geom.*, 32(1):235–268, 1990.

[18] Herbert Lange. Zur Klassifikation von Regelmannigfaltigkeiten. *Math. Ann.*, 262(4):447–459, 1983.

[19] Stacy G. Langton. Valuative criteria for families of vector bundles on algebraic varieties. *Ann. Math.*, 101:88–110, 1975.

[20] Joseph Le Potier. *Lectures on vector bundles*, volume 54 of *Cambridge Studies in Advanced Mathematics*. Cambridge University Press, 1997. Translated by A. Maciocia.

[21] M. S. Narasimhan and S. Ramanan. Moduli of vector bundles on a compact Riemann surface. *Ann. Math.*, 89:14–51, 1969. doi:10.2307/1970807.

[22] C. S. Seshadri. Vector bundles on curves. In *Linear algebraic groups and their representations (Los Angeles, CA, 1992)*, volume 153 of *Contemp. Math.*, pages 163–200. American Mathematical Society, 1993.

[23] The Stacks Project Authors. *The Stacks project*. https://stacks.math.columbia.edu.

4

Boundedness of semistable sheaves

Haoyang Guo
Max-Planck-Institut für Mathematik

Sanal Shivaprasad
University of Michigan, Ann Arbor

Dylan Spence
University of Wisconsin-Whitewater

Yueqiao Wu
University of Michigan, Ann Arbor

Abstract In this expository article, we follow Langer's work in [6] to prove the boundedness of the moduli space of semistable torsion-free sheaves over a projective variety, in any characteristic.

4.1 Introduction

Background and the main result The study of moduli is one of the oldest branches of algebraic geometry and forms a central pillar of our modern understanding. Historically, such questions date back to Riemann, who, in his pursuit to understand what we now call Riemann surfaces, determined that a complex projective curve of genus g depends on exactly $3g-3$ parameters (which he referred to as "moduli" – hence the name). The central question of moduli theory has not evolved much; we are interested in whether or not various algebraic objects can be "parametrized" (in some sense) by some other algebraic object. For example, one might be interested in forming a moduli space of algebraic curves with some fixed genus, or, in our case, a moduli space of torsion-free sheaves with fixed Hilbert polynomial on a fixed variety X.

To construct the moduli space of torsion-free sheaves $\{E_\alpha\}$ with fixed Hilbert polynomial P over a given variety X, one of the first and the most fundamental problems is the *boundedness* of the collection $\{E_\alpha\}$. This property is equivalent to the moduli space, if exists, being a finite type scheme over the base field, and thus a reasonable geometric object that one can work with. However it was quickly discovered that even if the Hilbert polynomial is fixed, certain pathological examples prevent such a moduli space from being well behaved. Mumford, in the case of curves, introduced the notion of *semistability* for a torsion-free sheaf as a solution for this problem, and to demonstrate its efficacy he showed that the collection of semistable vector bundles of a fixed rank and degree on a fixed curve is bounded. This boundedness is an important ingredient in a modern proof of the projectivity of the moduli space of semistable vector bundles on a curve, discussed in [1], Chapter 3 in this volume.

As the notion of semistability can be generalized to torsion-free sheaves over general projective varieties, it is natural to ask about the boundedness for semistable sheaves on higher dimensional varieties. Following Langer's work in [6], the goal of this chapter is to give a positive answer to the boundedness problem. Precisely, we prove the following:

Theorem 4.1.1 *Let $\mathcal{M}^P(X)$ be the moduli space of semistable torsion-free sheaves with fixed Hilbert polynomial P. Then it is of finite type.*

Idea of the proof Let us now discuss the idea of the proof. Given any (semistable) torsion-free sheaf E in our collection, we write $E|_H$ for the restriction of E to a general hypersurface H, and $\mu_{\max}(E|_H)$ for the maximal slope in the Harder–Narasimhan polygon of $E|_H$.

Our most important tool in the proof is Kleiman's criterion (Theorem 4.2.12) and an accompanying inequality (Lemma 4.2.13), which, when combined, state that to prove the boundedness of the family of torsion-free sheaves $\mathcal{M}^P(X)$, it suffices to give a uniform bound of $\mu_{\max}(E|_H)$. Because of this fact, most of the work in proving boundedness lies in producing such a uniform bound, which we obtain in Corollary 4.5.1 and Theorem 4.5.2. Explicitly, it can be understood as follows:

Theorem 4.1.2 *Let E be a torsion-free sheaf over X, with Hilbert polynomial P, and denote the maximal slope of its Harder–Narasimhan polygon be μ_0. Then we have*

$$\mu_{\max}(E|_H) \leq C_1(P) + C_2(P)\mu_0,$$

where H is a general hypersurface of X and $C_i(P)$ are constants determined only by P and X.

The key point is that this result allows us to bound the slopes $\mu_{\max}(E|_H)$ of the restrictions $E|_H$ by the slopes of the original E, plus extra terms and coefficients that are controlled only by the Hilbert polynomial of E.

The major contents of this chapter are thus devoted to obtaining the inequality above, following [6, Section 3]. The strategy can be understood as a double induction on two different results. The first result, Theorem 4.4.1, is a special case of Theorem 4.1.2 above, considering the invariants before and after the restriction of E to a hypersurface, where E is torsion-free of rank $\leq r$ (cf. [6, Theorem 3.1]). We will also refer to this result as Theorem Res(r). It relates the slopes of the Harder–Narasimhan filtration of the restriction $E|_H$ to the discriminant and Harder–Narasimhan filtration of E itself.

The second result is a collection of several various numerical inequalities, which we refer to as *Bogomolov's inequalities* (see Theorem 4.3.1, cf. [6, Theorems 3.2–3.4, and $T^5(r)$]), owing to their similarity to the well-known inequality in characteristic zero of the same name. We will refer to this result as Theorem BI(r). Up to an extra term, it says that the discriminant $\Delta(E)$ of E is nonnegative when the torsion-free sheaf E is (strongly) semistable, where E is of rank $\leq r$. Using the notation above, Langer's induction schema can be summarized as the following two implications (see Section 4.4):

$$\begin{cases} \mathrm{BI}(r) \Rightarrow \mathrm{Res}(r); \\ \mathrm{Res}(r) + \mathrm{BI}(r) \Rightarrow \mathrm{BI}(r+1). \end{cases} \tag{1}$$

We note that both results are true automatically for rank $r = 1$. In this way, the two technical results are proved together, thus so is the inequality in Theorem 4.1.2 and the boundedness in Theorem 4.1.1.

We also mention that in the positive characteristic case, a Frobenius pullback of a semistable sheaf may no longer be semistable. So we define semistable sheaves whose Frobenius pullbacks are all semistable as being *strongly semistable*. A priori, for a given sheaf, the Harder–Narasimhan filtrations of its Frobenius pullbacks could be unrelated to each other. One of Langer's key observations is that their differences are controllable: the Harder–Narasimhan filtrations eventually stabilize with respect to the Frobenius (Theorem 4.2.28, [6, Theorem 2.7]). This allows us to pass between semistable and strongly semistable sheaves in the proof.

Lastly, we briefly mention the history preceding Langer's result; a more detailed introduction can be found in [6]. Theorem 4.1.1 was known in characteristic zero, this was proven by Barth, Spindler, Maruyama, Forster, Hirschowitz, and Schneider. In positive characteristic, however, only the cases of curves and surfaces were known. At that time, it was understood that the numerical quantities associated to E before and after restricting E to a hypersurface (often called Grauert–Mülich type results) should allow one to prove boundedness for a given family, but prior to [6] only a coarser result by Mehta and Ramanathan was available. On the other hand, Bogomolov showed the nonnegativity of the discriminant of E whenever E is semistable, assuming X is in characteristic zero. It was unknown as well whether the result is completely true in positive characteristic, but it was clear that some version would be needed. Thus Langer's key contributions were in developing positive characteristic versions of the above and also in combining them in a fruitful way.

Chapter outline We start with Section 4.2 on necessary preliminaries. The section contains three subsections, including basics on the stability of coherent sheaves and additional results in positive characteristic. Moreover, as the proofs of the main theorems require working with polarizations consisting of nef divisors, we also include a subsection on how to approximate nef polarizations by ample polarizations. In Section 4.3 we introduce several forms of Bogomolov's inequalities in positive characteristic, and show that they are equivalent in Theorem 4.3.1(BI(r)). For the reader's convenience, this section corresponds to the implications $T^5(r) \Rightarrow T^3(r) \Rightarrow T^4(r) \Rightarrow T^2(r) \Rightarrow T^5(r)$ in [6, Sections 3.6–3.8]. In Section 4.4, we prove Theorem 4.4.1, and complete the major technical part of the chapter by proving the induction schema as in (4.1). This corresponds to [6, Sections 3.5 and 3.9] and [5]. Eventually, in Section 4.5, we combine the main technical results above to show the boundedness of the moduli space $\mathcal{M}^P(X)$, thus finishing the proof of Theorem 4.1.1.

4.2 Preliminaries

In this section, we provide the preliminaries for the chapter.

4.2.1 Stability of coherent sheaves

In this section we recall some useful facts on the (semi)stability of coherent sheaves. We refer to [4] for details and proofs. Let X be a smooth projective variety of dimension n over an algebraically closed field. Fix $n-1$ ample divisors $D_1, \ldots, D_{n-1}$ on X. Let E be a coherent sheaf on X. Recall that a sheaf E is said to be pure (of dimension $d \leq \dim X$) if, for all nontrivial subsheaves $f \subset E$, $\dim(F) := \dim \operatorname{supp}(F) = d$.

Definition 4.2.1 The *slope* of E, with respect to the polarization

$(D_1, \ldots, D_{n-1})$, is defined by

$$\mu(E) := \frac{\deg(E)}{\mathrm{rk}(E)} := \frac{c_1(E) \cdot D_1 \cdots D_{n-1}}{\mathrm{rk}(E)}$$

where $\mathrm{rk}(E)$ is the rank of E at the generic point. If $\mathrm{rk}(E) = 0$, then $\mu(E) := \infty$.

As indicated, the slope of a coherent sheaf does depend on the choice of ample (or nef) divisors defining a polarization. For the purpose of clarity, we will not reference this choice of polarization when discussing slopes unless there is a serious risk of confusion.

Definition 4.2.2 We say that E is *(semi)stable* if E is pure and, for any nonzero proper subsheaf $F \subset E$, $\mu(F) < \mu(E)$ (resp. $\mu(F) \leq \mu(E)$).

This notion of (semi)stability is more generally known as slope-(semi)-stability or μ-(semi)stability. An equivalent formulation of the definition can also be stated in terms of quotient sheaves.

Lemma 4.2.3 *[4, Proposition 1.2.6] Let E be a coherent sheaf. Then E is semistable if and only if, for all the quotient sheaves $E \to G$, $\mu(E) \leq \mu(G)$.*

This formulation of the definition gives a quick proof of the following very useful fact.

Lemma 4.2.4 *[4, Proposition 1.2.7] Let E_1 and E_2 be semistable sheaves with slopes μ_1 and μ_2 such that $\mu_1 > \mu_2$. Then* $\mathrm{Hom}(E_1, E_2) = 0$.

The definition of semistability also behaves well with regards to exact sequences, as the following two lemmas indicate.

Lemma 4.2.5 *Let $E' \to E \to E''$ be a short exact sequence of torsion-free coherent sheaves, and let μ', μ, μ'' and r', r, r'' be the slopes and the ranks of E', E, E'' respectively. Then, $r\mu = r'\mu' + r''\mu''$. In particular, if the slopes of two of E', E, E'' are the same, then this is also the case for the third.*

Lemma 4.2.6 *If $E' \to E \to E''$ is a short exact sequence and μ', μ, μ'' denote the slopes of E', E, E'' respectively, then*

- *If E is semistable and either $\mu' = \mu$ or $\mu'' = \mu$, then E' and E'' are also semistable.*
- *If E' and E'' are semistable and $\mu' = \mu''$, then E is semistable with $\mu = \mu' = \mu''$.*

One of the more important technical tools is the Harder–Narasimhan filtration, which is defined below.

Definition 4.2.7 A *Harder–Narasimhan filtration* for E is an increasing filtration

$$0 = E_0 \subset E_1 \subset \cdots \subset E_d = E$$

such that the factors $F_i := E_i/E_{i-1}, i = 1, \ldots, d$ are semistable sheaves with slopes μ_i satisfying

$$\mu_{\max}(E) := \mu_1 > \mu_2 > \cdots > \mu_d =: \mu_{\min}(E).$$

Proposition 4.2.8 *[4, Theorem 1.3.4] Every torsion-free sheaf E has a unique Harder–Narasimhan filtration.*

Remark 4.2.9 Just as for the slope, the Harder–Narasimhan filtration is also dependent on the choice of the polarization $(D_1, \ldots, D_{n-1})$, and could be different for different polarizations.

Remark 4.2.10 We should also remark here that in the construction of the Harder–Narasimhan filtration for a torsion-free sheaf, an important step is establishing the existence and uniqueness of a *maximal destabilizing subsheaf*. We will use this notion a few times, so we give its definition here. Given a torsion-free sheaf E, then the *maximal destabilizing subsheaf* $F \subset E$ is a semistable coherent subsheaf such that $\mu(F) \geq \mu(G)$ for all other subsheaves $G \subset E$, and moreover if $\mu(F) = \mu(G)$ then $F \supset G$.

Next, we wish to introduce the key tools of the chapter, Kleiman's criterion and a related inequality. For the conclusion of Kleiman's criterion to make sense, however, we should remind the reader of the technical property of boundedness.

Definition 4.2.11 Let M be a set of coherent sheaves on X. Then M is said to be *bounded* if there is a scheme B of finite type and a coherent sheaf F on $X \times B$ with $M \subset \{F_b \mid b \in B \text{ closed}\}$. Here F_b is the pullback of F along $X \times \{b\} \to X \times B$.

Kleiman's criterion provides a very convenient way of determining whether or not a given family is bounded.

Theorem 4.2.12 (Kleiman's criterion) *[4, Theorem 1.7.8] Let $\{E_\alpha\}$ be a family of coherent sheaves over X with the same Hilbert polynomial P. Then the family is bounded if and only if there are constants C_i, for*

$i = 0, \ldots, \deg(P)$, such that for every E_α there exists an E_α-regular sequence of hyperplanes $H_1, \ldots, H_{\deg(P)}$ satisfying

$$h^0(E_\alpha|_{\cap_{j\leq i} H_j}) \leq C_i, \quad \forall i.$$

Here we recall that a hyperplane $s \in H^0(X, \mathcal{O}_X(1))$ is said to be *E-regular* if the map

$$E(-1) \xrightarrow{\cdot s} E$$

is injective. A sequence $\{s_1, \ldots, s_l\} \subset H^0(X, \mathcal{O}_X(1))$ is E-regular if s_i is $(E/(s_1, \ldots, s_{i-1})E)(-1)$-regular for all $1 \leq i \leq l$.

Lemma 4.2.13 *[4, Lemma 3.3.2] Let E be a torsion-free sheaf of rank r. Then for any E-regular sequence of hyperplane sections $H_1, \ldots, H_n$, the following inequality holds for $i = 1, \ldots, n$:*

$$\frac{h^0(X_i, E_i)}{r \deg(X)} \leq \frac{1}{i!}\left[\frac{\mu_{\max}(E_1)}{\deg(X)} + i\right]_+^i,$$

where $X_i \in |H_1| \cap \cdots \cap |H_{n-i}|$, $E_i = E|_{X_i}$, and $[x]_+ = \max\{0, x\}$ for any real number x.

Combining the previous two results, provided we can get a uniform bound on $\mu_{\max}$, will prove boundedness for semistable sheaves with a fixed Hilbert polynomial. To get such estimates, we will often use the following lemma.

Lemma 4.2.14 *If $0 \to A \to B \to C \to 0$ is an exact sequence of torsion-free sheaves, then $\mu_{\max}(B) \leq \max\{\mu_{\max}(A), \mu_{\max}(C)\}$.*

As we end the subsection, we introduce the following definition that helps visualize the Harder–Narasimhan filtration.

Definition 4.2.15 (Harder–Narasimhan polygon) Consider a torsion-free sheaf E on X with its Harder–Narasimhan filtration $0 = E_0 \subset \cdots \subset E_m = E$. Let $p(E_i) = (\mathrm{rk}(E_i), \deg(E_i))$. We define the *Harder–Narasimhan polygon* of E, denoted as $\mathrm{HNP}(E)$, to be the convex hull of the points $p(E_0), \ldots, p(E_m)$ in $\mathbb{R}^2$. See Figure 4.1.

In fact, the Harder–Narasimhan polygon is a convex polygon with vertices $p(E_0), \ldots, p(E_m)$, and the slope of the line segment $\overline{p(E_i)p(E_{i+1})}$ is $\mu(E_i/E_{i+1})$.

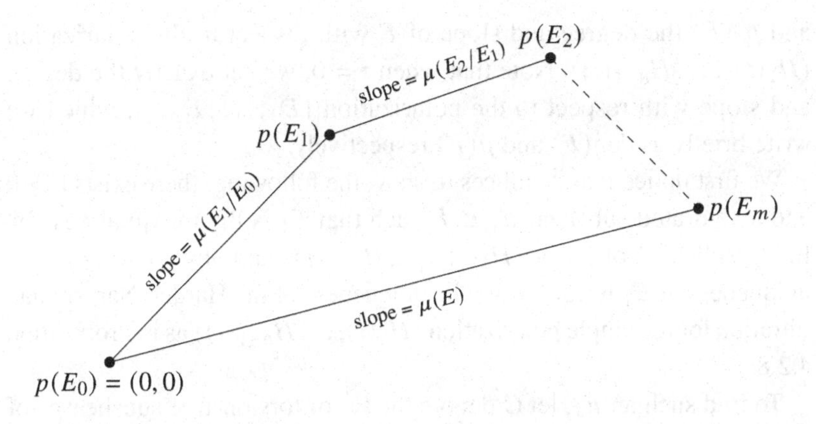

Figure 4.1 An illustration of the Harder–Narasimhan polygon.

Remark 4.2.16 If $F \subset E$ is a torsion-free subsheaf F inside a semistable sheaf E, then the point $p(F) = (\mathrm{rk}(F), \deg(F))$ lies below the polygon $\mathrm{HNP}(E)$, i.e., for any $x \in [0, \mathrm{rk}(F)]$, we have

$$\sup\{y \mid (x, y) \in \mathrm{HNP}(F)\} \leq \sup\{y \mid (x, y) \in \mathrm{HNP}(E)\}.$$

4.2.2 Approximation of nef polarizations

In this subsection, we approximate the Harder–Narasimhan filtration for a polarization consisting of *nef divisors*; this will be used later. As this is a nonstandard setup and does not appear in most of the literature, we provide the full details of the proof. We assume the existence and uniqueness of the (absolute) Harder–Narasimhan filtration with respect to an *ample polarization* as in the last subsection.

Our first main result in this subsection is the following approximation result.

Theorem 4.2.17 *Let $(D_1, \ldots, D_{n-1})$ be a set of nef divisors, and H an ample divisor. Assume E is a torsion-free coherent sheaf over X. Then there exists a positive number ϵ and a (unique) filtration of saturated subsheaves*

$$0 = E_0 \subset E_1 \subset \cdots E_l = E$$

such that for any $t \in (0, \epsilon)$ the above is the Harder–Narasimhan filtration for the ample polarization $(D_1 + tH, \ldots, D_{n-1} + tH)$.

Proof Denote by $H_i(t)$ the $\mathbb{R}$-divisor $D_i + tH$, which is ample [7, Corollary 1.4.10]. For a torsion-free sheaf F, we denote by $\deg_t(F)$

and $\mu_t(F)$ the degree and slope of F with respect to the polarization $(H_1(t), \ldots, H_{n-1}(t))$. Note that when $t = 0$, we get exactly the degree and slope with respect to the polarization $(D_1, \ldots, D_{n-1})$, which we write briefly as $\deg(F)$ and $\mu(F)$ respectively.

We first notice that it suffices to show the following: there exists $\epsilon > 0$ and a saturated subsheaf $E_1 \subset E$ such that E_1 is the maximal destabilizing subsheaf of E for $(H_1(t), \ldots, H_{n-1}(t))$ and any $t \in (0, \epsilon)$. The uniqueness of E_1 follows from the uniqueness of the Harder–Narasimhan filtration for the ample polarization $(H_1(t), \ldots H_{n-1}(t))$ as in Proposition 4.2.8.

To find such an E_1, let C denote the set of torsion-free subsheaves of E and consider the following map of sets:

$$\mu_t(-)\colon C \longrightarrow \{\text{degree } n-1 \text{ polynomials in } t\},$$
$$E' \longmapsto \mu_t(E') = \frac{c_1(E')H_1(t)\cdots H_{n-1}(t)}{\mathrm{rk}(E')}.$$

For each $E' \subset E$, the image $\mu_t(E')$ is the polynomial $a_0 + a_1 t + \cdots + a_{n-1}t^{n-1}$, where we have

$$a_0(E') = \mu(E');$$
$$a_1(E') = \sum_{1 \le i \le n-1} \mu_{(H, D_1, \ldots, \hat{D}_i, \ldots, D_{n-1})}(E').$$

For general $j \le n$, the coefficient a_j is the sum of the slopes of E' with respect to all possible choice of polarizations $(H, \ldots, H, D_{i_1}, \ldots, D_{i_{n-1-j}})$ such that each of the first j entries of the polarization are all equal to H. Moreover, each polynomial $\mu_t(E')$ has coefficients in $\frac{1}{r!}\mathbb{Z}$, whose coefficients are finite linear combinations of the slopes of E' for a fixed, finite number of choices of polarization. In particular, for each $0 \le i \le n$, the collection of coefficients $\{a_i(E') \mid E' \subset E\}$ is bounded above. [1] Here we denote the subset of polynomials consisting of the image of the map $\mu_t(-)$ by $\mathcal{D}$.

We then define a lexicographic order on the elements in $\mathcal{D}$ as follows:

$$W_1(t) = \sum^{d} a_i t^i < W_2(t) = \sum^{d} b_i t^i$$

[1] To see this, we first note that since $a_i(E')$ is a finite positive linear combination of $\mu_{H^i, D_{j_1}, \ldots, D_{j_{n-1-i}}}(E')$, it suffices to bound each slope for the given nef polarization. The latter can be proved by induction on ranks of E as in the classical case. To check the case when E is a line bundle, it suffices to show the set of slopes of sub line bundles is bounded by the slope of E itself, which is true again by approximation via adding tH to each D_{j_l} and letting t approach zero. Here we observe that the inequality holds for any $t > 0$, so by continuity we get the upper bound. The general case follows from the induction hypothesis on E' and E'' in a short exact sequence of vector bundles $E' \to E \to E''$.

if $a_0 = b_0, \ldots, a_{i-1} = b_{i-1}, a_i < b_i$ for some i.

Since the coefficients of polynomials in $\mathcal{D}$ are bounded above, we can find the maximum polynomial $P(t)$ in $\mathcal{D}$. Moreover, since the degree of a sheaf is a discrete quantity, we can find a subsheaf E_1 in the preimage of $\mu_t^{-1}(P(t))$ whose rank is maximal.

Finally, we prove that E_1 is exactly the maximal rank destabilizing subsheaf of E with respect to $(H_1(t), \ldots, H_{n-1}(t))$ for t small enough. To show this, it suffices to show the following, whose complete details are left to the reader.

Claim 4.2.18 Let $\mathcal{D}$ be a set of degree-n polynomials with coefficients bounded above and whose coefficients are in $\frac{1}{N}\mathbb{Z}$ for some positive integer N. Then a polynomial $P(t) \in \mathcal{D}$ is maximum with respect to the lexicographic order if and only if, for small enough $t > 0$, $P(t) > Q(t)$ (in $\mathbb{R}$) for all other $Q(t) \in \mathcal{D}$.

The idea of the claim is the following: as t approaches zero, higher-power terms are dominated by those with lower powers, so the polynomial that has the largest first few terms under lexicographic order will also have the maximum value in $\mathbb{R}$ for t small enough. □

The above filtration of subsheaves $\{E_i\}$ in general fails to be the Harder–Narasimhan filtration with respect to $(D_1, \ldots, D_{n-1})$ for $t = 0$. However, the filtration is in fact a *weak Harder–Narasimhan filtration* with respect to $(D_1, \ldots, D_{n-1})$, in the sense that each E_i/E_{i-1} is semistable with respect to $(D_1, \ldots, D_{n-1})$, and we have inequalities

$$\mu(E_{i+1}/E_i) \geq \mu(E_i/E_{i-1}), \quad 1 \leq i \leq l-1.$$

Note that we do not have strict inequalities here, as in the definition of the Harder–Narasimhan filtration.

The semistability of the subsheaves E_i can be seen as follows:

Lemma 4.2.19 *Let E' be a subsheaf of the torsion-free sheaf E such that for small enough $t > 0$, E' is semistable with respect to the polarization $(H_1(t), \ldots, H_{n-1}(t))$. Then E' is semistable with respect to $(D_1, \ldots, D_{n-1})$.*

Proof Assume E' admits a subsheaf E_1' whose slope with respect to $(D_1, \ldots, D_{n-1})$ is strictly larger than that of E'. By the continuity of the polynomial μ_t in t, we get the inequality $\mu_t(E_1') > \mu_t(E')$ for small enough $t > 0$, which contradicts our assumption. □

The same strategy above in fact implies the existence and uniqueness of the Harder–Narasimhan filtration for nef divisors, generalizing the result for ample divisors.

Corollary 4.2.20 *Let $(D_1, \ldots, D_{n-1})$ be a set of nef divisors of X. Then any torsion-free coherent sheaf E over X admits a unique Harder–Narasimhan filtration with respect to the polarization $(D_1, \ldots, D_{n-1})$.*

Proof Let H be a fixed ample divisor over X. By Theorem 4.2.17, there exists a unique filtration of saturated subsheaves E_i of E such that, when $t > 0$ is very small, E_i is the Harder–Narasimhan filtration of E with respect to $(H_1(t), \ldots, H_{n-1}(t))$. Each E_i is semistable with respect to $(D_1, \ldots, D_{n-1})$ by Lemma 4.2.19, and the slopes of E_i/E_{i-1} are non-strictly decreasing. Thus by picking the subsheaves in this filtration in such a way that the slopes of the factors become strictly decreasing, we get the Harder–Narasimhan filtration of E with respect to $(D_1, \ldots, D_{n-1})$. Thus the uniqueness of the Harder–Narasimhan filtration follows from the uniqueness of the maximal destabilizing subsheaf with respect to $(D_1, \ldots, D_{n-1})$, where the latter can be checked via the continuity of μ_t. So we are done. □

Now we give a simple example illustrating the degeneration of the Harder–Narasimhan filtration.

Example 4.2.21 Let $X = \mathbb{P}^1 \times \mathbb{P}^1$ be the product of two projective lines over a field k. Let D be the trivial divisor, let H be the ample divisor of bidegree $(1, 2)$ over X, and let $H(t) := D + tH$ be the sum. Denote by $L_1 = (0, 1)$ and $L_2 = (1, 0)$ the two line bundles over X, and let E be the the direct sum $L_1 \oplus L_2$. Then, for any $t > 0$, the divisor $H(t)$ is ample in X as it can be written as a sum of the pullback of ample divisors on two separate factors of X. Moreover, for each $t > 0$, we have $L_1 \cdot H(t) = 2t > t = L_2 \cdot H(t)$, and thus

$$0 \subset L_1 \subset E \qquad (*)$$

is the Harder–Narasimhan filtration of E for the ample polarization $H(t)$, for any $t > 0$. However, when $t = 0$, as $H(0) = D = 0$ is the trivial divisor, any subsheaf of E has the same slope 0. In particular, the Harder–Narasimhan filtration of E is the trivial filtration, and the filtration $(*)$ becomes a filtration of subsheaves whose graded pieces have the same slopes.

4.2.3 Positive characteristic

The main difference in characteristic p is that we need to work with the notion of *strong semistability*, as the Frobenius pullbacks of semistable sheaves are not necessarily semistable.

The Frobenius morphism First, we recall a few basic notions from algebraic geometry in positive characteristic. Let k be an algebraically closed field of characteristic $p > 0$. Let X be a smooth projective k-variety. The *absolute Frobenius* morphism $F_X : X \to X$ is the map on X given locally on an open subset $\operatorname{Spec} R \subset X$ by $a \mapsto a^p$. For simplicity of notation, we just denote F_X by F. Note that F is not a map of k-schemes.

We also fix nef divisors $D_1, \ldots, D_{n-1}$ on X, and we will compute slopes with respect to the polarization $(D_1, \ldots, D_{n-1})$.

Definition 4.2.22 (Strong semistability) A coherent sheaf E on X is said to be *strongly semistable* if $(F^e)^*E$ is a semistable sheaf on X for all $e \geq 0$.

Remark 4.2.23 A coherent sheaf that is semistable but not strongly semistable can be found for example in [3, Corollary 2].

In positive characteristic, instead of just keeping track of the values of $\mu_{\min}$ and $\mu_{\max}$ for a coherent sheaf, we also keep track of a few other invariants related to the Frobenius pullbacks. We note first that Frobenius pullback alters the slope of a sheaf, indeed we have that $\mu((F^e)^*E) = p^e \mu(E)$. For this reason, we define the related quantity

$$L_{\max}(E) := \lim_{e \to \infty} \frac{\mu_{\max}((F^e)^*E)}{p^e}.$$

Note that the sequence $\frac{\mu_{\max}((F^e)^*E)}{p^e}$ is increasing in e, thus the limit $L_{\max}(E)$ exists in $\mathbb{R} \cup \{\infty\}$. We will show later that this limit is indeed finite. Similarly, we can define $L_{\min}$. By definition, we have $L_{\max}(E) \geq \mu_{\max}(E)$ and $L_{\min}(E) \leq \mu_{\min}(E)$. It immediately follows from the definition that if E is strongly semistable, then

$$L_{\max} = \mu_{\max} = \mu(E) = \mu_{\min}(E) = L_{\min}(E).$$

Let us set

$$\alpha(E) := \max\{L_{\max}(E) - \mu_{\max}(E), \mu_{\min}(E) - L_{\min}(E)\}.$$

We would like to find an upper estimate for $\alpha(E)$. To do this, we first state a theorem on how to detect instability of the Frobenius pullback of a semistable sheaf. For details, refer to [6, Section 2]. The idea is to make use of the canonical connection $F^*E \to F^*E \otimes \Omega_X$ on the Frobenius pullback.

Theorem 4.2.24 *Let E be a semistable sheaf on X such that F^*E is not a semistable sheaf on X. Let $0 = E_0 \subset E_1 \subset \cdots \subset E_m = F^*E$ be the Harder–Narasimhan filtration of F^*E. Then, the natural $\mathcal{O}_X$ homomorphisms $E_i \to (E/E_i) \otimes \Omega_X$ induced by the canonical connection are nonzero.*

We now estimate the the minimum and maximum slopes of the Frobenius pullback of a semistable sheaf. The idea is to use the nonzero maps above, along with Theorem 4.2.24, to get an estimate of how far apart the slopes can be.

Lemma 4.2.25 *Let A be a nef divisor such that $T_X(A)$ is globally generated (or equivalently such that $\Omega \hookrightarrow \mathcal{O}_X(A)^{\oplus l}$ for some l) and let E be a torsion-free semistable sheaf on X. Then,*

$$\mu_{\max}(F^*E) - \mu_{\min}(F^*E) \leq (\mathrm{rk}(E) - 1)A \cdot D_1 \cdot \cdots \cdot D_{n-1}.$$

Proof Let $0 = E_0 \subset E_1 \subset \cdots \subset E_m = F^*E$ be the Harder–Narasimhan filtration of F^*E. Using Theorem 4.2.24, we have that the $\mathcal{O}_X$-homomorphism $E_i \to F^*E/E_i \otimes \Omega_X$ is nonzero. Thus, we get that

$$\mu(E_i/E_{i-1}) = \mu_{\min}(E_i) \leq \mu_{\max}(F^*E/E_i \otimes \Omega_X).$$

Since $\Omega \hookrightarrow \mathcal{O}_X(A)^{\oplus l}$, we obtain

$$\begin{aligned}&\mu_{\max}(F^*E/E_i \otimes \Omega_X)\\ &\qquad \leq \mu_{\max}(F^*E/E_i \otimes \mathcal{O}_X(A)) = \mu(E_{i+1}/E_i) + A \cdot D_1 \cdot \cdots \cdot D_{n-1}.\end{aligned}$$

Thus, we find that

$$\mu(E_i/E_{i-1}) - \mu(E_{i+1}/E_i) \leq A \cdot D_1 \cdot \cdots \cdot D_{n-1}.$$

Summing this inequality, we get the result. □

Proposition 4.2.26 *If A is a nef divisor such that $T_X(A)$ is globally generated and E is a torsion-free sheaf on X, then*

$$\frac{\mu_{\max}(F^*E)}{p} \leq \mu_{\max}(E) + \frac{(\mathrm{rk}(E) - 1)}{p} A \cdot D_1 \ldots D_{n-1}$$

Proof Let $0 = E_0 \subset E_1 \subset \cdots \subset E_m = E$ be the Harder–Narasimhan filtration of E. Applying the previous proposition to $F^*(E_i/E_{i-1})$, we find that

$$\frac{\mu_{\max}(F^*(E_i/E_{i-1}))}{p} \leq \mu(E_i/E_{i-1}) + \frac{(\mathrm{rk}(E)-1)}{p} A \cdot D_1 \ldots D_{n-1}.$$

Note that the F^*E_i form a filtration of F^*E and thus by Lemma 4.2.14 we have $\mu_{\max}(F^*E) \leq \max_i\{\mu_{\max}(F^*(E_i/E_{i-1}))\}$. Using this, we obtain the required result. □

We have the following bound on $\alpha(E)$ (which in particular shows that $L_{\max}(E)$ and $L_{\min}(E)$ are finite).

Proposition 4.2.27 *If A is a nef divisor such that $T_X(A)$ is globally generated, then*

$$\alpha(E) \leq \frac{(\mathrm{rk}(E)-1)}{p-1} A \cdot D_1 \cdots D_{n-1}.$$

Proof Applying induction to the previous proposition, we see that

$$\frac{\mu_{\max}((F_X^e)^*E)}{p^e} \leq \mu_{\max}(E) + (\mathrm{rk}(E)-1)\left(\frac{1}{p} + \frac{1}{p^2} + \cdots + \frac{1}{p^e}\right) A \cdot D_1 \cdots D_{n-1}.$$

Letting $e \to \infty$ gives the required result. □

Finite determinancy of the Harder–Narasimhan filtration We will now prove the following theorem.

Theorem 4.2.28 *For every torsion-free sheaf E, there exists a non-negative integer e_0 such that all the factors in the Harder–Narasimhan filtration of $(F^{e_0})^*E$ are strongly semistable.*

The proof uses the Harder–Narasimhan polygon (see Definition 4.2.15). For any sheaf G over a smooth projective variety X, let us define $p(G) := (\mathrm{rk}G, \deg G) \in \mathbb{R}^2$. We also define

$$\mathrm{HNP}_e(E) := \{(x,y) \in \mathbb{R}^2 \mid (x, p^e y) \in \mathrm{HNP}((F^e)^*E)\}.$$

Note that $\mathrm{HNP}_e(E)$ forms an increasing sequence of convex subsets of $\mathbb{R}^2$, and we define $\mathrm{HNP}_\infty(E) := \overline{\bigcup_e \mathrm{HNP}_e(E)}$.

Proposition 4.2.29 $\mathrm{HNP}_\infty(E)$ *is a bounded convex polygon.*

Proof It is clear that $\mathrm{HNP}_\infty(E)$ is convex as each of the $\mathrm{HNP}_e(E)$ are convex. To see that it is bounded, note that the rank coordinates of $\mathrm{HNP}_e(E)$ lie in in the interval $[0, \mathrm{rk}(E)]$. The fact that $\alpha(E)$ is finite also tells us that there is a uniform bound on the degree coordinates of $\mathrm{HNP}_e(E)$. Thus we see that $\mathrm{HNP}_\infty(E)$ is a bounded subset of $\mathbb{R}^2$.

To show that $\mathrm{HNP}_\infty(E)$ is a polygon, we claim that it is a convex hull of the set $\{(r, q_{\infty,r}) \mid r \in \{0, \ldots, \mathrm{rk}(E)\}$, where $q_{\infty,r} = \sup\{d \mid (r,d) \in \mathrm{HNP}_\infty(E)\}$. Note that $q_{\infty,0} = p(0) = (0,0)$ and $q_{\infty,\mathrm{rk}(E)} = p(E) = (\mathrm{rk}(E), \deg(E))$.

It is clear that the convex hull of these points is contained in $\mathrm{HNP}_\infty(E)$. To see the converse, first note that all $\mathrm{HNP}_k(E)$ and thus $\mathrm{HNP}_\infty(E)$ lie above the line segment joining the points $p(E_0) = (0,0)$ and $p(E) = (\mathrm{rk}(E), \deg(E))$. It is thus enough to show that every vertex of $\mathrm{HNP}_e(E)$ lies in the convex hull for all e. Pick any vertex (r,d) of $\mathrm{HNP}_e(E)$ for some e and some $r \in \{0, \ldots, \mathrm{rk}(E)\}$. Then, since (r,d) lies above the line segment $\overline{p(0)p(E)}$ and below the point (r, q_r), one can easily see that (r,d) lies in the convex hull of $p(0)$, $p(E)$ and (r, q_r). □

Before proving the finite determinancy of the Harder–Narasimhan filtration, let us introduce some notation. Let $0 = E_{0,e} \subset \cdots \subset E_{m_e,e}$ denote the Harder–Narasimhan filtration of $(F^e)^*E$. Let $p_{i,e} = (\mathrm{rk}(E_{i,e}), \deg(E_{i,e})/p^e)$ denote the vertices of $\mathrm{HNP}_e(E)$.

Let $(0,0) = p_{0,\infty}, \ldots, p_{s,\infty} = p(E)$ denote the vertices of the $\mathrm{HNP}_\infty(E)$ and denote the coordinates of $p_{j,\infty}$ as $p_{j,\infty} = (r_{j,\infty}, d_{j,\infty})$. Let us denote the slope of the line segment $\overline{p_{(j-1),\infty}p_{j,\infty}}$ as $\mu_{j,\infty}$. For every j, there exists a sequence $p_{i_j,e} \to p_{j,\infty}$ for some sequence i_j as $e \to \infty$.

Proof of Theorem 4.2.28 We prove that there exists some e_0 such that $\mathrm{HNP}_{e_0}(E) = \mathrm{HNP}_\infty(E)$. We show this by induction on the rank. In the rank-1 case, there is nothing to show.

Pick $0 < \epsilon \ll 1$. Then we have the Euclidean distance $\|p_{i_j,e} - p_{j,\infty}\| < \epsilon$ for all j and all $e \gg 0$. Replacing E by $(F^e)^*E$ for some large e, we may assume that $\|p_{i_j,e} - p_{j,\infty}\| < \epsilon$ holds for all $e \geq 0$. Since the rank coordinates of $p_{i_j,e}$ and $p_{j,\infty}$ can only take integer values, it must be the case that $r_{j,\infty} = \mathrm{rk}(E_{i_j,e})$.

We first show that there exists an integer e_0 such that $E_{i_1,e} = (F^{e-e_0})^*E_{i_1,e_0}$ for all $e \geq e_0$. Consider the first line segment s of the polygon $\mathrm{HNP}_\infty(E_{i_1,0})$. Note that s cannot lie above the line segment $\overline{p_{0,\infty}p_{1,\infty}}$ (see Remark 4.2.16). We will show that s actually lies on the line segment $\overline{p_{0,\infty}p_{1,\infty}}$. Assuming that is the case, we apply the induction

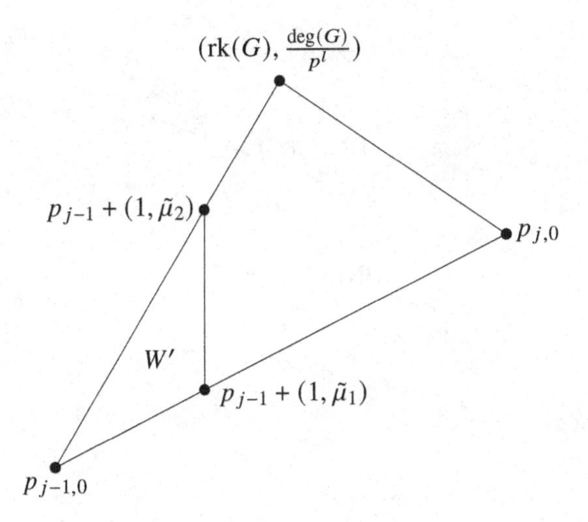

Figure 4.2 The triangles W and W' appearing in the proof of Theorem 4.2.28. The larger triangle is W and the smaller triangle inside it is W'.

hypothesis to $E_{i_1,0}$ to obtain the following: there exists a subsheaf G of $(F^l)^*E_{i_1,0}$ such that $(\mathrm{G}, \frac{\deg G}{p^l}) \in s \subset \overline{p_{0,\infty}p_{1,\infty}}$, i.e., G is strongly semistable. Once again, we may replace E by $(F^l)^*E$ to assume that $l = 0$. By the induction hypothesis, since the theorem holds for E/G and since $\mu(G) = \mu_{\max}(E) = \frac{d_{1,\infty}}{r_{1,\infty}}$, the theorem also holds for E.

Now suppose that s lies strictly below the line segment $\overline{p_{0,\infty}p_{1,\infty}}$. We will deduce a contradiction from this. Since $p_{i_1,l} \to p_{i_1,\infty}$ as $l \to \infty$, we can find an integer l such that $\overline{p_{0,l}p_{i_1,l}}$ lies above s. Thus, $\mu_{\max}((F^l)^*E) > \mu_{\max}((F^l)^*E_{i_1,0})$ for such an integer l.

Since

$$\{(F^l)^*E_{i_1,0}, (F^l)^*E_{i_1+1,0}, \ldots, (F^l)^*E\}$$

form a filtration of $(F^l)^*E$, using Lemma 4.2.14, we obtain that there exists an integer $j > i_1$ such that

$$\mu_{\max}((F^l)^*(E_{j,0}/E_{j-1,0})) > \mu_{\max}((F^l)^*E_{i_1,0}).$$

Now consider a saturated subsheaf $G \subset (F^l)^*(E_{j,0})$ such that

$$\mu(G/(F^l)^*(E_{j-1,0})) = \mu_{\max}((F^l)^*(E_{j,0}/E_{j-1,0})).$$

Then $(\mathrm{rk}(G), \frac{\deg(G)}{p^l}) \in \mathrm{HNP}_l(E)$.

Let us try to estimate the difference in the areas of $\mathrm{HNP}_l(E)$ and $\mathrm{HNP}(E)$ in Figure 4.2. There is a lower bound on the difference in areas

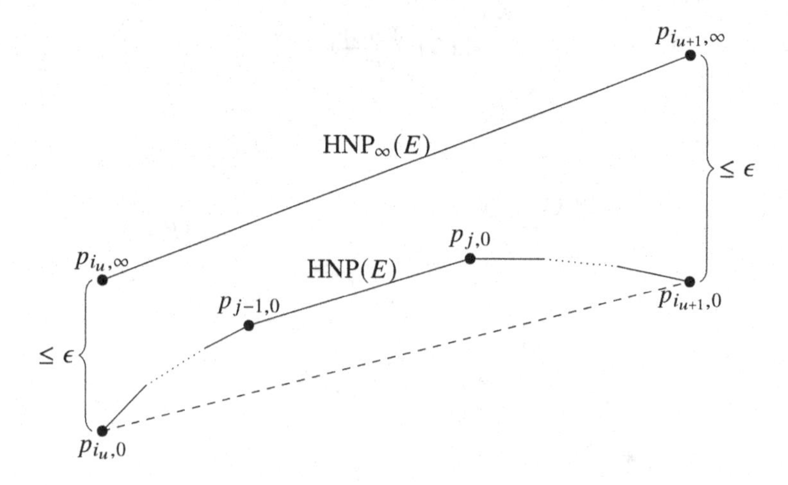

Figure 4.3 Estimating $\mu(E_{j,0}/E_{j-1,0})$. The lines show the boundaries of $\mathrm{HNP}_\infty(E)$ and $\mathrm{HNP}(E)$.

found by considering the triangle W joining the points $p_{j-1,0}$, $p_{j,0}$, and $(\mathrm{rk}(G), \frac{\deg(G)}{p^l})$.

Let

$$\tilde{\mu}_1 := \frac{\mu((F^l)^*(E_{j,0}/E_{j-1,0}))}{p^l},$$

$$\tilde{\mu}_2 := \frac{\mu(G/(F^l)^* E_{j-1,0})}{p^l}$$

denote the slopes of the sides of W that contain $p_{j-1,0}$. Using the fact that the rk-coordinates of these points can only take integer values, we can get a get a lower bound on the area of the triangle W by considering the triangle W' joining $p_{j-1,0}$, $p_{j-1,0} + (1, \tilde{\mu}_1)$, $p_{j-1,0} + (1, \tilde{\mu}_2)$ as in Figure 4.2:

$$\begin{aligned}
&\mathrm{Area}(\mathrm{HNP}_l(E)) - \mathrm{Area}(\mathrm{HNP}(E)) \\
&\geq \mathrm{Area}(W') \\
&\geq \frac{1}{2}(\tilde{\mu}_2 - \tilde{\mu}_1) \\
&= \frac{1}{2}\left(\frac{\mu(G/(F_X^l)^*(E_{j-1,0}))}{p^l} - \frac{\mu((F_X^l)^*(E_{j,0}/E_{j-1,0}))}{p^l}\right) \\
&\geq \frac{1}{2}\left(\mu(E_{i_1,0}) - \mu(E_{j,0}/E_{j-1,0})\right) \\
&\geq \frac{1}{2}((\mu_{1\infty} - \epsilon) - \mu(E_{j,0}/E_{j-1,0})).
\end{aligned}$$

To finish the estimate, we need to get an upper bound on $\mu(E_{j,0}/E_{j-1,0})$. To do this, let u be such that $i_u < j \leq i_{u+1}$. Then the line segment $\overline{p_{j-1,0}p_{j,0}}$ is sandwiched between the line segments $\overline{p_{i_u,0}p_{i_{u+1},0}}$ and $\overline{p_{i_u,\infty}p_{i_{u+1},\infty}}$ (see Figure 4.3).

Using the fact that $\|p_{i_u,\infty} - p_{i_u 0}\| \leq \epsilon$ and estimating the slope of the line segment $\overline{p_{j-1,0}p_{j,0}}$, we obtain that $|\mu(E_{j,0}/E_{j-1,0}) - \mu_{u+1,\infty}| \leq 3\epsilon$. Thus, we have

$$\mathrm{Area}(\mathrm{HNP}_l(E)) - \mathrm{Area}(\mathrm{HNP}(E)) \geq \frac{1}{2}(\mu_{1\infty} - \mu_{2\infty} - 4\epsilon).$$

But the difference in areas between $\mathrm{HNP}(E)$ and $\mathrm{HNP}_\infty(E)$ is at most $r\epsilon$, which gives us a contradiction for a small enough choice of ϵ. □

4.3 Bogomolov's inequality

In this section, we give several equivalent statements on Bogomolov's inequality concerning the positivity of $\Delta(E) \cdot D_2 \cdots D_{n-1}$.

Fix a positive integer r. Let X be a smooth projective variety of dimension $n \geq 2$ over an algebraically closed field k in any characteristic. Note that in the characteristic zero case, we define strong semistability to be the same as semistability. We fix a nef divisor A over X such that $T_X(A)$ is globally generated. Define the constant $\beta_r = \beta_r(A; D_1, \ldots, D_{n-1})$ for a choice of divisors $(D_1, \ldots, D_{n-1})$ as below

$$\beta_r(A; D_1, \ldots, D_{n-1}) := \begin{cases} 0, & \text{if } \mathrm{char}(k) = 0, \\ \left(\frac{r(r-1)}{p-1} A D_1 \cdots D_{n-1}\right)^2, & \text{if } \mathrm{char}(k) = p. \end{cases}$$

We also reference the *discriminant* of a rank-r torsion-free sheaf E, denoted as $\Delta(E)$, which is defined as

$$\Delta(E) := 2r \cdot c_2(E) - (r-1)c_1(E)^2.$$

When G is a nontrivial subsheaf of rank s in E, we set

$$\xi_{G,E} = \frac{c_1(G)}{s} - \frac{c_1(E)}{r}.$$

We also use K^+ to denote an open cone in $\mathrm{Num}(X)$, defined as in the beginning of Section 4.3.2.

The main theorem is the following.

Theorem 4.3.1 (BI(r)) *Let $(D_1, \ldots, D_{n-1})$ be a collection of nef line bundles over X, and let $d = D_1^2 \cdot D_2 \cdots D_{n-1} \geq 0$. Fix a positive integer r. Then the following statements are equivalent.*

(i) *Assume each* D_i *is ample, and* E *is a strongly* $(D_1,\dots,D_{n-1})$*-semistable torsion-free sheaf of rank* $r' \le r$. *Then we have*

$$\Delta(E)D_2\cdots D_{n-1} \ge 0.$$

(ii) *Let* E *be a* $(D_1,\dots,D_{n-1})$*-semistable torsion-free sheaf of rank* r' *for some* $r' \le r$. *Then we have*

$$d\cdot\Delta(E)D_2\cdots D_{n-1} + \beta_{r'} \ge 0.$$

(iii) *Let* E *be a torsion-free sheaf of rank* r' *for some* $r' \le r$ *over* X, *and assume* $d\cdot\Delta(E)D_2\cdots D_{n-1} + \beta_{r'} < 0$. *Then there exists a saturated subsheaf* E' *of* E *such that* $\xi_{E',E} \in K^+$.

(iv) *Let* E *be a strongly* $(D_1,\dots,D_{n-1})$*-semistable torsion-free sheaf of rank* $r' \le r$ *over* X. *Then*

$$\Delta(E)D_2\cdots D_{n-1} \ge 0.$$

Remark 4.3.2 In terms of the notation as in [6], the above corresponds to $T^5(r) \Rightarrow T^3(r) \Rightarrow T^4(r) \Rightarrow T^2(r) \Rightarrow T^5(r)$. Note that the last implication from $T^2(r)$ to $T^5(r)$ is trivial, as the statement (i) is the same as (iv) but with a stronger assumption.

Remark 4.3.3 In characteristic zero, since $\beta_r = 0$ and strong semistability is the same as the semistability, the statement above can be more or less combined into one single statement about the nonnegativity of $\Delta(E)D_2\cdots D_{n-1} \ge 0$. This was first proved by Bogomolov in [2], thus the name.

4.3.1 BI(r)(i) $\Rightarrow$ BI(r)(ii)

In this subsection, our goal is to prove the implication (i) $\Rightarrow$ (ii) in Theorem BI(r) for a torsion-free sheaf E of rank r, and we follow [6, Section 3.6].

To prove the statement, we start with the following preparation. Let H be an ample line bundle over X, and let $(D_1,\dots,D_{n-1})$ be as in Theorem BI(r)(ii). Then $(H_1(t),\dots,H_{n-1}(t)) := (D_1+tH,\dots,D_{n-1}+tH)$ is a polarization of ample line bundles.

Lemma 4.3.4 *Assume Theorem* BI(r)(i). *Then for* $t > 0$ *we have*

$$H_1(t)^2H_2(t)\cdots H_{n-1}(t)\cdot\Delta(E)H_2(t)\dots H_{n-1}(t) \\ + r^2(L_{\max,t}(E)-\mu_t(E))(\mu_t(E)-L_{\min,t}(E)) \ge 0.$$

Here we follow the notation $L_{\max}$ and $L_{\min}$ as in Section 4.2.3, and denote $L_{\max,t}$, $L_{\min,t}$, and μ_t as those defined using the polarization $(H_1(t), \ldots, H_{n-1}(t))$.

Remark 4.3.5 Before we prove the lemma, it is worth noting that the above inequality together with its proof, which only uses Theorem BI(r)(i), applies also to a general polarization $(D_1, \ldots, D_{n-1})$ for ample D_i, in place of $(H_1(t), \ldots, H_{n-1}(t))$. More explicitly, Theorem BI(r)(i) will imply the inequality

$$D_1^2 D_2 \cdots D_{n-1} \cdot \Delta(E) D_2 \cdots D_{n-1} + r^2(L_{\max} - \mu)(L_{\min} - \mu) \geq 0,$$

where the D_i are all ample.

Proof We first notice that by Theorem 4.2.28, there exists $k \in \mathbb{N}$ such that all the factors in the Harder–Narasimhan filtration of $(F^k)^*E$ are strongly semistable. Denote by $0 = E_0 \subset \cdots E_m = (F^k)^*E$ the corresponding filtration of $(F^k)^*E$, and let F_i be the quotient E_i/E_{i-1}, let $r_i = \mathrm{rk}(F_i)$, and let $\mu_{i,t} = \mu_t(F_i)$. Then by the Hodge index theorem,[2] we have

$$\begin{aligned} &\frac{\Delta((F^k)^*E) H_2(t) \cdots H_{n-1}(t)}{r} \\ &= \sum_i \frac{\Delta(F_i) H_2(t) \cdots H_{n-1}(t)}{r_i} \\ &\quad - \frac{1}{r} \sum_{i<j} r_i r_j \left(\frac{c_1(F_i)}{r_i} - \frac{c_1(F_j)}{r_j} \right)^2 H_2(t) \cdots H_{n-1}(t) \\ &\geq \sum_i \frac{\Delta(F_i) H_2(t) \cdots H_{n-1}(t)}{r_i} \\ &\quad - \frac{1}{rd} \sum_{i<j} r_i r_j (\mu_{i,t} - \mu_{j,t})^2. \end{aligned}$$

The assumption of Theorem BI(r)(i) and the ampleness of $H_i(t)$ provide us with the inequality

$$\Delta(F_i) H_2(t) \cdots H_{n-1}(t) \geq 0.$$

On the other hand, we have the following elementary inequality on r_i

[2] Here we are using the form of the Hodge index theorem that $(D^2 D_2 \cdots D_{n-1}) \cdot (D_1^2 D_2 \cdots D_{n-1}) \leq (D D_1 D_2 \cdots D_{n-1})^2$, where the D_i are nef divisors satisfying $D_1^2 D_2 \cdots D_{n-1} > 0$. This can be proved using the standard Hodge index theorem for ample divisors together with the approximation technique as in Section 4.2.2.

and μ_i (with $r = \sum_i r_i$ and $r\mu = \sum_i r_i\mu_i$):

$$\sum_{i<j} r_i r_j(\mu_i - \mu_j)^2 \le r^2(\mu_1 - \mu)(\mu - \mu_m). \tag{1}$$

Combining with the inequality above, we get the inequality in the statement of the lemma. □

With the above lemma together with the approximation technique as in Section 4.2.2, we are ready to prove the aforementioned implication.

Proof of Theorem BI(r), (i) ⇒ (ii) Let H be a fixed ample line bundle. By Theorem 4.2.17, we can find a filtration of torsion-free coherent subsheaves $0 = E_0 \subset E_1 \subset \cdots \subset E_m = E$ of E such that it is equal to the Harder–Narasimhan filtration with respect to the polarizations $(H_1(t), \ldots, H_{n-1}(t))$, where $t > 0$ is very close to 0. As E is $(D_1, \ldots, D_{n-1})$-semistable, we have

$$\mu(E) \ge \mu(E_1) = \lim_{t\to 0^+} \mu_t(E_1) \ge \lim_{t\to 0^+} \mu_t(E) = \mu(E).$$

By the choice of E_1, the above implies the equality

$$\mu(E) = \lim_{t\to 0^+} \mu_{\max,t}(E).$$

Similarly, we have

$$\mu(E) = \lim_{t\to 0^+} \mu_{\min,t}(E).$$

On the other hand, Proposition 4.2.27 states that for a nef divisor A such that $T_X(A)$ is globally generated, we have the two inequalities

$$L_{\max,t}(E) - \mu_{\max,t}(E),\ \mu_{\min,t}(E) - L_{\min,t}(E) \le \frac{r-1}{p-1} A H_1(t) \cdots H_{n-1}(t).$$

In particular, the limit of the right-hand side of the inequalities above is equal to $\frac{r-1}{p-1} A D_1 \cdots D_{n-1}$, whose square is $\frac{1}{r^2}\beta_r$. In this way, applying the inequality of Lemma 4.3.4 for each $t > 0$, we get

$$\begin{aligned} D_1^2 D_2 \cdots D_{n-1} &\cdot \Delta(E) D_2 \cdots D_{n-1} + \beta_r \\ &= \lim_{t\to 0^+} \Bigg(D_1^2 D_2 \cdots D_{n-1} \cdot \Delta(E) D_2 \cdots D_{n-1} \\ &\quad + r^2 \left(\frac{r-1}{p-1} A H_1(t) \cdots H_{n-1}(t) \right)^2 \Bigg) \end{aligned}$$

$$\geq \lim_{t\to 0^+} \Big(H_1(t)^2 H_2(t)\cdots H_{n-1}(t)\cdot \Delta(E) H_2(t)\cdots H_{n-1}(t)\Big)$$
$$+ r^2(L_{\max,t}(E)-\mu_t(E))(\mu_t(E)-L_{\min,t}(E))$$
$$\geq 0. \qquad \square$$

4.3.2 BI(r)(ii) $\Rightarrow$ BI(r)(iii)

In this subsection, we show the implication (ii) $\Rightarrow$ (iii), following [6, Section 3.7]

As a preparation, we define an open cone in Num(X) as follows.

Definition 4.3.6 Let K^+ be the following open cone in Num(X):

$$K^+ = \{D \in \mathrm{Num}(X) \mid D^2 D_2\cdots D_{n-1} > 0, \text{ and } DD'D_2\cdots D_{n-1} \geq 0 \text{ for all nef } D'\}.$$

The following description is used in the proof.

Lemma 4.3.7 *A divisor $D \in K^+$ if and only if it satisfies the inequalities $DLD_2\cdots D_n > 0$ for all $L \in \bar{K}^+ \setminus \{0\}$.*

Before starting, we also mention several elementary computational results whose proofs are left as exercises to the reader.

Lemma 4.3.8 *Let $0 \to E' \to E \to E'' \to 0$ be a short exact sequence of coherent sheaves. Let r' (resp. r'') be the rank of E' (resp. E''), and assume they are positive. We have*

$$\frac{\Delta(E)D_2\cdots D_{n-1}}{r} + \frac{rr'}{r''}\xi^2_{E',E}D_2\cdots D_{n-1} = \frac{\Delta(E')D_2\cdots D_{n-1}}{r'} + \frac{\Delta(E'')D_2\cdots D_{n-1}}{r''}.$$

Lemma 4.3.9 *Let $0 \to E' \to E \to E'' \to 0$ be a short exact sequence of coherent sheaves. Let r' (r'') be the rank of E' (E'') respectively.*

- *If G is a nontrivial subsheaf of E', then*

$$\xi_{G,E} = \xi_{E',E} + \xi_{G,E'}.$$

- *If $G'' \subset E''$ is a proper subsheaf of rank s, and we denote by G the kernel of the map $E \to E''/G''$, then*

$$\xi_{G,E} = \frac{r'(r''-s)}{(r'+s)r''}\xi_{E',E} + \frac{s}{r'+s}\xi_{G'',E''}.$$

Proof of Theorem BI(r), (ii) $\Rightarrow$ (iii) We will prove (iii) inductively on $\mathrm{rk}(E) = r$, assuming (ii). As a starting point, we notice that, when E is of rank 1, as $\Delta(E) = 0$ and $\beta_1 \geq 0$, (iii) is automatically true.

In general, the assumption of Theorem BI(r)(ii) tells us that E is not semistable with respect to $(D_1, \ldots, D_{n-1})$. Thus we have the maximal destabilizing subsheaf $E' \subset E$ with respect to the polarization. Let $E'' = E/E'$, and let r', r'' be the ranks of E', E'' respectively. Then by Lemma 4.3.8 we have

$$\frac{\Delta(E)D_2 \cdots D_{n-1}}{r} + \frac{rr'}{r''}\xi_{E',E}^2 D_2 \cdots D_{n-1} = \frac{\Delta(E')D_2 \cdots D_{n-1}}{r'} + \frac{\Delta(E'')D_2 \cdots D_{n-1}}{r''}.$$

We also have

$$\frac{\beta_r}{r} \geq \frac{\beta_{r'}}{r'} + \frac{\beta_{r''}}{r''}.$$

So, by setting $d = D_1^2 D_2 \cdots D_{n-1}$, either $\xi_{E',E}^2 D_2 \cdots D_{n-1} > 0$, or one of $d\Delta(E')D_2 \cdots D_{n-1} + \beta_{r'}$ or $d\Delta(E'')D_2 \cdots D_{n-1} + \beta_{r''}$ must be negative.

In the first case, we claim that $\xi_{E',E} \in K^+$. To see this, by the definition of K^+ and the assumption on $\xi_{E',E}$ it suffices to show that for any $L \in K^+ \backslash \{0\}$ we have $\xi_{E',E} L D_2 \cdots D_{n-1} > 0$.

Note first that the assumption of E' and E implies that the inequality is true for $L = D_1$, and it reduces to showing the sign of the function $L \mapsto \xi_{E',E} L D_2 \cdots D_{n-1}$ does not change on K^+. This then follows from the continuity of the function, the connectivity of K^+, and the Hodge index theorem, that

$$(\xi_{E',E} L D_2 \cdots D_{n-1})^2 \geq \xi_{E',E}^2 D_2 \cdots D_{n-1} \cdot L^2 D_2 \cdots D_{n-1} > 0,$$

for $L \in K^+ \backslash \{0\}$.

Now suppose that $\xi_{E',E}^2 D_2 \cdots D_{n-1} \leq 0$ and we are in the second case above. Note that since both E' and E'' are of smaller ranks, we can apply the induction hypotheses to get the following dichotomy:

- There is a saturated subsheaf $G \subset E'$ such that $\xi_{G,E'} \in K^+$, or
- There is a saturated subsheaf $G'' \subset E''$ such that $\xi_{G'',E''} \in K^+$.

Applying Lemma 4.3.9, we get in either case that $\xi_{G,E}$ is a positive linear combination of $\xi_{E',E}$ and some element L in K^+.

To proceed, define the open subcone

$$C(\xi) := \{D \in \overline{K^+} \setminus \{0\} : \xi \cdot D D_2 \cdots D_{n-1} > 0\} \subset \overline{K^+}$$

for a given $\xi \in \mathrm{Num}(X)$. We observe that by replacing E' with G in either case, we get strictly larger cones than before: $C(\xi_{E',E}) \subsetneq C(\xi_{G,E})$. Here the inclusion part is clear, as $\xi_{G,E}$ is the sum of $\xi_{E',E}$ with some $L \in K^+$, where the second term contributes only positive values in the product $\xi_{G,E} D D_2 \cdots D_{n-1}$. To see that the inclusion is strict, as $\xi_{E',E}$ is not in K^+, then by Lemma 4.3.7 there exists some $L' \in \overline{K^+}\setminus\{0\}$ such that $\xi_{E',E} L' D_2 \cdots D_{n-1} \leq 0$ but $LL'D_2 \cdots D_{n-1} > 0$. On the other hand, as E' is the maximal destabilizing subsheaf of E with respect to the polarization $(D_1, \ldots, D_{n-1})$, we have $\xi_{E',E} D_1 \cdots D_{n-1} > 0$ and $LD_1 \cdots D_{n-1} > 0$, where the latter follows as $L \in K^+$. In this way, there exists some real number $t \in [0, 1)$ such that

$$\xi_{E',E}(tD_1 + (1-t)L')D_2 \cdots D_{n-1} = 0,$$

and thus

$$\begin{aligned}
\xi_{G,E}&(tD_1 + (1-t)L')D_2 \cdots D_{n-1} \\
&= \xi_{E',E}(tD_1 + (1-t)L')D_2 \cdots D_{n-1} \\
&\quad + L(tD_1 + (1-t)L')D_2 \cdots D_{n-1} \\
&= 0 + L(tD_1 + (1-t)L')D_2 \cdots D_{n-1} \\
&> 0.
\end{aligned}$$

So the element $tD_1 + (1-t)L'$ is in $C(\xi_{G,E})$ but not in $C(\xi_{E',E})$. In this way, we get a sequence of strictly increasing subcones of $\overline{K^+}$ by replacing E' by G until we reach the situation where $\xi^2_{E',E} D_2 \cdots D_{n-1} > 0$.

It remains to show that this process terminates within finite replacements. First note that by choosing an ample $\mathbb{R}$-basis $H_1, \ldots, H_\rho$, the H_i being contained in $C(\xi_{E',E})$ for $\mathrm{Num}(X)$,[3] we have

$$\xi_{G,E} \in \frac{1}{r!}(\mathbb{Z}H_1 + \cdots + \mathbb{Z}H_\rho).$$

Furthermore, we also have

$$0 < \xi_{G,E} H_j D_2 \cdots D_{n-1} < \mu^j_{\max}(E) - \mu^j(E),$$

for all $j = 1, 2, \ldots, \rho$, where the μ^j denote the slopes with respect to $(H_j, D_2, \ldots, D_{n-1})$. Thus $\xi_{G,E}$ is in fact contained in a bounded discrete, hence finite, subset of $\mathrm{Num}(X)$. □

[3] Note that since $C(\xi_{E'E})$ is an open subcone in $K^+ \subset \mathrm{Num}(X)$ that is nonempty (it contains D_1), such a basis exists.

4.3.3 BI(r)(iii) $\Rightarrow$ BI(r)(iv)

In this subsection, we follow [6, Section 3.8] to show the implication (iii) $\Rightarrow$ (iv).

We follow the definition of β_r as in the beginning of this section, and the definition of the cone K^+ with respect to the polarization $(D_1, \ldots, D_{n-1})$ as in the last subsection.

By the definitions of $\xi_{E',E}$ and K^+, (iii) implies that E is not semistable with respect to $(D_1, \ldots, D_{n-1})$.

Remark 4.3.10 The implication (iii) $\Longrightarrow$ (iv) is easy in $\mathrm{char}(k) = 0$, by the definition of K^+ applying at $\xi_{E',E} D_1 \cdots D_{n-1}$.

Proof of Theorem BI(r), (iii) $\Rightarrow$ (iv) Assume E is a torsion-free sheaf such that $\Delta(E) D_2 \cdots D_{n-1} < 0$. We will deduce that E is not strongly semistable. We would like to apply Theorem BI(r)(iii) to the Frobenius pullback $(F^l)^* E$ of E. Since $\Delta((F^l)^* E) = p^{2l} \Delta(E)$, we get that

$$
\begin{aligned}
& D_1^2 D_2 \cdots D_{n-1} \cdot \Delta((F^l)^* E) D_2 \cdots D_{n-1} + \beta_r \\
& \qquad = D_1^2 D_2 \cdots D_{n-1} \cdot p^{2l} \cdot \Delta(E) D_2 \cdots D_{n-1} + \beta_r
\end{aligned}
$$

is negative when l is large enough. By the assumption of Theorem BI(r)(iii), we can find a saturated subsheaf E' of $(F^l)^* E$ with $\xi_{E',(F^l)^* E} \in K^+$. By the definition of K^+ as in Definition 4.3.6, we have

$$
\xi_{E',(F^l)^* E} D_1 \cdots D_{n-1} > 0;
$$

in particular, the pullback $(F^l)^* E$ is not semistable. Hence the torsion-free sheaf E itself is not strongly semistable. □

4.4 Restriction to hypersurfaces and Bogomolov's inequality

In this section, we prove one of the main technical ingredients, Theorem 4.4.1, which controls the change in various numerical invariants as one passes to hypersurfaces. Following this, we complete the induction schema sketched in the introduction of this chapter. We follow the proof as in [6, Sections 3.5 and 3.9] and [5]. Combining the results of the section with the equivalent statements in Theorem BI(r), we finish the proofs of all the theorems in [6, Section 3].

Fix a positive integer r, and let X be a smooth projective variety over k. Consider the following statement.

Theorem 4.4.1 (Res(r)) *Let E be a torsion-free sheaf of rank r over X, and let $(D_1, \ldots, D_{n-1})$ be a collection of nef divisors over X with $d := D_1^2 D_2 \cdots D_{n-1} \geq 0$.*

Assume that D_1 is very ample and the restriction of E to a very general divisor $D \in |D_1|$ is not semistable (with respect to $(D_2|_D, \ldots, D_{n-1}|_D)$). Let μ_i (r_i) denote the slopes (ranks) of the Harder–Narasimhan filtration of $E|_D$. Then

$$\sum_{i<j} r_i r_j (\mu_i - \mu_j)^2 \leq d\Delta(E) D_2 \cdots D_{n-1} + 2r^2 (L_{\max} - \mu)(\mu - L_{\min}).$$

Our goal this section is to show the following two implications:

$$\mathrm{Res}(r) + \mathrm{BI}(r-1) \Longrightarrow \mathrm{BI}(r) \Longrightarrow \mathrm{Res}(r+1).$$

Note that since both Theorem Res(r) and Theorem BI(r) are empty and thus automatically true for $r = 1$, the above induction process proves that both Theorems Res(r) and BI(r) hold true for all $r \geq 1$.

4.4.1 Res(r) + BI($r - 1$) ⇒ BI(r)

In this subsection, we consider the implication Res(r)+BI($r-1$) ⇒ BI(r), following [6, Section 3.5] and [5]. This is achieved via an inductive argument on the dimension of X. We first prove the case when X is a surface. Further, we do not need to assume BI($r - 1$) for the base case.

Proposition 4.4.2 *Let X be a surface. Assume Theorem Res(r) holds for all torsion-free sheaves of rank $\leq r$, then Theorem BI(r) holds for all torsion-free sheaves of rank $\leq r$ and all ample divisors D_1.*

Proof By the equivalence of the statements in Theorem BI(r), it suffices to show that the assumption implies the statement in Theorem BI(r), (i), i.e., the inequality $\Delta(E) \geq 0$, for a strongly D_1-semistable torsion-free sheaf E of rank $\leq r$ where D_1 is ample, holds.

Let E be a rank-r vector bundle that is strongly D_1-semistable. Suppose by way of contradiction that $\Delta(E) < 0$. We may further assume that E is locally free, by the inequality

$$\Delta(E^{**}) = \Delta(E) - 2r \operatorname{length}(E^{**}/E),$$

where the reflection E^{**} is locally free on X. We first claim that under the assumption, the restriction to a general curve $C \in |D_1|$ is also strongly semistable. To see this, since E is strongly semistable, $\mu = L_{\max} = L_{\min}$, and since (by assumption) $\Delta(E) < 0$, from the assumption we would get a contradiction,

$$0 \leq \sum_{i<j} r_i r_j (\mu_i - \mu_j)^2 < 0,$$

where we use the notation in Theorem Res(r). Similarly, as $(F^k)^*E$ is semistable, its restriction to a general curve C is also semistable. Thus, the restriction of E to a very general curve C is strongly semistable. For the remainder of the proof, fix such a very general C.

Now consider the symmetric power $S^{kr}E|_C$. The strategy is to compute $\chi(S^{kr}E)$ in two different ways and use them to deduce a contradiction. Since symmetric powers of strongly semistable sheaves on curves are strongly semistable (see [4, Corollary 3.2.10] for the characteristic zero case and [8, Sections 3 and 5] for the positive characteristic case), it follows that $S^{kr}E|_C$ is strongly semistable. Consider the short exact sequence arising from C:

$$0 \to S^{kr}E(-kc_1(E) - C) \to S^{kr}E(-kc_1(E)) \\ \to S^{kr}E(-kc_1(E))|_C \to 0$$

This gives us the estimate

$$h^0(S^{kr}E(-kc_1(E))) \\ \leq h^0(S^{kr}E(-kc_1(E) - C)) + h^0(S^{kr}E(-kc_1(E))|_C).$$

Consider the first term on the right-hand side. This quantity is equal to $\dim\left(\mathrm{Hom}(\mathcal{O}_X(kc_1(E) + C), S^{kr}E)\right)$, and we claim that this number is zero. Note that the slope of $\mathcal{O}_X(kc_1(E) + C)$ is given by $kr\mu(E) + D_1^2$, and, by the splitting principle, the slopes of the symmetric powers can be seen to be

$$\mu(S^{kr}(E)) = \frac{\binom{r+kr-1}{kr} kc_1(E)D_1}{\binom{r+kr-1}{kr}} = kr\mu(E).$$

Since D_1 is ample, $D_1^2 > 0$, so the claim follows as there are no morphisms from semistable sheaves of higher slope to those of lower slope. Thus we get $h^0(S^{kr}E(-kc_1(E))) \leq h^0(S^{kr}E(-kc_1(E))|_C)$.

Given any semistable vector bundle G on a curve, we have the further estimate $h^0(G) \leq \max\{0, \deg G + \mathrm{rk}\, G\}$. Using the sequence

$$0 \longrightarrow G(-lP) \longrightarrow G \longrightarrow G|_{lP} \longrightarrow 0,$$

the above inequality can be seen from the estimate $h^0(G) \leq h^0(G \otimes \mathcal{O}_C(-lP)) + \mathrm{rk}(G)l = 0 + \mathrm{rk}(G)l \leq \max\{0, \mathrm{rk}G + \deg G\}$, where P is a point on C, $l = \max\{0, \lceil \mu(G) \rceil\}$, and $h^0(G \otimes \mathcal{O}_C(-lP)) = 0$ since $\mu(\mathcal{O}_C(lP)) > \mu(G)$.

Thus, we have $h^0(S^{kr}E(-kc_1(E))) = O(k^r)$. By Serre duality, we have the same order of magnitude estimate for $h^2(S^{kr}E(-kc_1(E)))$. On the other hand, the splitting principle for Chern classes and the Hirzebruch–Riemann–Roch theorem can be used to get the following estimate (for example, see [2, Section 10]):

$$\chi(X, S^{kr}E(-kc_1(E))) = -\frac{r^r \Delta(E)}{2(r+1)!} k^{r+1} + O(k^r).$$

Since $\Delta(E) < 0$, this polynomial is eventually positive and is of order k^{r+1}, but this is an obvious contradiction as h^0 and h^2 are of order k^r. □

To show the implication for general X, we induct on the dimension of X, using the above surface case as the base case. However, differently from the assumption in Proposition 4.4.2, for this to work we require the additional assumption that Theorem BI$(r-1)$ holds. Precisely, we want to show that, assuming

- Theorems BI$(r-1)$ and Res(r) hold for X of any dimension,
- Theorem BI(r) holds for varieties of dimension $< n$,

then we have that Theorem BI(r) holds for any variety X of dimension n.

Proof of Theorem BI(r) Similarly to the proposition above, we aim to prove Theorem BI(r)(i) for X of dimension n, namely the inequality $\Delta(E)D_2 \cdots D_{n-1} \geq 0$ for a strongly $(D_1, \ldots, D_{n-1})$-semistable torsion-free sheaf E over X, where the D_i are all ample.

Assume to the contrary that $\Delta(E)D_2 \cdots D_{n-1} < 0$. For a general divisor in $|D_2|$ (which we also denote as D_2), we have $\Delta(E|_{D_2})D_3 \cdots D_{n-1} = \Delta(E)D_2 \cdots D_{n-1} < 0$. Consider the polarization $(D_2^2, D_3, \ldots, D_{n-1}) := (D_2, D_2, D_3, \ldots, D_{n-1})$.

We first assume that E is strongly $(D_2^2, D_3, \ldots, D_{n-1})$-semistable. Applying Theorem Res(r) to the Frobenius twists $(F^k)^*E$ of E with respect to $(D_2^2, \ldots, D_{n-1})$, then if $(F^k)^*E|_{D_2}$ were not semistable we would get the inequality

$$\sum_{i<j} r_i r_j (\mu_i - \mu_j)^2 \leq dp^k \Delta(E) D_2 \cdots D_{n-1} + 0 < 0,$$

which is impossible. Thus $E|_{D_2}$ is strongly semistable for ample divisors $(D_2, \ldots, D_{n-1})$, and by the induction hypothesis, since $E|_{D_2}$ is defined over a hypersurface of X we have $\Delta(E|_{D_2})D_3 \cdots D_{n-1} = \Delta(E)D_2 \cdots D_{n-1} \geq 0$, a contradiction.

For the rest, we assume E is not strongly $(D_2^2, D_3, \ldots, D_{n-1})$-semistable. The trick is to interpolate between the polarizations given by $(D_1, D_2, \ldots, D_{n-1})$ and $(D_2^2, D_3, \ldots, D_{n-1})$. Let B_t be given by the product $((1-t)D_1 + tD_2)D_2 \cdots D_{n-1}$. By assumption, E is strongly B_0-semistable while it is not strongly B_1-semistable. Since not being strongly semistable is an open condition with respect to t, there exists a number $t_k \in [0, 1)$ such that E is strongly B_{t_k}-semistable but not strongly B_t-semistable for all $t_k < t \leq 1$.

Let k be a positive integer such that $(F^k)^*E$ is not semistable for B_1. Using Theorem 4.2.17, we see that the Harder–Narasimhan filtration of E with respect to B_t remains constant for all $t_k < t < t_k + \epsilon$ for some $\epsilon > 0$ small enough. We let E' denote the maximal destabilizing sheaf of $(F^k)^*E$ with respect to the polarization B_t for $t_k < t < t_k + \epsilon$. By the continuity of the slope with respect to t, it follows that E' and $(F^k)^*E$ have the same B_{t_k}-slope (see Lemma 4.2.19). Since E' is a subsheaf of a semistable sheaf with the same slope, E' is also semistable with respect to B_{t_k}. It also follows that the quotient $E'' = (F^k)^*E/E'$ is B_{t_k}-semistable as well. Let r, r', and r'' denote the ranks of E, E', and E'' respectively.

Now we apply Lemma 4.3.8 to the short exact sequence $E' \to E \to E''$ to get that

$$\frac{\Delta((F^k)^*E)D_2 \cdots D_{n-1}}{r} = \frac{\Delta(E')D_2 \cdots D_{n-1}}{r'} + \frac{\Delta(E'')D_2 \cdots D_{n-1}}{r''} - \frac{r'r''}{r}\xi_{E',E''}^2 D_2 \cdots D_{n-1}.$$

Note that by the Hodge index theorem, we have

$$\xi_{E',E''}^2 D_2 \cdots D_{n-1} \cdot (t_k D_1 + (1-t_k)D_2)^2 D_2 \cdots D_{n-1} \leq (\xi_{E',E''} B_{t_k})^2.$$

Since E' and E'' have the same B_{t_k}-slope, $(\xi_{E',E''} B_{t_k})^2 = 0$ and, by the ampleness of $D_1, \ldots, D_{n-1}$, we have that

$$d(t_k) := (t_k D_1 + (1-t_k)D_2)^2 D_2 \cdots D_{n-1} > 0.$$

Moreover by Theorem BI(r)(ii) but with $(r-1)$ in place of r, we get the inequalities below:

$$\frac{\Delta((F^k)^*E)D_2 \cdots D_{n-1}}{r} \geq -\frac{1}{d(t_k)}\left(\frac{\beta_{r'}(t_k)}{r'} + \frac{\beta_{r''}(t_k)}{r''}\right) \geq -\frac{\beta_r(t_k)}{r d(t_k)},$$

where we denote $\beta_r(t) := \beta_r((1-t)D_1, +tD_2, D_2, \ldots, D_{n-1})$. In this way, we get

$$\Delta(E)D_2 \cdots D_{n-1} \geq -\frac{\beta_r(t_k)}{d(t_k)p^{2k}}.$$

Notice that the function $\frac{\beta_r(t)}{d(t)}$ for $t \in [0,1]$ is defined and continuous, thus bounded. Taking the limit as k approaches infinity, we get the inequality

$$\Delta(E)D_2 \cdots D_{n-1} \geq 0,$$

which is a contradiction, so we are done. □

Remark 4.4.3 As pointed out by the referee, the last paragraph of the proof above can be argued as follows. As E' and E'' are of the same B_{t_k}-slope as the strongly B_{t_k}-semistable sheaf E, both of them are strongly B_{t_k}-semistable. Thus applying the induction hypothesis of Theorem BI(r)(i) to E' and E'' for the polarization B_{t_k}, we see that

$$\frac{\Delta(E')D_2 \cdots D_{n-1}}{r'} + \frac{\Delta(E'')D_2 \cdots D_{n-1}}{r''} \geq 0.$$

In this way, by the vanishing of $\xi_{E',E''}D_2 \cdots D_{n-1}$ proved above, we get the positivity of $\Delta(E)D_2 \cdots D_{n-1}$.

4.4.2 BI(r) ⇒ Res($r+1$)

We now consider the implication BI(r) ⇒ Res($r+1$), following [6, Section 3.9].

Let r be a nonnegative integer, and $(D_1, \ldots, D_{n-1})$ be a nef polarization over X with $d = D_1^2 D_2 \cdots D_{n-1} \geq 0$.

Proof Let Π be the projective space associated to the linear system $|D_1|$. Let $Z \subset \Pi \times X$ be the incident scheme, defined as the locus $\{(D,x) \in \Pi \times X \mid x \in D\}$, where $p\colon Z \to \Pi$ and $q\colon Z \to X$ are the two natural maps induced by the projections. For each $s \in \Pi$, we let Z_s be its fiber along the map q, which is a divisor in X. Let $0 = E_0 \subset E_1 \subset \cdots \subset E_m = q^*E$ be the Harder–Narasimhan filtration of q^*E associated to the polarization

$$(p^*\mathcal{O}_\Pi(1)^{\dim \Pi}, q^*D_2, \ldots, q^*D_{n-1})$$
$$= (p^*\mathcal{O}_\Pi(1), \ldots, p^*\mathcal{O}_\Pi(1), q^*D_2, \ldots, q^*D_{n-1})$$

over Z. Denote by F_i thc subquotient E_i/E_{i-1}. Here we observe that the pullback of the filtration $\{E_i\}$ at a general fiber Z_s, for $s \in \Pi$, is the Harder–Narasimhan filtration of $E|_{Z_s}$ at the hypersurface $Z_s \in |D|$. Indeed, the slope of q^*E with respect to $(p^*\mathcal{O}_\Pi(1)^{\dim \Pi}, q^*D_2, \ldots, q^*D_{n-1})$ is the same as the slope of $q^*E|_{Z_s}$. This shows that the filtration $\{E_i\}$ is also the relative Harder–Narasimhan filtration for q^*E with respect to the map p. In particular, we have the equalities $r_i = \text{rk}F_i$ and $\mu_i =$

$\mu(F_i)$, where the latter is with respect to the polarization given by $(p^*\mathcal{O}_\Pi(1)^{\dim \Pi}, q^*D_2, \ldots, q^*D_{n-1})$.

Now let $\mathbb{P}^1 \cong \Lambda \subset \Pi$ be a sufficiently general pencil, corresponding to a linear subsystem of $|D|$ in X parametrized by a line $\mathbb{P}^1 \subset \Pi$. Let $B \subset X$ be the base of Λ, which is the intersection of Z_s for any two (hence all) $s \in \Lambda$, and is of codimension 2 in X. Denote by Y the incident scheme for Λ, which is equal to the closed subscheme $p^{-1}\Lambda$ inside Z. Here, by the construction of the incident scheme for a pencil, the projection map $Y = p^{-1}\Lambda \to X$ coincides with the blowup of X at the base locus B. Moreover, depending on the dimension of X, we can write the first Chern class of the restriction $F_i|_Y$ in terms of the following:

- If $\dim(X) = 2$, then B consists of a union of a finite amount of points, and we may write each $c_1(F_i|_Y)$ as $M_i + \sum_j b_{ij}N_j$, with $M_i \in \mathrm{Pic}(X)$ and N_j the jth exceptional divisor for $1 \le j \le l$. We define b_i to be the number $\frac{\sum_j b_{ij}}{l}$.
- If $\dim(X) \ge 3$, then by Bertini's theorem the blowup center B is a smooth connected closed subscheme of codimension 2 in X, and we may write each $c_1(F_i|_Y)$ as $M_i + b_iN$, for $M_i \in \mathrm{Pic}(X)$ and N the exceptional divisor.

In any of the above cases, with respect to the polarization

$$(p^*\mathcal{O}_\Lambda(1), q^*D_2, \ldots, q^*D_{n-1})$$

over Y, we can write the slope μ_i as the following:

$$\mu_i = \frac{M_iD_1 \cdots D_{n-1} + b_id}{r_i}. \tag{0}$$

Here, as the collection of torsion-free sheaves $\{F_i\}$ forms the graded pieces of a filtration of q^*E, we have

$$\sum_i b_i = c_1(q^*E)p^*\mathcal{O}_\Lambda(1)q^*D_2 \cdots q^*D_{n-1} = 0. \tag{1}$$

Furthermore, since $(q|_Y)_*(E_i|_Y) \subset E$, we have the inequalities

$$\frac{\sum_{j\le i} M_jD_1 \cdots D_{n-1}}{\sum_{j\le i} r_j} \le \mu_{\max}, \quad \text{for all } i,$$

which is equivalent to the inequalities

$$\sum_{j\le i} b_jd \ge \sum_{j\le i} r_j(\mu_j - \mu_{\max}), \quad \text{for all } i. \tag{2}$$

Finally, we are ready to prove the implication. As each F_i is of rank $\leq r-1$, by Theorem BI(r)(i) for the polarization

$$(p^*\mathcal{O}_\Pi(1)^{\dim \Pi}, q^*D_2, \ldots, q^*D_{n-1})$$

over Z, we have

$$\begin{aligned}\Delta(F_i)p^*\mathcal{O}_\Pi(1)^{\dim\Pi}q^*D_2\cdots q^*D_{n-1}\\ = \Delta(F_i|_Y)p^*\mathcal{O}_\Lambda(1)q^*D_2\cdots q^*D_{n-1} \geq 0, \quad \text{for all } i.\end{aligned}$$

Applying this, we get

$$\begin{aligned}&\frac{d\Delta(E)D_2\cdots D_{n-1}}{r}\\ &= \sum_i \frac{d\Delta(F_i|_Y)p^*\mathcal{O}_\Lambda(1)q^*D_2\cdots q^*D_{n-1}}{r_i}\\ &\quad - \frac{d}{r}\sum_{i<j} r_ir_j\left(\frac{c_1(F_i|_Y)}{r_i} - \frac{c_1(F_j|_Y)}{r_j}\right)^2 (q|_Y)^*(D_2\cdots D_{n-1})\\ &\geq -\frac{d}{r}\sum_{i<j} r_ir_j\left(\frac{c_1(F_i|_Y)}{r_i} - \frac{c_1(F_j|_Y)}{r_j}\right)^2 (q|_Y)^*(D_2\cdots D_{n-1})\\ &= \frac{d}{r}\sum_{i<j} r_ir_j\left(d\left(\frac{b_i}{r_i} - \frac{b_j}{r_j}\right)^2 - \left(\frac{M_i}{r_i} - \frac{M_j}{r_j}\right)^2 D_2\cdots D_{n-1}\right)\\ &\geq \frac{1}{r}\sum_{i<j} r_ir_j\left(d^2\left(\frac{b_i}{r_i} - \frac{b_j}{r_j}\right)^2 - \left(\frac{M_iD_1\cdots D_{n-1}}{r_i} - \frac{M_jD_1\cdots D_{n-1}}{r_j}\right)^2\right),\end{aligned}$$

where the first inequality is an application of Theorem BI(r)(i), the equality after it is a rearrangement using the formula (0), and the last inequality follows from the Hodge index theorem applied to $\frac{M_i}{r_i} - \frac{M_j}{r_j}$ and the polarization $D_1, \ldots, D_{n-1}$. Moreover, using the formula (0), we may rewrite the last expression of the inequalities above as

$$\frac{2d}{r}\sum_{i<j}(\mu_i - \mu_j)(b_ir_j - b_jr_i) - \frac{1}{r}\sum_{i<j} r_ir_j(\mu_i - \mu_j)^2.$$

Using (1), this can be further simplified to

$$2\sum_i db_i\mu_i - \frac{1}{r}\sum_{i<j} r_ir_j(\mu_i - \mu_j)^2.$$

Here, more concretely, the simplification above can be seen via an induction on m. If $m = 1$, this is obvious. Suppose it is true for $m - 1$. Then

$$
\begin{aligned}
&\frac{2d}{r}\sum_{i<j}(\mu_i - \mu_j)(b_i r_j - b_j r_i)\\
&= 2\sum_{i<j\le m-1}(\mu_i - \mu_j)\left(\left(b_i + \frac{b_m}{m}\right)r_j - \left(b_j + \frac{b_m}{m}\right)r_i\right)\\
&\quad + \frac{b_m}{m}\sum_{i<j\le m-1}(\mu_i - \mu_j)(r_i - r_j) + \frac{2d}{r}\sum_{i\le m-1}(\mu_i - \mu_m)(b_i r_m - b_m r_i)\\
&= \frac{2(r - r_m)}{r}\sum_{i\le m-1} d\left(b_i + \frac{b_m}{m}\right)\mu_i\\
&\quad + \frac{2d}{r}\left(r_m\sum_{i\le m-1} b_i\mu_i - b_m(\mu r - \mu_m r_m) + b_m\mu_m r_m + b_m\mu_m(r - r_m)\right)\\
&\quad + \frac{b_m}{m}\sum_{i<j\le m-1}(\mu_i - \mu_j)(r_i - r_j)\\
&= 2\sum_i d b_i\mu_i
\end{aligned}
$$

where the second equality follows from the induction hypothesis. Furthermore, using formula (2) together with an elementary equality

$$\sum_i a_i b_i = \sum_i\left(\sum_{j\le i} a_j\right)(b_i - b_{i+1}),$$

we get

$$
\begin{aligned}
\sum_i d b_i\mu_i &= \sum_i\left(\sum_{j\le i} d b_j\right)(\mu_i - \mu_j)\\
&\ge \sum_i\left(\sum_{j\le i} r_j(\mu_j - \mu_{\max})(\mu_i - \mu_{i+1})\right)\\
&= \sum_i (r_i\mu_i^2 - r\mu\mu_{\max})\\
&\ge \sum_i r_i\mu_i^2 - r\mu^2 + r(\mu - \mu_{\max})(\mu - \mu_{\min})\\
&= \sum_{i<j}\frac{r_i r_j}{r}(\mu_i - \mu_j)^2 + r(\mu - \mu_{\max})(\mu - \mu_{\min}).
\end{aligned}
$$

As a consequence, we get

$$\frac{d\Delta(E)D_2\cdots D_{n-1}}{r} \geq \sum_{i<j} \frac{r_i r_j}{r}(\mu_i - \mu_j)^2 + 2r(\mu - \mu_{\max})(\mu - \mu_{\min}).$$

Finally, by moving the second term on the right-hand side and the inequalities $L_{\max} \geq \mu_{\max}, \mu_{\min} \geq L_{\min}$, we get the expression in Theorem Res(r). □

4.5 Boundedness of torsion-free sheaves

Now we use the ingredients in the last two sections to show the boundedness, following [6, Section 4].

We start with the following result combining the inequalities in the last two sections into a uniform inequality, following [6, Corollary 3.11].

Corollary 4.5.1 *Let E be a torsion-free sheaf on X and let D_1 be very ample, let $D_2, \ldots, D_{n-1}$ be ample, and let $D \in |D_1|$ be a general divisor. Then, we have that*

$$\frac{r}{2}(L_{\max}(E|_D) - L_{\min}(E|_D))^2$$
$$\leq d\Delta(E)D_2\cdots D_{n-1} + 2r^2(L_{\max} - \mu)(\mu - L_{\min}).$$

Proof First consider the case when $E|_D$ is not strongly semistable. Then, the inequality follows from Theorem Res(r) as well as the elementary inequality [6, Lemma 1.3] below:

$$\sum_{i<j} r_i r_j(\mu_i - \mu_j)^2 \geq \frac{r_1 r_m}{r_1 + r_m} r(\mu_1 - \mu_m)^2 \geq \frac{r}{2}(\mu_1 - \mu_m)^2$$

where $r_1, \ldots, r_m$ are positive real numbers, $\mu_1, \ldots, \mu_m$ are real numbers, and $r = r_1 + \cdots + r_m$.

Now consider the case when $E|_D$ is strongly semistable. Then, the left-hand side of the inequality is just zero and the result follows from the inequality in Remark 4.3.5 (where the latter is proved using Theorem BI(r)(i)). □

The boundedness is deduced from the following result. A more precise version is given in [6].

Theorem 4.5.2 *Let $H_1, \ldots, H_{n-1}$ be very ample divisors on X and let $X_l = |H_1| \cap \cdots \cap |H_l|$, $1 \leq l \leq n-1$, be very general complete intersections. Pick a nef divisor A such that $T_{X_l}(A)$ is globally generated* *for all* $0 \leq l \leq n-1$. *Set* $\beta_r = \beta(r; A, H_1, \ldots, H_{n-1})$. *Let* $\mu_{\max,l}, \mu_{\min,l}$ *denote the maximal and minimal slopes of the Harder–Narasimhan filtration of* $E|_{X_l}$. *Then we have the following inequality:*

$$\begin{aligned}&\mu_{\max,l} - \mu_{\min,l}\\&\quad\leq C(r,n)\left(\sqrt{\max\{d\Delta(E)H_2\cdots H_{n-1}, 0\}} + \sqrt{\beta_r} + (\mu_{\max} - \mu_{\min})\right),\end{aligned}$$

where $C(r,n)$ *is a constant depending only on* r *and* n.

Remark 4.5.3 Before we move on to the proof, it is worth mentioning that, except for $\mu_{\max} - \mu_{\min}$, the rest of the terms on the right-hand side together with $\mu_{\min,l}$ can all be bounded using constants depending only on X and the Hilbert polynomial of E.

Proof For $l = 1$, we use Corollary 4.5.1, and get

$$\begin{aligned}(L_{\max,1} - L_{\min,1})^2 &\leq \frac{2}{r} d\Delta(E)H_2\cdots H_{n-1} + 4r(L_{\max} - \mu)(\mu - L_{\min})\\&\leq \frac{2}{r}\max\{d\Delta(E)H_2\cdots H_{n-1}, 0\} + 4r(L_{\max} - L_{\min})^2\\&\leq 4r\left(\max\{d\Delta(E)H_2\cdots H_{n-1}, 0\} + (L_{\max} - L_{\min})^2\right).\end{aligned}$$

Now using the definition and Proposition 4.2.27, which implies that

$$\mu_{\max} - \mu_{\min} \leq L_{\max} - L_{\min} \leq \mu_{\max} - \mu_{\min} + \frac{2\sqrt{\beta_r}}{r},$$

we get

$$\begin{aligned}&(\mu_{\max,1} - \mu_{\min,1})^2\\&\quad\leq 4r\left(\max\{d\Delta(E)H_2\cdots H_{n-1}, 0\} + (\mu_{\max} - \mu_{\min} + 2\sqrt{\beta_r})^2\right).\end{aligned}$$

Using $\sqrt{a+b} \leq \sqrt{a} + \sqrt{b}$ for $a, b \geq 0$, we get

$$\begin{aligned}&\mu_{\max,1} - \mu_{\min,1}\\&\quad\leq 4\sqrt{r}\left(\sqrt{\max\{d\Delta(E)H_2\cdots H_{n-1}, 0\}} + (\mu_{\max} - \mu_{\min}) + \sqrt{\beta_r}\right),\end{aligned}$$

which gives the required inequality for $l = 1$. For higher l, we use induction. □

Now we are ready to prove the main theorem of the chapter, which states that the moduli space parametrizing torsion-free sheaves with a fixed Hilbert polynomial and with a upper bound for the slopes of their Harder–Narasimhan filtrations is bounded.

Theorem 4.5.4 *Let X be a projective variety over an algebraically closed field with an ample line bundle $O_X(1)$. Let P be a polynomial of degree d and let $\mu_0 \in \mathbb{R}$. Then, the family of torsion-free sheaves whose Hilbert polynomials are P and whose $\mu_{\max}$ are at most μ_0 is bounded.*

Proof We use the same notation as in Theorem 4.5.2. According to Kleiman's criteria (Theorem 4.2.12) and Lemma 4.2.13, to prove the boundedness it is enough to give an upper bound of $\mu_{\max,l}$ (or more specifically for $l = 1$). Note that $\mu_l = \mu(E|_{X_l})$ is independent of E and depends only on P for all $0 \leq l \leq n$. To get such an upper bound, we use Theorem 4.5.2 to get

$$\begin{aligned}\mu_{\max,l}(E) &\leq \mu_l + (\mu_{\max,l}(E) - \mu_{\min,l}(E))\\ &\leq C_1 + C_2(\mu_{\max}(E) - \mu_{\min}(E))\\ &\leq C_1 + C_2 r(\mu_0 - \mu),\end{aligned}$$

where C_1, C_2 are constants independent of E. Here the last inequality follows from $\mu_{\max} \leq \mu_0$ and the observation that $r\mu \leq \mu_{\min}(E) + (r - 1)\mu_{\max}(E)$. □

A more general version of the above theorem for pure-dimensional sheaves can be proved by imitating the proof in [9, Theorem 1.1].

Acknowledgements We would like to express our gratitude to the organizers of the Stacks Project Workshop for running the event online during the pandemic. We thank Alex Perry for kindly guiding us through the paper during the week of the workshop and beyond. We also thank Faidon Andriopoulos for helpful discussions during the workshop, and Adrian Langer for answering technical questions. Finally, we thank the referee for reading the draft carefully and proposing various comments to help improve the chapter.

References

[1] Jarod Alper, Pieter Belmans, Daniel Bragg, Jason Liang, and Tuomas Tajakka. Projectivity of the moduli space of vector bundles on a curve. In *Stacks project expository collection*, number 480 in London Mathematical Society Lecture Note Series, Chapter 3. Cambridge University Press, 2022.

[2] F. A. Bogomolov. Holomorphic tensors and vector bundles on projective manifolds. *Izv. Akad. Nauk SSSR Ser. Mat.*, 42(6):1227–1287; 1439, 1978.

[3] Holger Brenner. On a problem of Miyaoka. In *Number fields and function fields – two parallel worlds*, volume 239 of *Progr. Math.*, pages 51–59. Birkhäuser, 2005. doi:10.1007/0-8176-4447-4_3.

[4] Daniel Huybrechts and Manfred Lehn. *The geometry of moduli spaces of sheaves*. Cambridge Mathematical Library. Cambridge University Press, second edition, 2010.

[5] Adrian Langer. Addendum to: Semistable sheaves in positive characteristic [Ann. Math. **159**(1), 251–276, 2004]. *Ann. Math.*, 160(3):1211–1213, 2004.

[6] Adrian Langer. Semistable sheaves in positive characteristic. *Ann. Math.*, 159(1):251–276, 2004.

[7] Robert Lazarsfeld. *Positivity in algebraic geometry. I*, volume 48 of *Ergebnisse der Mathematik und ihrer Grenzgebiete. 3. Folge*. Springer, 2004. doi:10.1007/978-3-642-18808-4.

[8] Yoichi Miyaoka. The Chern classes and Kodaira dimension of a minimal variety. In *Algebraic geometry, Sendai, 1985*, volume 10 of *Adv. Stud. Pure Math.*, pages 449–476. North-Holland, 1987. doi:10.2969/aspm/01010449.

[9] Carlos T. Simpson. Moduli of representations of the fundamental group of a smooth projective variety. I. *Inst. Hautes Études Sci. Publ. Math.*, 79:47–129, 1994.

5

Theorem of the Base

Raymond Cheng
Columbia University

Lena Ji
University of Michigan

Matt Larson
Stanford University

Noah Olander
Columbia University

Abstract We explain a proof of the Theorem of the Base: the Néron–Severi group of a proper variety is a finitely generated abelian group. We discuss, quite generally, the Picard functor and its torsion and identity components. We study representability and finiteness properties of the Picard functor, both absolutely and in families. Along the way, we streamline the original proof by using alterations, and we discuss some examples of peculiar Picard schemes.

Introduction

The Theorem of the Base is a fundamental finiteness result on the Picard group of a proper variety X. A line bundle $\mathcal{L}$ on X is *algebraically trivial* if it is possible to deform $\mathcal{L}$ to $\mathcal{O}_X$; it is *numerically trivial* if it has degree 0 on every curve. See Definition 5.2.6. The *Néron–Severi group* $\mathrm{NS}(X)$ of X is the Picard group of X modulo algebraic triviality. We have the following result, see Proposition 5.4.3 and Theorem 5.7.4:

Theorem B *Let X be a proper scheme over a field. Then*

(i) $\mathrm{NS}(X)$ *is finitely generated, and*
(ii) *its torsion subgroup consists of numerically trivial classes.*

In fact, we prove a stronger version of Theorem B that gives a uniform bound on the rank and size of the torsion subgroup in families over a Noetherian base; see Theorem 5.7.7.

The proof of Theorem B for a smooth projective variety X over $\mathbf{C}$ is simple. The first Chern class gives an injection $\mathrm{NS}(X) \hookrightarrow H^2(X, \mathbf{Z})$, so the finite generation of $\mathrm{NS}(X)$ follows from the finite generation of the topological cohomology group $H^2(X, \mathbf{Z})$. The Lefschetz $(1,1)$-theorem and the Hard Lefschetz theorem imply that the Hodge conjecture holds for curves on X. Therefore, by Poincaré duality, any line bundle $\mathcal{L}$ that is numerically trivial has torsion first Chern class, which implies (ii).

Severi first proved Theorem B using transcendental methods in [17, 18], extending work of Picard in [15]. Néron proved (i) in arbitrary characteristic and gave the first algebraic proof by reinterpreting the question using rational points of an abelian variety related to the relative Jacobian of a curve fibration; see [14]. Lang and Néron later simplified this proof when they proved the Lang–Néron theorem in [10]. The proof of (ii) in arbitrary characteristic was first done by Matsusaka in [12].

The modern approach, developed by Kleiman in [9] and [1, Exposé XIII], proves the finiteness of the rank of $\mathrm{NS}(X)$ and the torsion subgroup of $\mathrm{NS}(X)$ separately. The proof that the rank of $\mathrm{NS}(X)$ is finite is similar to the simple proof over the complex numbers except that it uses a Weil cohomology theory, such as étale cohomology, to work in arbitrary characteristic. In order for the Weil cohomology theory to have the desired finiteness properties, we need to reduce to the case of smooth projective varieties. We use the existence of regular alterations [5] to do so. Alternatively, one could reduce to the case of surfaces, and then use that the resolution of singularities for surfaces is known in arbitrary characteristic; see [1, Exp. XIII, Section 5].

There is no Weil cohomology theory with integral coefficients in positive characteristic, so this approach does not show the finiteness of the torsion subgroup. In order to show the finiteness of the torsion subgroup, we reduce to the case of projective varieties using Proposition 5.1.4, a theorem of Raynaud. We then show that all numerically trivial line bundles are parametrized by a single finite type Quot scheme, from which we deduce the finiteness of the torsion subgroup and (ii).

The Theorem of the Base is used throughout algebraic geometry, and is required for the formulation of many fundamental results. It is frequently useful to study the *numerical* properties of line bundles, i.e., properties of line bundles that depend only the image of the line bundle in $\mathrm{NS}(X) \otimes \mathbf{R}$. Many properties of line bundles, such as whether they are big [11, Theorem 2.2.26] or ample [9], can be shown to be numerical properties. Furthermore, the locus in $\mathrm{NS}(X) \otimes \mathbf{R}$ of line bundles having a given numerical property often has nice topological properties. For example, the cone of ample line bundles is open in $\mathrm{NS}(X) \otimes \mathbf{R}$ [9], which allows one to deform the polarization of a variety. The formulation of the openness of the ample cone requires the Theorem of the Base: otherwise, when $\dim(X) > 2$, there is no natural topology on $\mathrm{NS}(X) \otimes \mathbf{R}$.

The Theorem of the Base is used very frequently in birational geometry. The *Picard number* $\rho(X)$, defined as the rank of $\mathrm{NS}(X)$, is a basic measure of the complexity of a variety, and it is frequently used to show the termination of algorithms. For example, the minimal model program for surfaces X consists of repeatedly contracting curves on X. After each contraction, one shows that the Picard number has dropped. As $\rho(X) < \infty$, this implies that eventually our variety will have no more curves that can be contracted. Many of the deepest results in birational geometry rely on the study of the ample cone and its closure in $\mathrm{NS}(X)\otimes\mathbf{R}$.

Our chapter is organized as follows. In Sections 5.1 and 5.2, we discuss some fundamental results on the Picard functor and its components. In Sections 5.3 and 5.4, we prove the finiteness of the torsion subgroup of $\mathrm{NS}(X)$ and (ii) using projective geometry. The finiteness of the Picard number and hence the finite generation of $\mathrm{NS}(X)$ is proved in Sections 5.5–5.7. We close in Section 5.8 with some examples of Picard schemes.

Conventions Throughout, k is a field. A *variety* is a separated integral scheme of finite type over a field k. Unadorned fiber products are taken over k.

5.1 The Picard functor

In this section, we recall the definition of the Picard functor in the Stacks project, and summarize some of its main properties. We will use these definitions in the remainder of the chapter. For a more detailed treatment, see [6, Part 5] and [4, Chapter 8].

Let $f\colon X \to S$ be a morphism of schemes. The Picard functor, restricting the general definition given in [19, Situation 0D25],

$$\mathrm{Pic}_{X/S}\colon (\mathrm{Sch}/S)^{\mathrm{opp}} \to \mathrm{Sets}$$

is the fppf sheafification of the functor sending a scheme T over S to the group $\mathrm{Pic}(X_T)$ of isomorphism classes of invertible $\mathcal{O}_{X_T}$-modules; here, $f_T\colon X_T \to T$ is the base change of f along $T \to S$. The basic representability result is the following:

Theorem 5.1.1 *Let $f\colon X \to S$ be a morphism of schemes. If*

(i) *f is proper, flat, of finite presentation, and*
(ii) *the formation of $f_* \mathcal{O}_X$ commutes with all base changes,*

then $\mathrm{Pic}_{X/S}$ *is an algebraic space. The morphism* $\mathrm{Pic}_{X/S} \to S$ *is quasi-separated and locally of finite presentation.*

Proof In the case where $\mathcal{O}_T \to f_{T,*}\mathcal{O}_{X_T}$ is an isomorphism for all schemes T over S, this is [19, Proposition 0D2C] and [19, Lemma 0DNI]. For the general case, see [3, Theorem 7.3]. □

Let $f: X \to S$ and $g: Y \to S$ be morphisms of schemes, and let $h: Y \to X$ be a morphism over S. Then there exists a morphism of group functors

$$h^*: \mathrm{Pic}_{X/S} \to \mathrm{Pic}_{Y/S}$$

obtained as the fppf sheafification of the natural pullback map of invertible modules. The basic finiteness result for pullbacks is the following:

Theorem 5.1.2 *Let $f: X \to S$ and $g: Y \to S$ be morphisms of schemes, and let $h: Y \to X$ be a morphism over S. Assume that*

(i) *S is integral and Noetherian,*
(ii) *f and g are proper, and*
(iii) *h is surjective.*

Then there exists a nonempty open subscheme S° of S such that

$$h^*|_{S^\circ}: \mathrm{Pic}_{X/S}|_{S^\circ} \to \mathrm{Pic}_{Y/S}|_{S^\circ}$$

is representable by a quasi-affine morphism of finite presentation.

Proof See [2, Exposé XII, Théorème 1.1]. □

For simplicity, we will mostly restrict ourselves to the situation where our base scheme S is the spectrum of a field k. In this case, since separated group algebraic spaces locally of finite type over fields are actually schemes, the basic representability result reads:

Proposition 5.1.3 *Let X be a proper scheme over k. Then* $\mathrm{Pic}_{X/k}$ *is a separated scheme locally of finite type over k.*

Proof Theorem 5.1.1 already shows that $\mathrm{Pic}_{X/k}$ is a quasi-separated algebraic space locally of finite type over k. By [19, Lemma 08BH], $\mathrm{Pic}_{X/k}$ is actually separated, so by [19, Lemma 0B8F], it is a scheme. See also the discussion of [19, Tag 06E9]. □

Similarly, Theorem 5.1.2 implies:

Proposition 5.1.4 *Let $h\colon Y \to X$ be a surjective morphism of schemes which are proper over k. Then $h^*\colon \mathrm{Pic}_{X/k} \to \mathrm{Pic}_{Y/k}$ is a finite type quasi-affine morphism of schemes.* □

Finally, although the value of the Picard functor on general schemes T may be rather subtle, its geometric points are as expected. This follows from the following very general consideration:

Lemma 5.1.5 *Let U be an object of a site C. Assume there exists a cofinal system of coverings of U of the form $\{V \to U\}$ such that each $V \to U$ admits a section. Then $\mathcal{F}(U) = \mathcal{F}^\#(U)$ for any presheaf $\mathcal{F}$ on C.*

Proof Here, $\mathcal{F}^\# := \mathcal{F}^{++}$ is the sheafification of $\mathcal{F}$, and $\mathcal{F}^+$ is

$$\mathcal{F}^+(U) := \mathrm{colim}_{\mathcal{U}}\, H^0(\mathcal{U}, \mathcal{F})$$

where the colimit is taken over all coverings $\mathcal{U}$ of U; see [19, Section 00W1]. Thus the result will follow if we can show that $\mathcal{F}(U) \to H^0(\mathcal{U}, \mathcal{F})$ is bijective for every covering $\mathcal{U} = \{p\colon V \to U\}$ in our cofinal system where each p admits a section σ. For injectivity, simply observe that

$$\mathcal{F}(U) \to H^0(\mathcal{U}, \mathcal{F}) \subset \mathcal{F}(V) \xrightarrow{\sigma^*} \mathcal{F}(U)$$

is the identity. For surjectivity, let $s \in H^0(\mathcal{U}, \mathcal{F})$ and set $t := \sigma^*(s)$. We claim $p^*(t) = s$. Indeed, writing $\mathrm{pr}_1, \mathrm{pr}_2\colon V \times_U V \to V$ for the projections, we have $\mathrm{pr}_1^*(s) = \mathrm{pr}_2^*(s)$, as s is a section on $\mathcal{U}$. Pulling this identity back along $(\mathrm{id}_V, \sigma \circ p) : V \to V \times_U V$ yields

$$\begin{aligned} s = \mathrm{id}_V^*(s) &= (\mathrm{id}_V, \sigma \circ p)^* \mathrm{pr}_1^*(s) \\ &= (\mathrm{id}_V, \sigma \circ p)^* \mathrm{pr}_2^*(s) = p^* \sigma^*(s) = p^*(t). \end{aligned}$$ □

Lemma 5.1.6 *Let X be a proper scheme over an algebraically closed field k. Then $\mathrm{Pic}_{X/k}(k) = \mathrm{Pic}(X)$.*

Proof The Hilbert Nullstellensatz, [19, Theorem 00FV], implies that any fppf covering $T \to \mathrm{Spec}(k)$ admits a section. Therefore Lemma 5.1.5 applies with $U = \mathrm{Spec}(k)$ and $\mathcal{F}$ the fppf presheaf $T \mapsto \mathrm{Pic}(X_T)$ defining the Picard functor. □

5.2 Components of the Picard functor

The Picard functor as a whole is almost never of finite type as it generally has countably infinitely many connected components. Thus to make sense of the finiteness properties for the Picard functor, it is helpful to consider the subgroup functors $\mathrm{Pic}^0_{X/k}$ and $\mathrm{Pic}^{\tau}_{X/k}$ giving the connected component of the identity and all torsion components, respectively.

Definition 5.2.1 Let G be a group scheme over k. The *identity component* G^0 of G is the connected component of the identity. The subscheme of *torsion components* G^{τ} of G is

$$G^{\tau} := \bigcup_{n>0} (g \mapsto g^n)^{-1} G^0,$$

the union of the preimage of G_0 under all nth-power maps with $n > 0$.

Lemma 5.2.2 *Let G be a group scheme locally of finite type over k. Then*

(i) *the formation of G^0 and G^{τ} commutes with extending k;*
(ii) *G^0 is an open and closed group subscheme of finite type;*
(iii) *G^0 is geometrically irreducible;*
(iv) *G^{τ} is an open group subscheme; and*
(v) *if G^{τ} is quasi-compact, then it is closed and of finite type.*

Proof For the statements about G^0; see [19, Proposition 0B7R]. Now (i) for G^{τ} and (iv) follow from the corresponding properties of G^0. For (v), if G^{τ} is quasi-compact, then there exists $N > 0$ such that

$$G^{\tau} = \bigcup_{n=1}^{N} (g \mapsto g^n)^{-1} G^0.$$

Thus G^{τ} is closed and of finite type since the same is true for G^0. □

The following is the observation that a finite type morphism between two group schemes must respect torsion components in a strong way.

Lemma 5.2.3 *Let $f\colon H \to G$ be a finite type morphism of group schemes over k, which are locally of finite type over k. Then $f^{-1}(G^{\tau}) = H^{\tau}$.*

Proof Since $f(H^0) \subseteq G^0$, we have

$$f(H^{\tau}) = \bigcup_{n>0} f((h \mapsto h^n)^{-1}(H^0)) = \bigcup_{n>0} (g \mapsto g^n)^{-1}(f(H^0)) \subseteq G^{\tau}$$

so that $f^{-1}(G^{\tau}) \supseteq H^{\tau}$. Conversely, let $f(h) \in G^{\tau}$. Replace h by h^n for $n > 0$ to assume $f(h) \in G^0$. By Lemma 5.2.2(ii), G^0 is of finite type; since f is of finite type, the same goes for $f^{-1}(G^0)$ as the composition of finite type morphisms is of finite type; see [19, Lemma 01T3]. In particular, it has finitely many components by [19, Lemma 0BA8].

So there exist $n > m > 0$ such that h^n and h^m are in the same component of $f^{-1}(G^0)$. Then $h^{n-m} \in H^0$, showing that $h \in H^\tau$. □

Applying this to the inclusion of a finite type subgroup scheme shows:

Corollary 5.2.4 *Let G be a group scheme locally of finite type over k. If $H \subseteq G$ is a subgroup scheme of finite type over k, then $H \subseteq G^\tau$.* □

Thus if G^τ is quasi-compact, then it is the largest subgroup scheme of G of finite type over k.

Now we apply the above notions to obtain subgroup schemes

$$\mathrm{Pic}^0_{X/k} \subseteq \mathrm{Pic}^\tau_{X/k} \subseteq \mathrm{Pic}_{X/k}.$$

These components make sense by the basic finiteness in the representability result Proposition 5.1.3. Already, we can show that formation of the torsion component commutes with certain pullbacks.

Lemma 5.2.5 *Let $f\colon Y \to X$ be a surjective morphism of proper schemes over k. Then $f^{*,-1}(\mathrm{Pic}^\tau_{Y/k}) = \mathrm{Pic}^\tau_{X/k}$.*

Proof The finiteness of the representability result of Proposition 5.1.4 allows us to apply Lemma 5.2.3. □

The structure sheaf $\mathcal{O}_X$ is the unit in the group of invertible $\mathcal{O}_X$-modules. So we may attempt to characterize the points of $\mathrm{Pic}^0_{X/k}$ and $\mathrm{Pic}^\tau_{X/k}$ by relating invertible modules with $\mathcal{O}_X$.

Definition 5.2.6 Let X be a proper scheme over k. An invertible $\mathcal{O}_X$-module $\mathcal{L}$ is called

(i) *numerically trivial* if $\deg(\mathcal{L}|_C) := \chi(C, \mathcal{L}|_C) - \chi(C, \mathcal{O}_C) = 0$ for every closed integral curve C in X;
(ii) *algebraically trivial* if there exists a connected scheme T of finite type over k, an invertible $\mathcal{O}_{X\times T}$-module $\mathcal{M}$, and geometric points t_0 and t_1 of T such that $\mathcal{M}|_{X\times t_0} \cong \mathcal{O}_X$ and $\mathcal{M}|_{X\times t_1} \cong \mathcal{L}$; and
(iii) *τ-trivial* if $\mathcal{L}^{\otimes n}$ is algebraically trivial for some integer $n \neq 0$.

These notions characterize the points of components of $\mathrm{Pic}_{X/k}$:

Lemma 5.2.7 *Let X be a proper scheme over k. Let $\mathcal{L}$ be an invertible $\mathcal{O}_X$-module corresponding to a point $[\mathcal{L}] \in \mathrm{Pic}_{X/k}(k)$. Then*

(i) *$\mathcal{L}$ is algebraically trivial if and only if $[\mathcal{L}] \in \mathrm{Pic}^0_{X/k}(k)$; and*
(ii) *$\mathcal{L}$ is τ-trivial if and only if $[\mathcal{L}] \in \mathrm{Pic}^\tau_{X/k}(k)$.*

Proof If $\mathcal{L}$ is algebraically trivial, then the witnessing data $(T, \mathcal{M}, t_0, t_1)$ is a morphism $[\mathcal{M}]: T \to \mathrm{Pic}_{X/k}$ such that $t_0 \mapsto [\mathcal{O}_X]$ and $t_1 \mapsto [\mathcal{L}]$. Since T is connected, $[\mathcal{L}]$ is in the connected component of the identity.

Suppose $[\mathcal{L}] \in \mathrm{Pic}^0_{X/k}(k)$. Let $f: T' \to \mathrm{Pic}^0_{X/k}$ be an fppf cover with an invertible sheaf $\mathcal{M}'$ on $X_{T'}$ representing the inclusion $\mathrm{Pic}^0_{X/k} \subseteq \mathrm{Pic}_{X/k}$. For $i = 0, 1$, let $T'_i \subseteq T'$ be irreducible components with geometric points t_i such that $f(t_0) = [\mathcal{O}_X]$ and $f(t_1) = [\mathcal{L}]$. Since $\mathrm{Pic}^0_{X/k}$ is irreducible and fppf morphisms are open, see Lemmas 5.2.2(iii) and [19, Tag 01UA], $f(T'_0) \cap f(T'_1)$ is a nonempty open subset of $\mathrm{Pic}^0_{X/k}$. Let s be a geometric point therein and let s_i be geometric points in the T'_i lying over it, so that, by Lemma 5.1.6, $\mathcal{M}'|_{X \times s_0} \cong \mathcal{M}'|_{X \times s_1}$. Up to replacing T' by a further fppf covering, we may assume that the images of s_i in T'_i are closed points with the same residue field. Then [19, Lemma 0B7M] allows us to glue the T'_i together along these closed points to obtain a scheme T. Furthermore, the sheaves $\mathcal{M}'|_{X \times T'_i}$ glue to an invertible $\mathcal{O}_{X \times T}$-module $\mathcal{M}$. Then $(T, \mathcal{M}, t_0, t_1)$ witness the algebraic triviality of $\mathcal{L}$. □

It is not *a priori* clear that numerical triviality characterizes the points of some component of $\mathrm{Pic}_{X/k}$, but we will see in Proposition 5.4.3 that it is actually equivalent to τ-triviality. The proof of that proposition will use some finiteness properties of $\mathrm{Pic}^\tau_{X/k}$. In any case, we can already prove that numerical triviality is implied by both algebraic and τ-triviality.

Lemma 5.2.8 *Let X be a proper scheme over k and $\mathcal{L}$ an invertible $\mathcal{O}_X$-module. If $\mathcal{L}$ is either algebraically trivial or τ-trivial, then $\mathcal{L}$ is numerically trivial.*

Proof If $\mathcal{L}$ is algebraically trivial, then there is a connected scheme T over k, an invertible $\mathcal{O}_{X \times T}$-module $\mathcal{M}$, and geometric points t_0 and t_1 of T such that $\mathcal{M}|_{X \times t_0} \cong \mathcal{O}_X$ and $\mathcal{M}|_{X \times t_1} \cong \mathcal{L}$. Let C be any closed integral curve in X. Since Euler characteristics are locally constant in flat proper families, as in [19, Lemma 0B9T], we have

$$
\begin{aligned}
\deg(\mathcal{L}|_C) &= \chi(C, \mathcal{L}|_C) - \chi(C, \mathcal{O}_C) \\
&= \chi(C \times t_1, \mathcal{M}|_{C \times t_1}) - \chi(C, \mathcal{O}_C) \\
&= \chi(C \times t_0, \mathcal{M}|_{C \times t_0}) - \chi(C, \mathcal{O}_C) \\
&= \chi(C, \mathcal{O}_C) - \chi(C, \mathcal{O}_C) = 0,
\end{aligned}
$$

so $\mathcal{L}$ is numerically trivial.

If $\mathcal{L}$ is τ-trivial, let n be a positive integer such that $\mathcal{L}^{\otimes n}$ is algebraically trivial. Then we have just proven that $\mathcal{L}^{\otimes n}$ is then numerically trivial.

But if C is now any integral closed curve in X, the additivity of degrees as from [19, Lemma 0AYX] gives

$$\deg(\mathcal{L}|_C) = \frac{1}{n}\deg(\mathcal{L}^{\otimes n}|_C) = 0,$$

so $\mathcal{L}$ itself is numerically trivial. □

The various notions of triviality behave well under pullback.

Lemma 5.2.9 *Let $f: Y \to X$ be a morphism of proper schemes over k. Let $\mathcal{L}$ be an invertible O_X-module.*

(i) *If $\mathcal{L}$ is numerically trivial, then $f^*\mathcal{L}$ is numerically trivial.*
(ii) *If $\mathcal{L}$ is algebraically trivial, then $f^*\mathcal{L}$ is algebraically trivial.*
(iii) *If $\mathcal{L}$ is τ-trivial, then $f^*\mathcal{L}$ is τ-trivial.*

Suppose furthermore that f is surjective.

(iv) *If $f^*\mathcal{L}$ is numerically trivial, then $\mathcal{L}$ is numerically trivial.*
(v) *If $f^*\mathcal{L}$ is τ-trivial, then $\mathcal{L}$ is τ-trivial.*

Proof For (i), if D is any integral curve in Y, its image $C := f(D)$ is an integral curve in X. Thus by the compatibility of degrees on curves with pullbacks, [19, Lemma 0AYZ],

$$\deg(f^*\mathcal{L}|_D) = \deg(D \to C)\deg(\mathcal{L}|_C) = 0.$$

For (ii), let $(T, \mathcal{M}, t_0, t_1)$ be the data witnessing the algebraic triviality of $\mathcal{L}$ on X. Let $f_T : Y \times T \to X \times T$ be the base change of f to T, and let $\mathcal{M}' = f_T^*\mathcal{M}$. Then $(T, \mathcal{M}', t_0, t_1)$ witness algebraic triviality of $f^*\mathcal{L}$. Now (iii) follows directly from (ii).

Assume $f: Y \to X$ is surjective. To see (iv), let C be an integral closed curve in X. The closure of a height-1 generic point in $f^{-1}(C)$ yields an integral closed curve D in Y with image C. Thus, again,

$$\deg(\mathcal{L}|_C) = \deg(D \to C)^{-1}\deg(f^*\mathcal{L}|_D) = 0.$$

Finally, (v) follows from Lemma 5.2.7(ii) and Lemma 5.2.5. □

5.3 Castelnuovo–Mumford regularity

When X is projective over k, the finiteness of $\mathrm{Pic}^\tau_{X/k}$ comes by exhibiting all its points as a quotient of a fixed finite locally free O_X-module: that is, $\mathrm{Pic}^\tau_{X/k}$ will be realized as an open subscheme of a Quot scheme. In this section, we use projective techniques to study the cohomology of numerically trivial invertible modules, the main result being Proposition 5.3.5. This is formulated using the following notion; see [19, Definition 08A3].

Definition 5.3.1 A coherent sheaf $\mathcal{F}$ on $\mathbf{P}^n_k$ is said to be *m-regular* if

$$H^i(\mathbf{P}^n_k, \mathcal{F}(m-i)) = 0 \quad \text{for } 1 \le i \le n.$$

Let Λ be a set of coherent sheaves on $\mathbf{P}^n_k$. The *Castelnuovo–Mumford regularity* of Λ is the smallest integer m, if it exists, such that each $\mathcal{F} \in \Lambda$ is m-regular.

Regularity has many consequences for the cohomology of sheaves; see [19, Section 089X] and [13, Lecture 14], for example. Most important for us, however, is that if $\mathcal{F}$ is m-regular, then $\mathcal{F}(m)$ is globally generated; see [19, Lemma 08A8]. Moreover, the definition is designed to be robust under passing to hyperplane sections; see [19, Lemma 08A5]. Conversely, and crucially for inductive arguments, we now show that regularity upon passing to a divisor yields the vanishing of cohomology:

Lemma 5.3.2 *Let $\mathcal{F}$ be a coherent sheaf on $\mathbf{P}^n_k$. Let $\iota\colon H \hookrightarrow \mathbf{P}^n_k$ be an effective Cartier divisor of degree d. If*

(i) *H avoids all associated points of $\mathcal{F}$, and*
(ii) *$\iota_*\mathcal{F}|_H$ is b-regular,*

then $H^i(\mathbf{P}^n_k, \mathcal{F}(\nu)) = 0$ for all $i \ge 2$ and $\nu \ge b - d$.

Proof Let $\sigma \in H^0(\mathbf{P}^n_k, \mathcal{O}_{\mathbf{P}^n_k}(d))$ be a section defining H. Multiplication by σ is injective on $\mathcal{F}$ as H avoids all its associated points. Thus, twisting the ideal sheaf sequence for H by $\mathcal{F}(\nu)$ gives sequences

$$0 \to \mathcal{F}(\nu - d) \xrightarrow{\sigma} \mathcal{F}(\nu) \to \iota_*\mathcal{F}(\nu)|_H \to 0 \quad \text{for all } \nu \in \mathbf{Z}.$$

Now $\iota_*\mathcal{F}|_H$ is ν-regular for all $\nu \ge b$; see [19, Lemma 08A6]. Therefore, the long exact sequence in cohomology gives

$$H^i(\mathbf{P}^n_k, \mathcal{F}(\nu - d)) \cong H^i(\mathbf{P}^n_k, \mathcal{F}(\nu)) \quad \text{for all } i \ge 2 \text{ and } \nu \ge b.$$

Serre vanishing, [19, Lemma 01YS], shows that these vanish for large ν, so

$$H^i(\mathbf{P}^n_k, \mathcal{F}(\nu - d)) = 0 \quad \text{for all } i \ge 2 \text{ and } \nu \ge b. \qquad \square$$

With the language of regularity, the goal of this section is to show that, for each closed subscheme $j\colon X \hookrightarrow \mathbf{P}^n_k$, the set

$$\Lambda(j) := \{ j_*\mathcal{L} \mid \mathcal{L} \text{ a numerically trivial } \mathcal{O}_X\text{-module}\}$$

has finite Castelnuovo–Mumford regularity which can be bounded by an integer depending only on the Hilbert polynomial of X. We deduce this from the following induction principle. Compare with [13, pp. 101–103].

Proposition 5.3.3 (Induction principle) *Let Λ be a set of coherent sheaves on $\mathbf{P}^n_k$. Assume that there exists*

(i) *a positive integer s such that* $\dim \operatorname{Supp}(\mathcal{F}) = s$,

(ii) *a positive number a such that*

$$\dim_k H^0(\mathbf{P}^n_k, \mathcal{F}(j)) \le a\binom{\nu + s}{s} \quad \textit{for every } \nu \in \mathbf{Z}, \textit{ and}$$

(iii) *a polynomial $P(t)$ which is the Hilbert polynomial of $\mathcal{F}$*

for every $\mathcal{F} \in \Lambda$. Also assume that there exists

(iv) *an integer b such that, for every $\mathcal{F} \in \Lambda$, $\iota_*\mathcal{F}|_H$ is b-regular for some hyperplane $\iota\colon H \hookrightarrow \mathbf{P}^n_k$ not containing any associated points of $\mathcal{F}$.*

Then there exists an integer m depending only on a, b, and $P(t)$ such that every sheaf in Λ is m-regular.

Proof Consider any $\mathcal{F} \in \Lambda$. Choose a hyperplane $\iota\colon H \hookrightarrow \mathbf{P}^n_k$ for $\mathcal{F}$ as in (iv). Twisting the ideal sheaf sequence of H by $\mathcal{F}(\nu)$ gives sequences

$$0 \to \mathcal{F}(\nu - 1) \to \mathcal{F}(\nu) \to \iota_*\mathcal{F}(\nu)|_H \to 0 \quad \text{for every } \nu \in \mathbf{Z}.$$

Note that multiplication by an equation for H is injective since it avoids all associated points of $\mathcal{F}$. Apply Lemma 5.3.2 with $d = 1$ to get

$$H^i(\mathbf{P}^n_k, \mathcal{F}(\nu - 1)) = 0 \quad \text{for all } i \ge 2 \text{ and } \nu \ge b.$$

So when $\nu \ge b$, the long exact sequence in cohomology reduces to

$$\begin{aligned} 0 &\longrightarrow H^0(\mathbf{P}^n_k, \mathcal{F}(\nu - 1)) \longrightarrow H^0(\mathbf{P}^n_k, \mathcal{F}(\nu)) \xrightarrow{\rho_\nu} H^0(\mathbf{P}^n_k, \iota_*\mathcal{F}(\nu)|_H) \\ &\longrightarrow H^1(\mathbf{P}^n_k, \mathcal{F}(\nu - 1)) \longrightarrow H^1(\mathbf{P}^n_k, \mathcal{F}(\nu)) \longrightarrow 0. \end{aligned}$$

Either ρ_ν is surjective, or else

$$\dim_k H^1(\mathbf{P}^n_k, \mathcal{F}(\nu)) < \dim_k H^1(\mathbf{P}^n_k, \mathcal{F}(\nu - 1)).$$

But observe: if ρ_ν is surjective, then $\rho_{\nu+1}$ is surjective. Indeed, consider the commutative square

$$\begin{array}{ccc} H^0(\mathbf{P}^n_k, \mathcal{F}(\nu)) \otimes_k H^0(\mathbf{P}^n_k, \mathcal{O}_{\mathbf{P}^n_k}(1)) & \longrightarrow & H^0(\mathbf{P}^n_k, \mathcal{F}(\nu+1)) \\ \downarrow{\scriptstyle \rho_\nu \otimes \mathrm{id}} & & \downarrow{\scriptstyle \rho_{\nu+1}} \\ H^0(\mathbf{P}^n_k, \iota_*\mathcal{F}(\nu)|_H) \otimes_k H^0(\mathbf{P}^n_k, \mathcal{O}_{\mathbf{P}^n_k}(1)) & \longrightarrow & H^0(\mathbf{P}^n_k, \iota_*\mathcal{F}(\nu+1)|_H) \end{array}$$

where the horizontal arrows are given by the multiplication of sections. By assumption, the map on the left given by ρ_ν is a surjection; since $\iota_*\mathcal{F}|_H$ is ν-regular, [19, Lemma 08A7] implies that the map on the bottom is a surjection. Commutativity of the square then implies $\rho_{\nu+1}$ is a surjection.

Thus the sequence $\{\dim_k H^1(\mathbf{P}^n_k, \mathcal{F}(\nu)) \mid \nu \geq b\}$ strictly decreases until it reaches 0. From this we see that $\mathcal{F}$ is m-regular for

$$m = b + \dim_k H^1(\mathbf{P}^n_k, \mathcal{F}(b)).$$

It remains to see that the latter quantity is bounded in terms of a and $P(t)$. By the vanishing property from Lemma 5.3.2,

$$\chi(\mathcal{F}(b)) = \dim_k H^0(\mathbf{P}^n_k, \mathcal{F}(b)) - \dim_k H^1(\mathbf{P}^n_k, \mathcal{F}(b)).$$

By (iii), $\chi(\mathcal{F}(b)) = P(b)$. Applying (ii) and rearranging then yields

$$\dim_k H^1(\mathbf{P}^n_k, \mathcal{F}(b)) \leq a\binom{b+s}{s} - P(b).$$

Thus $\mathcal{F}$ is m-regular for

$$m := b + a\binom{b+s}{s} - P(b)$$

and this depends only on a, b, and $P(t)$, as claimed. □

The following gives the uniform bound on twists of global sections of numerically trivial invertible modules. Compare with [6, Lemma 9.6.5]

Lemma 5.3.4 *Assume the base field k is infinite. Let $X \hookrightarrow \mathbf{P}^n_k$ be a closed subscheme. For every coherent $\mathcal{O}_X$-module $\mathcal{F}$, there exists a positive number $a(\mathcal{F})$ such that*

$$\dim_k H^0(X, \mathcal{F} \otimes \mathcal{L}(\nu)) \leq a(\mathcal{F})\binom{\nu+s}{s} \quad \text{where } s := \dim \operatorname{Supp}(\mathcal{F}),$$

for every numerically trivial $\mathcal{O}_X$-module $\mathcal{L}$ and $\nu \in \mathbf{Z}$. Moreover, $a(\mathcal{O}_X)$ can be chosen to depend only on the degree of X.

Proof We proceed via dévissage, in the form of [19, Lemma 01YM]. There are three conditions to check. The first is that for every short exact sequence of coherent $\mathcal{O}_X$-modules

$$0 \to \mathcal{F}_1 \to \mathcal{F} \to \mathcal{F}_2 \to 0$$

in which $a(\mathcal{F}_1)$ and $a(\mathcal{F}_2)$ exist, then $a(\mathcal{F})$ also exists. The cohomology sequence shows we may take $a(\mathcal{F}) := a(\mathcal{F}_1) + a(\mathcal{F}_2)$.

The second condition is that if $\mathcal{F}$ is a $\mathcal{O}_X$-module and $a(\mathcal{F}^{\oplus r})$ exists for some $r \geq 1$, then $a(\mathcal{F})$ also exists. Additivity shows that we may take $a(\mathcal{F}) := a(\mathcal{F}^{\oplus r})/r$.

The third and final condition concerns passing to closed subschemes $Z \hookrightarrow X$. Let $Y \hookrightarrow X$ be another closed subscheme not containing Z. Let $\mathcal{I} \subseteq \mathcal{O}_Z$ be the ideal sheaf of $Y \cap Z$ in Z. Then we must show there exists a quasi-coherent subsheaf of ideals $\mathcal{J} \subseteq \mathcal{I}$ for which $a(\mathcal{J})$ exists. Since the restriction $\mathcal{L}|_Z$ is a numerically trivial invertible module on Z, we may replace X by Z and $\mathcal{L}$ by its restriction. Next, every closed subscheme $Y \hookrightarrow X$ is contained in some hypersurface section; hence we may find $\mathcal{J} \subseteq \mathcal{I}$ of the form $\mathcal{O}_X(-d)$ for some nonnegative integer d. So, to complete the dévissage, it suffices to construct $a(\mathcal{O}_X(-d))$. In fact, if $d > 0$, then multiplication by a section of $\mathcal{O}_X(d)$ yields an injection

$$H^0(X, \mathcal{L}(-d)) \hookrightarrow H^0(X, \mathcal{L})$$

so we may take $a(\mathcal{O}_X(-d)) = a(\mathcal{O}_X)$.

Thus we are reduced to the special case of the lemma in which $\mathcal{F} = \mathcal{O}_X$. Proceed by induction on the dimension s of X. If $s = 0$, take $a(\mathcal{O}_X) = \dim_k H^0(X, \mathcal{O}_X)$; note this is the degree of X. Assume $s > 1$ and the statement holds for all closed subschemes of $\mathbf{P}^n_k$ of dimension $s - 1$. Since k is infinite, we may find a hyperplane section $\iota \colon H \hookrightarrow X$ not containing any associated points of X; see [19, Lemma 08A0]. This gives a short exact sequence

$$0 \to \mathcal{L}(\nu - 1) \to \mathcal{L}(\nu) \to \iota_* \mathcal{L}(\nu)|_H \to 0.$$

Taking global sections and applying induction on $\mathcal{L}(\nu)|_H$ gives the following inequalities, for all $\nu \in \mathbf{Z}$,

$$\begin{aligned} \dim_k H^0(X, \mathcal{L}(\nu)) &\leq \dim_k H^0(X, \mathcal{L}(\nu - 1)) + \dim_k H^0(H, \mathcal{L}(\nu)|_H) \\ &\leq \dim_k H^0(X, \mathcal{L}(\nu - 1)) + a(\mathcal{O}_{X \cap H}) \binom{\nu + s - 1}{s - 1}. \end{aligned}$$

Now note that $\mathcal{L}(-1)$ has negative degree on all integral curves in X, so $H^0(X, \mathcal{L}(-1)) = 0$: indeed, if there were a nonzero section, restricting it to a curve C not contained in its zero locus would yield the contradiction $\deg(\mathcal{L}(-1)|_C) > 0$; see [19, Lemma 0B40]. Therefore we may iterate the above inequalities to obtain

$$\dim_k H^0(X, \mathcal{L}(\nu)) \leq a(\mathcal{O}_{X\cap H}) \sum_{\mu=0}^{\nu} \binom{\mu+s-1}{s-1} = a(\mathcal{O}_{X\cap H}) \binom{\nu+s}{s},$$

so we may take $a(\mathcal{O}_X) = a(\mathcal{O}_{X\cap H})$. Since the degrees of X and $X \cap H$ are the same, this depends only on the degree of X. □

We now come to the crucial boundedness result: any numerically trivial invertible sheaf has the same Hilbert polynomial as the structure sheaf. The following argument largely follows [6, Lemma 9.6.6] and works for any projective scheme. It requires a careful induction on dimension that simultaneously computes the Hilbert polynomial and proves the boundedness of Castelnuovo–Mumford regularity. A much easier argument in the smooth case can be made via the Hirzebruch–Riemann–Roch theorem together with Lemma 5.5.2. One could also reduce to the smooth case using the existence of regular alterations and Proposition 5.1.4.

Proposition 5.3.5 *Assume the base field k is infinite. Let $j\colon X \hookrightarrow \mathbf{P}_k^n$ be a projective scheme. Then every member of the set*

$$\Lambda(j) := \{j_*\mathcal{L} \mid \mathcal{L} \text{ a numerically trivial } \mathcal{O}_X\text{-module}\}$$

has the same Hilbert polynomial $P(t)$, namely, that of $\mathcal{O}_X$ with respect to the embedding j. The Castelnuovo–Mumford regularity of $\Lambda(j)$ is bounded by an integer m depending only on $P(t)$.

Proof Proceed by induction on $s := \dim(X)$. When $s = 0$, the set in question consists only of the structure sheaf so the conclusion follows. So, assume $s \geq 1$. First we show that every member of $\Lambda(j)$ has the same Hilbert polynomial. For that, it suffices to show that if $\mathcal{L}$ is any numerically trivial $\mathcal{O}_X$-module, then the Hilbert polynomial of $\mathcal{L}^{\otimes q}$ is independent of $q \in \mathbf{Z}$, as we may take $q = 0$. Choose d large enough that $\mathcal{L}(d)$ is very ample. Now choose effective divisors H and D determined by short exact sequences

$$0 \to \mathcal{O}_X(-d) \to \mathcal{O}_X \to \mathcal{O}_H \to 0,$$
$$0 \to \mathcal{L}^\vee(-d) \to \mathcal{O}_X \to \mathcal{O}_D \to 0.$$

Twisting both sequences by $\mathcal{L}^{\otimes q}(d+\nu)$ yields the sequences

$$0 \to \mathcal{L}^{\otimes q}(\nu) \to \mathcal{L}^{\otimes q}(d+\nu) \to \iota_{H,*}\mathcal{L}^{\otimes q}(d+\nu)|_H \to 0,$$

$$0 \to \mathcal{L}^{\otimes q-1}(\nu) \to \mathcal{L}^{\otimes q}(d+\nu) \to \iota_{D,*}\mathcal{L}^{\otimes q}(d+\nu)|_D \to 0.$$

Taking Euler characteristics and subtracting yields, for all $q, \nu \in \mathbf{Z}$,

$$\chi(\mathcal{L}^{\otimes q}(\nu)) - \chi(\mathcal{L}^{\otimes q-1}(\nu)) = \chi(\mathcal{L}^{\otimes q}(d+\nu)|_D) - \chi(\mathcal{L}^{\otimes q}(d+\nu)|_H).$$

Since the restriction of $\mathcal{L}^{\otimes q}$ to any closed subscheme remains numerically trivial, induction can be applied to show that the right-hand side is a polynomial depending only on ν. Therefore, as a function of q and ν,

$$\chi(\mathcal{L}^{\otimes q}(\nu)) = \varphi_1(\nu)q + \varphi_0(\nu)$$

for some polynomials φ_1 and φ_0.

We now show that $\varphi_1 = 0$. If not, choose ν_0 sufficiently large that

(i) $\varphi_1(\nu)$ is the same sign for all $\nu \geq \nu_0$, and
(ii) $\mathcal{L}^{\otimes q}|_H$ is b-regular for $b := \nu_0 + d$ and all $q \in \mathbf{Z}$,

where the induction hypothesis is used for the second condition. Then Lemma 5.3.2 applies, to show

$$\chi(\mathcal{L}^{\otimes q}(\nu)) = \dim_k H^0(X, \mathcal{L}^{\otimes q}(\nu)) - \dim_k H^1(X, \mathcal{L}^{\otimes q}(\nu))$$

for all $q \in \mathbf{Z}$ and $\nu \geq \nu_0$. Thus we see that

$$\dim_k H^0(X, \mathcal{L}^{\otimes q}(\nu)) \leq \chi(\mathcal{L}^{\otimes q}(\nu)) = \varphi_1(\nu)q + \varphi_0(\nu).$$

So, taking $q \to \pm\infty$, depending on whether $\varphi(\nu_0)$ is positive or negative, shows that $\dim_k H^0(X, \mathcal{L}^{\otimes q}(\nu)) \to \infty$. But this contradicts Lemma 5.3.4, which uniformly bounds this dimension independently of q. Therefore $\varphi_1 = 0$ and the Hilbert polynomials of $\mathcal{L}^{\otimes q}$ are independent of q. Hence all members of $\Lambda(j)$ have the same Hilbert polynomial.

Now, to show that $\Lambda(j)$ has bounded Castelnuovo–Mumford regularity depending only on $P(t)$, we apply the induction principle from Proposition 5.3.3. We verify the following hypotheses:

(i) Every member of $\Lambda(j)$ is supported on all of the s-dimensional scheme X;
(ii) The quantity $a := a(\mathcal{O}_X)$ from Lemma 5.3.4 bounds global sections;
(iii) We have just proven that every member of $\Lambda(j)$ has the same Hilbert polynomial $P(t)$;
(iv) Since k is infinite, we may choose a hyperplane $\iota\colon H \hookrightarrow \mathbf{P}^n_k$ that avoids the associated points of X; see [19, Lemma 08A0]. Then induction gives a b depending only on the Hilbert polynomial of $H \cap X$ such that the $\iota_*\mathcal{L}|_H$ are b-regular.

Thus Proposition 5.3.3 applies to give an integer m, depending only on a, b, and $P(t)$, such that all members of $\Lambda(j)$ are m-regular. It remains to see that both a and b depend only on $P(t)$. In fact, the second statement of Lemma 5.3.4 shows that $a = a(\mathcal{O}_X)$ depends only on the degree of the embedding $j\colon X \hookrightarrow \mathbf{P}^n_k$; this is simply the leading coefficient of $P(t)$. As for b, observe from the short exact sequence

$$0 \to \mathcal{O}_X(-1) \to \mathcal{O}_X \to \mathcal{O}_{X\cap H} \to 0$$

that the Hilbert polynomial of $X \cap H$ is $P(t) - P(t-1)$, and hence depends only on $P(t)$. Thus m depends only on $P(t)$. □

5.4 Boundedness

In this section, we give the finiteness result for $\mathrm{Pic}^{\tau}_{X/k}$.

Definition 5.4.1 Let X be a proper scheme over k. We say that a set Λ of invertible $\mathcal{O}_X$-modules is *bounded* if there exists

- a scheme T of finite type over k, and
- an invertible $\mathcal{O}_{X\times T}$-module $\mathcal{M}$,

such that for every $\mathcal{L} \in \Lambda$, there exists $t \in T$ such that $\mathcal{L} \cong \mathcal{M}|_{X\times t}$.

Lemma 5.4.2 *Let X be a projective scheme over an infinite field k. Then the set Λ of numerically trivial invertible $\mathcal{O}_X$-modules is bounded.*

Proof Fix a very ample invertible module $\mathcal{O}_X(1)$. Then, by Proposition 5.3.5, all numerically trivial invertible $\mathcal{O}_X$-modules have the same Hilbert polynomial $P(t)$ and Castelnuovo–Mumford regularity bounded by an integer m depending only on $P(t)$. So for every $\mathcal{L} \in \Lambda$, $\mathcal{L} \otimes \mathcal{O}_X(m)$ is globally generated by [19, Lemma 08A8] and its space of global sections has as its dimension a fixed integer M. Thus every such $\mathcal{L}$ is a quotient of

$$\mathcal{F} := \mathcal{O}_X(-m)^{\oplus M} \cong H^0(X, \mathcal{L} \otimes \mathcal{O}_X(m)) \otimes_k \mathcal{O}_X(-m).$$

Therefore Λ is parameterized by the open subscheme T of $\mathrm{Quot}^P_{\mathcal{F}/X/k}$ parameterizing locally free quotients. The latter is of finite type over k by [19, Lemma 0DPC], hence T is of finite type over k. □

The following characterizes the points of $\mathrm{Pic}^{\tau}_{X/k}$ numerically. See [1, Exp. XIII, Théorème 4.6] for more details.

Proposition 5.4.3 *Let X be a proper scheme over an infinite field k, and let $\mathcal{L}$ be an invertible $\mathcal{O}_X$-module. Then the following are equivalent.*

(i) *the set* $\{\mathcal{L}^{\otimes m} \mid m \in \mathbf{Z}\}$ *is bounded;*
(ii) $\mathcal{L}$ *is* τ*-trivial; and*
(iii) $\mathcal{L}$ *is numerically trivial.*

Then (i) $\Rightarrow$ (ii) $\Leftrightarrow$ (iii). *If X is, furthermore, projective over k, then all three statements are equivalent.*

Proof To see (i) $\Rightarrow$ (ii), assume $\{\mathcal{L}^{\otimes m} \mid m \in \mathbf{Z}\}$ is bounded. Then there exists a scheme T of finite type over k, a line bundle $\mathcal{M}$ on $X \times T$, and, for each $m \in \mathbf{Z}$, a geometric point t_m of T such that $\mathcal{L}^{\otimes m} \cong \mathcal{M}|_{X \times t_m}$. But T has only finitely many connected components, so there exists $m, n \in \mathbf{Z}$ such that t_m and t_n lie in the same connected component, and so $\mathcal{L}^{\otimes m-n}$ is algebraically trivial.

Now for (ii) $\Rightarrow$ (iii) use Lemma 5.2.8. For the converse (ii) $\Leftarrow$ (iii), Chow's lemma [19, Lemma 0200] and Lemma 5.2.9 together show that it is enough to consider the projective case. But when X is, furthermore, projective, Lemma 5.4.2 gives (iii) $\Rightarrow$ (i) and all three statements are then equivalent. □

Theorem 5.4.4 (Finiteness of $\mathrm{Pic}^{\tau}_{X/k}$) *Let X be a proper scheme over a field k. Then* $\mathrm{Pic}^{\tau}_{X/k}$ *is a quasi-compact closed subscheme of* $\mathrm{Pic}_{X/k}$.

Proof Using Chow's lemma [19, Lemma 0200], the functoriality of the torsion component from Lemma 5.2.5, and Proposition 5.1.4, it suffices to consider the case when X is projective and k is algebraically closed. Since $\mathrm{Pic}_{X/k}$ is locally of finite type over k by Proposition 5.1.3, it suffices to show that $\mathrm{Pic}^{\tau}_{X/k}$ is quasi-compact, from which everything else follows by Lemma 5.2.3. Now Proposition 5.4.3 shows that τ-triviality and numerical triviality are the same, and so Lemma 5.4.2 gives a scheme T of finite type over k and a morphism $[\mathcal{M}] : T \to \mathrm{Pic}_{X/k}$ whose image contains all geometric points corresponding to τ-trivial invertible $\mathcal{O}_X$-modules. In other words, by Lemma 5.2.7(ii) and Lemma 5.1.6,

$$[\mathcal{M}](T) \supseteq \mathrm{Pic}^{\tau}_{X/k}.$$

Being of finite type over k, T is a Noetherian topological space, and so are any of its subspaces. Thus $\mathrm{Pic}^{\tau}_{X/k}$ is Noetherian and hence quasi-compact. □

5.5 Finiteness of cycles modulo numerical equivalence

In this section, we show that the group of cycles modulo numerical equivalence is of finite rank, using a cycle-class map into a Weil cohomology theory, after which finiteness comes from the finiteness of Weil cohomology theories generally.

Throughout this section, we take X to be a smooth projective variety of dimension d over an algebraically closed field k. Then [19, Chapters 0AZ6 and 0FFG] on intersection theory and Weil cohomology theories apply.

Definition 5.5.1 Let $0 \le i \le d$ and let $\alpha \in \mathrm{CH}^i(X)$.

(i) We say that α is *numerically trivial* if $\deg(\alpha \cdot \beta) = 0$ for every $\beta \in \mathrm{CH}^{d-i}(X)$. Here, $\deg \colon \mathrm{CH}^*(X) \to \mathrm{CH}^*(\mathrm{Spec}(k)) = \mathbf{Z}$ is the degree map and $\alpha \cdot \beta$ is the intersection product from [19, Section 0B0G].

(ii) We say that α is *H^*-trivial* if it lies in the kernel of the cycle-class map $\gamma \colon \mathrm{CH}^i(X) \to H^{2i}(X)(i)$ associated with H^*.

Write $\mathrm{CH}^i(X)_{\mathrm{num}}$ and $\mathrm{CH}^i(X)_{H^*}$ for the subgroups of cycles which are numerically trivial and H^*-trivial, respectively, and let

$$\mathrm{Num}^i(X) := \mathrm{CH}^i(X)/\mathrm{CH}^i(X)_{\mathrm{num}}$$

be the *numerical group of codimension-i cycles.*

The goal of this section is to bound the rank of the numerical group in terms of the Betti numbers of a Weil cohomology theory. The relevance of this to our situation is given by the first Chern class homomorphism

$$c_1 \colon \mathrm{Pic}_{X/k}(k) \to \mathrm{CH}^1(X);$$

see [19, Section 02SI]. Since X is smooth, [19, Lemma 0BE9] shows that c_1 is an isomorphism. The following shows that the two notions of numerical triviality in Definitions 5.2.6 and 5.5.1 are compatible under c_1.

Lemma 5.5.2 *An invertible $\mathcal{O}_X$-module $\mathcal{L}$ is numerically trivial if and only if $c_1(\mathcal{L})$ is numerically trivial. Thus c_1 induces an isomorphism of abelian groups* $\mathrm{Pic}_{X/k}(k)/\mathrm{Pic}^{\tau}_{X/k}(k) \cong \mathrm{Num}^1(X)$.

Proof The invertible module $\mathcal{L}$ is numerically trivial if and only if for every integral curve C in X, $\deg(\mathcal{L}|_C) = 0$. By [19, Lemma 0BEY], this degree is the numerical intersection number on the left of

$$(\mathcal{L} \cdot C) = \deg(c_1(\mathcal{L}) \cdot [C]).$$

By [19, Lemma 0BFI], the numerical intersection number is compatible with the Chow-theoretic intersection number on the right, the vanishing of which is equivalent to the numerical triviality of $c_1(\mathcal{L})$. □

We now proceed to bound the ranks of the numerical groups.

Lemma 5.5.3 $\mathrm{CH}^i(X)_{H^*} \subseteq \mathrm{CH}^i(X)_{\mathrm{num}} \subseteq \mathrm{CH}^i(X)$.

Proof We need to show that any codimension-i cycle α that is H^*-trivial is numerically trivial. So consider any $\beta \in \mathrm{CH}^{d-i}(X)$. By the cycle-class axioms (C)(c) of [19, Section 0FHA] and [19, Lemma 0FHR],

$$\deg(\alpha \cdot \beta) = \int_X \gamma(\alpha) \cup \gamma(\beta) = 0.$$ □

In particular, this means that the numerical group is naturally a quotient of $\mathrm{CH}^i(X)/\mathrm{CH}^i(X)_{H^*}$. We use this observation to bound the rank over $\mathbf{Z}$ of $\mathrm{Num}^i(X)$ in terms of the dimension over F of $H^{2i}(X)$, by exploiting the compatibility between the intersection pairing on cycles and the perfect pairing on H^*. The situation is abstracted into the following technical result:

Lemma 5.5.4 *Suppose that we are given*

- *a field F containing a ring R,*
- *finite-dimensional vector spaces V_1 and V_2 over F,*
- *R-submodules $A_1 \subseteq V_1$ and $A_2 \subseteq V_2$, and*
- *an F-bilinear map $\langle -, - \rangle\colon V_1 \times V_2 \to F$.*

Let $\bar{A}_1 := \langle A_1, - \rangle \subseteq \mathrm{Hom}_F(V_2, F)$. If

(i) *the restriction of $\langle -, - \rangle$ to $A_1 \times A_2$ takes values in R, and*
(ii) *the F-span of A_2 is V_2,*

then $\mathrm{rank}_R(\bar{A}_1) \leq \dim_F(V_2)$.

Proof Consider the R-module $U := \{\varphi \in \mathrm{Hom}_F(V_2, F) \mid \varphi(A_2) \subseteq R\}$. Condition (i) means that $\bar{A}_1 \subseteq U$. Thus it suffices to show that U is an R-module of rank at most $d = \dim_F(V_2)$. Use condition (ii) to choose a R-submodule $M \subseteq A_2$ free of rank d and such that $M \otimes_R F = V_2$. Then

$$\mathrm{Hom}_F(V_2, F) = \mathrm{Hom}_F(M \otimes_R F, F) \cong \mathrm{Hom}_R(M, F),$$

and, since $M \subseteq A_2$, U is mapped under this isomorphism to a submodule of $R^d \cong \mathrm{Hom}_R(M, R) \subseteq \mathrm{Hom}_R(M, F)$. Thus U, and hence A_1, has rank over R at most d. □

Proposition 5.5.5 *Let X be a smooth projective variety of dimension d over an algebraically closed field k. Then, for every Weil cohomology theory H^* over k and every $0 \le i \le d$,*

$$\mathrm{rank}_{\mathbf{Z}}(\mathrm{Num}^i(X)) \le \dim_F(H^{2i}(X)).$$

Proof We apply Lemma 5.5.4. Set $R := \mathbf{Z}$ and F the characteristic 0 coefficient field of H^*. Set

$$A_1 := \mathrm{CH}^i(X)/\mathrm{CH}^i(X)_{H^*} \subset H^{2i}(X)(i) =: V_1$$

where the inclusion is given by the cycle-class map. Set

$$A_2 := \mathrm{CH}^{d-i}(X)/\mathrm{CH}^{d-i}(X)_{H^*}$$

and denote by V_2 the F-span of the image of A_2 inside $H^{2(d-i)}(X)(d-i)$ under the cycle-class map. Finally, let $\langle -, - \rangle \colon V_1 \times V_2 \to F$ be the restriction of the pairing on H^*. Condition 5.5.4(i) now follows from from the compatibility of intersection products with $\langle -, - \rangle$, and 5.5.4(ii) is true by construction.

It remains to observe that, thanks again to the compatibility between intersection products and the pairing on H^*, the image $\bar{A}_1$ of

$$\begin{aligned}\mathrm{CH}^i(X)/\mathrm{CH}^i(X)_{H^*} \subseteq H^{2i}(X)(i) &\xrightarrow{\cong} \mathrm{Hom}_F(H^{2(d-i)}(X)(d-i), F)\\ &\twoheadrightarrow \mathrm{Hom}_F(\mathrm{CH}^{d-i}(X)/\mathrm{CH}^{d-i}(X)_{H^*}, F) = \mathrm{Hom}_F(V_2, F)\end{aligned}$$

is exactly $\mathrm{Num}^i(X) = \mathrm{CH}^i(X)/\mathrm{CH}^i(X)_{\mathrm{num}}$: indeed, an element in the kernel above is represented by a cycle of $\mathrm{CH}^i(X)$ which intersects every cycle of $\mathrm{CH}^{d-i}(X)$ trivially. Thus the lemma gives the first inequality in

$$\mathrm{rank}_{\mathbf{Z}}(\mathrm{Num}^i(X)) \le \dim_F(V_2) \le \dim_F(H^{2(d-i)}(X)) = \dim_F(H^{2i}(X)),$$

where the last equality comes from Poincaré duality for H^*. □

5.6 Alterations in families

In this section, we formulate a version of de Jong's alterations theorem [5] in families. This will be used to study the behavior of the Picard rank in families in the following section.

Lemma 5.6.1 *Let X be a proper scheme over a perfect field k. Then there exists a smooth projective scheme X' over k and a surjective morphism $X' \to X$ over k.*

Proof Let $\{V_i\}_{i\in I}$ be the reduced irreducible components of X. For each i, apply de Jong's alterations theorem, [5, Theorem 4.1], to choose a smooth projective alteration $V_i' \to V_i$. Set $X' := \coprod_{i\in I} V_i'$ and $X' \to X$ as the composition $X' \to \coprod_{i\in I} V_i \to X$. □

Lemma 5.6.2 *Let A be a Noetherian domain with fraction field K; then the perfect closure K^{perf} is a filtered union of rings B, each of which contains a subring A_f with $0 \neq f \in A$ such that $A_f \subset B$ is finite free and a universal homeomorphism.*

Proof We may assume $\mathbf{F}_p \subset A$. We will show that every finitely generated A-algebra $C = A[x_1,\dots,x_n]$ with $A \subset C \subset K^{\mathrm{perf}}$ is contained in some B as in the statement of the lemma. Since $K \subset K^{\mathrm{perf}}$ is purely inseparable, there is a pth power q such that $x_i^q \in K$ for all i. Therefore we can find $0 \neq f \in A$ such that $x_i^q \in A_f$ for all i; thus each x_i is integral over A_f. Then $A_f[x_1,\dots,x_n]$ is a finitely generated A_f-module, so there is a further localization $(A_f[x_1,\dots,x_n])_{f'} = A_{ff'}[x_1,\dots,x_n]$, $0 \neq f' \in A$, which is free over $A_{ff'}$. Take this as our B. Then $A_{ff'} \subset B$ is finite free and a universal homeomorphism by [19, Lemma 0CNF]. □

Lemma 5.6.3 *Let S be an integral scheme. Let $X \to S$ be a proper morphism of schemes. Then there exists a nonempty open set $U \subset S$, a finite locally free universal homeomorphism $S' \to U$, a smooth projective morphism $X' \to S'$, and a surjective morphism $X' \to X \times_S S'$.*

Proof Let K be the function field of S. Then, by Lemma 5.6.1, there exists a smooth projective scheme Y over the perfect closure K^{perf} of K together with a surjective morphism $Y \to X \times_S \mathrm{Spec}(K^{\mathrm{perf}})$. Lemma 5.6.2 now gives an open $U \subset S$, a finite free universal homeomorphism $S' \to U$, and a projective scheme morphism $X' \to S'$ with a morphism $X' \to X \times_S S'$ such that the diagram

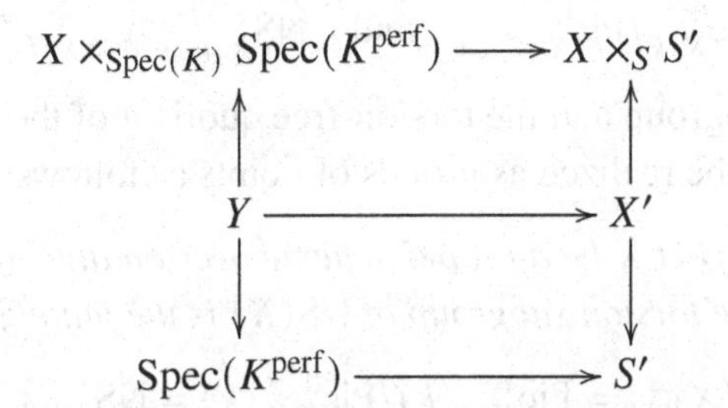

is commutative and the bottom square is Cartesian. By possibly further changing S', we may arrange for $X' \to S'$ to be smooth, see [19, Lemma 0C0C], and for $X' \to X \times_S S'$ to be surjective; see [19, Lemma 07RR]. □

5.7 Theorem of the Base

Let X be a proper scheme over an algebraically closed field k. The *Néron–Severi group* of X is the abelian group

$$\mathrm{NS}(X) := \mathrm{Pic}_{X/k}(k)/\mathrm{Pic}^0_{X/k}(k)$$

parameterizing the components of the Picard scheme; its rank $\rho(X)$ is the *Picard rank* of X. The goal of this section is to show in Theorem 5.7.4 that $\mathrm{NS}(X)$ is a finitely generated abelian group. We then study its behavior in families in Lemmas 5.7.5 and 5.7.6. Everything is put together in Theorem 5.7.7.

Two general observations before we begin. First, given a normal subgroup scheme H of a group scheme G locally of finite type over an algebraically closed field k, Lemma 5.1.5 shows that $(G/H)(k) = G(k)/H(k)$. Second, $Y_{\mathrm{red}}(k) = Y(k)$ for any scheme Y over any field k, since field-valued points are insensitive to nonreduced structure.

To formulate our results, we find it helpful to consider the following slight enrichment of the Néron–Severi group:

Definition 5.7.1 The *Néron–Severi scheme* of a proper scheme X over k is the commutative group scheme

$$\mathrm{NS}_{X/k} := \mathrm{Pic}_{X/k}/\mathrm{Pic}^0_{X/k,\mathrm{red}}.$$

This is a scheme of dimension 0 which is, by Proposition 5.1.3, locally of finite type over k. When k is algebraically closed, we recover the Néron–Severi group as the group of k-points:

$$\mathrm{NS}_{X/k}(k) = \mathrm{Pic}_{X/k}(k)/\mathrm{Pic}^0_{X/k}(k) = \mathrm{NS}(X).$$

The constructions of Definition 5.2.1 give subgroup schemes of $\mathrm{NS}_{X/k}$ closely related to the corresponding subgroup schemes of $\mathrm{Pic}_{X/k}$:

$$\mathrm{NS}^0_{X/k} \cong \mathrm{Pic}^0_{X/k}/\mathrm{Pic}^0_{X/k,\mathrm{red}} \quad \text{and} \quad \mathrm{NS}^\tau_{X/k} \cong \mathrm{Pic}^\tau_{X/k}/\mathrm{Pic}^0_{X/k,\mathrm{red}}.$$

The torsion subgroup and the torsion-free quotient of the Néron–Severi group can now be realized as groups of points as follows:

Lemma 5.7.2 *Let X be a proper scheme over an algebraically closed field k. Then the torsion subgroup of* $\mathrm{NS}(X)$ *is the finite group*

$$\mathrm{NS}(X)_{\mathrm{tors}} = \mathrm{Pic}^\tau_{X/k}(k)/\mathrm{Pic}^0_{X/k}(k) = \mathrm{NS}^\tau_{X/k}(k).$$

The torsion-free quotient is

$$\mathrm{NS}(X)_{\mathrm{tf}} := \mathrm{NS}(X)/\mathrm{NS}(X)_{\mathrm{tors}} = \mathrm{Pic}_{X/k}(k)/\mathrm{Pic}^\tau_{X/k}(k).$$

Proof The class $[\mathcal{L}]$ lies in $\mathrm{NS}(X)_{\mathrm{tors}}$ if and only if

$$n[\mathcal{L}] = [\mathcal{L}^{\otimes n}] = [O_X] \quad \text{for some integer } n \neq 0.$$

As $\mathrm{NS}(X)$ is the component group of $\mathrm{Pic}_{X/k}$, this means that $\mathcal{L}^{\otimes n} \in \mathrm{Pic}^0_{X/k}$, so $\mathcal{L} \in \mathrm{Pic}^{\tau}_{X/k}$. Hence the identification of $\mathrm{NS}(X)_{\mathrm{tors}}$. That this is finite then follows from Theorem 5.4.4: $\mathrm{NS}^{\tau}_{X/k}$ is a zero-dimensional scheme of finite type over k, so it has only finitely many k-points.

As for $\mathrm{NS}(X)_{\mathrm{tf}}$, consider the short exact sequence of group schemes

$$0 \to \mathrm{NS}^{\tau}_{X/k} \to \mathrm{NS}_{X/k} \to \mathrm{NS}_{X/k}/\mathrm{NS}^{\tau}_{X/k} \to 0.$$

The preceding discussion shows that $\mathrm{NS}_{X/k}/\mathrm{NS}^{\tau}_{X/k} \cong \mathrm{Pic}_{X/k}/\mathrm{Pic}^{\tau}_{X/k}$ and taking points identifies the torsion-free quotient. □

Theorem 5.7.3 (Theorem of the Base, smooth projective case) *Let X be a smooth projective scheme over an algebraically closed field k. Then*

(i) $\mathrm{NS}(X)$ *is a finitely generated abelian group, and*
(ii) $\rho(X) \leq \dim_F(H^2(X))$ *for any Weil cohomology theory H^* over k.*

Proof We have $X = \coprod_i X_i$ with X_i the connected components of X, and $\mathrm{Pic}_{X/k} = \prod_i \mathrm{Pic}_{X_i/k}$, so $NS(X) = \bigoplus_i NS(X_i)$. Since also $H^2(X) = \bigoplus_i H^2(X_i)$, it suffices to prove the theorem for each X_i. Thus we may assume X is a smooth projective variety. The short exact sequence

$$0 \to \mathrm{NS}(X)_{\mathrm{tors}} \to \mathrm{NS}(X) \to \mathrm{NS}(X)_{\mathrm{tf}} \to 0$$

implies that it suffices to show that $\mathrm{NS}(X)_{\mathrm{tors}}$ is finite and that $\mathrm{NS}(X)_{\mathrm{tf}}$ is of finite rank. Lemma 5.7.2 already gives finiteness of the torsion; it moreover identifies the quotient as the numerical group of invertible modules, giving the first isomorphism in

$$\mathrm{NS}(X)_{\mathrm{tf}} \cong \mathrm{Pic}_{X/k}(k)/\mathrm{Pic}^{\tau}_{X/k}(k) \cong \mathrm{Num}^1(X).$$

Since X is smooth projective, the first Chern class identifies the numerical group of invertible modules with the numerical group $\mathrm{Num}^1(X)$ of divisors on X, as in Lemma 5.5.2. Then Proposition 5.5.5 shows that the rank of this is at most the dimension of $H^2(X)$ for any Weil cohomology theory H^* over k. □

We can now deduce the finite generation of $\mathrm{NS}(X)$ for any proper scheme over k by reducing to the smooth projective case above:

Theorem 5.7.4 (Theorem of the Base, proper case) *Let X be a proper scheme over an algebraically closed field k. Then* $\mathrm{NS}(X)$ *is a finitely generated abelian group.*

Proof As before, it suffices to show that $\mathrm{NS}(X)$ has finite torsion and that its torsion-free quotient is of finite rank. Finiteness of the torsion is again handled by Lemma 5.7.2. As for the torsion-free quotient, observe that if $f: Y \to X$ is a surjective morphism, then Lemma 5.2.9 says that an $\mathcal{O}_X$-module $\mathcal{L}$ is τ-trivial if and only if $f^*\mathcal{L}$ is τ-trivial; thus pullback induces an injective homomorphism $f^*: \mathrm{NS}(X)_{\mathrm{tf}} \to \mathrm{NS}(Y)_{\mathrm{tf}}$. If the latter is of finite rank, then so is the former. Thus to prove that X has finite Picard rank, we may replace X by schemes surjecting onto it. But, by Lemma 5.6.1, there is a surjective morphism $Y \to X$ with Y a smooth projective scheme over k. We conclude by Theorem 5.7.3. □

Now we consider how the Néron–Severi group varies in families. More precisely, consider a proper morphism $f: X \to S$. For each point $s \in S$, let $\kappa(s)$ be its residue field and let

$$\bar{s}: \mathrm{Spec}(\overline{\kappa(s)}) \to S$$

be a geometric point lying above s. Let $X_{\bar{s}} := X \times_S \bar{s}$ be the geometric fiber of f over s. Then the Theorem of the Base allows us to define numerical functions

$$\begin{aligned} \mathrm{tors}_{X/S}: S &\to \mathbf{Z} & \rho_{X/S}: S &\to \mathbf{Z} \\ s &\mapsto \#\mathrm{NS}(X_{\bar{s}})_{\mathrm{tors}}, & s &\mapsto \rho(X_{\bar{s}}), \end{aligned}$$

giving the torsion size and Picard rank of geometric fibers.

In order to study the function $\mathrm{tors}_{X/S}$, we will need to make use of the subfunctor $\mathrm{Pic}^{\tau}_{X/S} \subset \mathrm{Pic}_{X/S}$ consisting of sections $\xi \in \mathrm{Pic}_{X/S}(T)$ such that $\xi|_{\bar{t}} \in \mathrm{Pic}^{\tau}_{X_{\bar{t}}/\bar{t}}(\bar{t})$ for every geometric point $\bar{t}$ of T. Note that for a geometric point $\bar{s}$ of S, we have $(\mathrm{Pic}^{\tau}_{X/S})_{\bar{s}} = \mathrm{Pic}_{X_{\bar{s}}/\bar{s}}$.

Lemma 5.7.5 (Bounded torsion) *Let $f: X \to S$ be a proper morphism. If S is Noetherian, then the function* $\mathrm{tors}_{X/S}$ *is bounded.*

Proof By Noetherian induction, it suffices to find a nonempty open subset of S on which $\rho_{X/S}$ is bounded. Since $\mathrm{tors}_{X/S}$ concerns geometric fibers, we may replace S by its reduction and assume S is reduced. We may then replace S by an irreducible component since a nonempty open subset of an irreducible component contains a nonempty open subset of S. Thus we may assume S is integral.

For a point $s \in S$, consider the base change map $\varphi(s): (f_*\mathcal{O}_X)_s \otimes_{\mathcal{O}_{S,s}} \kappa(s) \to H^0(X_s, \mathcal{O}_{X_s})$. When s is the generic point of S this map is an isomorphism by [19, Lemma 02KH], since $\mathrm{Spec}(\kappa(s)) \to S$ is flat. Therefore, there exists a nonempty open subset of S consisting of points

$s \in S$ such that $\varphi(s)$ is an isomorphism, and, after replacing S with this open subset, the formation of $f_*\mathcal{O}_X$ commutes with arbitrary base change by [7, Section 7.7]. Therefore by Theorem 5.1.1, we may assume that $\mathrm{Pic}_{X/S}$ is representable by an algebraic space. Then [1, Exposé XIII, Théorème 4.7] shows that $\mathrm{Pic}^{\tau}_{X/S}$ is an algebraic space of finite type over S. Since the generic fiber of $\mathrm{Pic}^{\tau}_{X/S} \to S$ is a scheme that is a finite type group algebraic space over a field, by [19, Lemma 07SR] we may assume, after replacing S by a nonempty open subset, that $\mathrm{Pic}^{\tau}_{X/S}$ is a scheme. For every $s \in S$,

$$\mathrm{tors}_{X/S}(s) = \#(\text{connected components of } (\mathrm{Pic}^{\tau}_{X/S})_{\bar{s}}).$$

There is a nonempty open subset of S on which the number of connected components of fibers is constant, see [19, Lemma 055H]; replacing S by such an open subset completes the proof. □

To bound Picard ranks in families, we assume that there exists a Weil cohomology theory that varies in families. That is, suppose we have the following situation. Fix a coefficient field F; for every scheme S and every geometric point

$$\bar{s}\colon \mathrm{Spec}(\overline{\kappa(s)}) \to S,$$

we have a Weil cohomology theory $H^*_{\bar{s}}$ over $\overline{\kappa(s)}$ with coefficients in F. The important hypothesis is as follows: for every smooth projective morphism $X \to S$, the function $s \mapsto \dim_F(H^2_{\bar{s}}(X_{\bar{s}}))$ is constructible on S.

Such a theory exists: for example, one may take ℓ-adic étale cohomology for suitable ℓ. The required constructibility result is then the Théorème de Finitude et Spécialisations of [2, Exposé XVI, Corollaire 2.2]. See also [19, Proposition 0GLI].

Lemma 5.7.6 (Bounded ranks) *Assume that there exists a Weil cohomology theory that varies in families as above. Let $f\colon X \to S$ be a proper morphism of schemes. If S is Noetherian, then $\rho_{X/S}$ is bounded.*

Proof By Noetherian induction, it suffices to find a nonempty open subset of S on which $\rho_{X/S}$ is bounded. Since $\rho_{X/S}$ concerns geometric fibers, we may replace S by its reduction to assume S is reduced. We may then replace S by an irreducible component since a nonempty open subset of an irreducible component contains a nonempty open subset of S. Thus we assume S is integral.

By the alterations theorem in families from Lemma 5.6.3, after replacing S by a nonempty open subset we may assume that there exist a finite surjective morphism $S' \to S$, a smooth projective morphism $X' \to S'$, and a surjective morphism $X' \to X \times_S S'$. Since $S' \to S$ is surjective and $X \to S$ and $X \times_S S' \to S'$ have the same geometric fibers, in order to show that $\rho_{X/S}$ is bounded it suffices to show that $\rho_{X \times_S S'/S'}$ is bounded. Thus we may replace S by S' and assume there exists a surjective morphism $X' \to X$ with X' smooth and projective over S. But then as observed in the proof of the Theorem of the Base, Lemma 5.2.9 implies that $\rho_{X/S} \leq \rho_{X'/S}$, so we may replace X with X'.

Thus we are in the situation where $X \to S$ is a smooth projective morphism with S an integral Noetherian scheme. The Theorem of the Base in the projective case, Theorem 5.7.3, implies that for, every $s \in S$, $\rho_{X/S}(s) \leq \dim_F(H^2_{\bar{s}}(X_{\bar{s}}))$. Our hypothesis about the Weil cohomology theory in families is that the latter function is constructible on S and so, since S is Noetherian, it is actually uniformly bounded on S. Therefore $\rho_{X/S}$ is also bounded on S. □

Putting the results of this section together yields the following result, originally due to [1, Exposé XIII, Théorème 5.1]:

Theorem 5.7.7 (Boundedness of the Néron–Severi group in families) *Let $X \to S$ be a proper morphism. If S is Noetherian, then* $\mathrm{NS}(X_{\bar{s}})$ *is a finitely generated abelian group for every geometric point $\bar{s}$ of S. Moreover, the order of torsion and ranks of these groups are bounded over S.*

Proof This follows from Theorem 5.7.4 together with Lemmas 5.7.5 and 5.7.6. □

5.8 Examples of Picard schemes

We close this chapter by giving three examples of Picard schemes.

Example 5.8.1 (Picard schemes of quotients) The Picard schemes of quotients by finite commutative group schemes G often contain the Cartier dual $G^\vee$; see [8]. To explain this, let S be a scheme and G a finite locally free commutative group scheme over S. The presheaf of abelian groups on the category Sch/S of schemes over S, given by

$$G^\vee : T \mapsto \mathrm{Hom}_T(G_T, (\mathbf{G}_m)_T),$$

the right-hand side being homomorphisms in the category of group schemes over T, is representable by a finite locally free commutative group scheme over S called the *Cartier dual of* G. The functor $G \mapsto G^\vee$ defines a duality on the category of finite locally free commutative group schemes over S, in that $G \mapsto (G^\vee)^\vee$ is isomorphic to the identity functor.

The Cartier dual often appears in the Picard scheme of a quotient:

Lemma 5.8.2 *Let S be a scheme, G a finite locally free commutative group scheme over S, and $\pi\colon Y \to X$ a G-torsor with X and Y algebraic spaces over S. Assume that $(Y \to S)_* \mathcal{O}_Y = \mathcal{O}_S$ holds universally. Then there is an exact sequence*

$$0 \to G^\vee \to \mathrm{Pic}_{X/S} \xrightarrow{\pi^*} \mathrm{Pic}_{Y/S}$$

of sheaves on $(\mathrm{Sch}/S)_{\mathrm{fppf}}$.

Proof It suffices to see that, for every $T \to S$, there is an exact sequence

$$0 \to G^\vee(T) \to \mathrm{Pic}(X_T) \to \mathrm{Pic}(Y_T).$$

By descent theory, the kernel of $\mathrm{Pic}(X_T) \to \mathrm{Pic}(Y_T)$ can be identified with the set of G_T-equivariant structures on $\mathcal{O}_{Y_T}$; see [19, Lemma 043U]. By definition, this is an isomorphism $\mathcal{O}_{G_T \times_T X_T} \to \mathcal{O}_{G_T \times_T X_T}$ satisfying two conditions. Such an isomorphism is given by a section of

$$H^0(G_T \times_T X_T, \mathcal{O}_{G_T \times_T X_T})^\times = H^0(G_T, \mathcal{O}_{G_T})^\times,$$

by the assumption on the pushforward of the structure sheaf. This is also the same as a morphism $G_T \to (\mathbf{G}_m)_T$ of schemes over T. The two conditions in the definition of an equivariant structure say exactly that this morphism is a homomorphism of group schemes. Thus the kernel is canonically identified with $G^\vee(T)$. □

Now for a particular example. Take $S := \mathrm{Spec}(k)$, for k an algebraically closed field of characteristic $p > 0$, and let G be the constant group scheme $\mathbf{Z}/p$. Then $G^\vee = \mu_p$. Fix $d \geq 2$. Then there exists a smooth complete intersection Y of dimension d in some projective space that carries a free G action; see [16, Proposition 15]. Then the fppf quotient sheaf $X := Y/G$ is a scheme and the quotient map $Y \to X$ is a G-torsor in the fppf topology; see [19, Lemma 07S7]. It follows from [19, Lemma 0BBM] that X is separated. Since $Y \to X$ is étale and surjective and Y is smooth over k, X is smooth over k by [19, Lemma 02K5]. Since $Y \to X$ is surjective, and Y is proper and irreducible, so is X; see [19, Lemma 03GN]. Thus Y is a smooth proper variety over k. Since the group scheme μ_p is connected, the exact sequence of Lemma 5.8.2 gives an exact sequence

$$0 \to \mu_p \to \mathrm{Pic}^0_{X/k} \to \mathrm{Pic}^0_{Y/k}.$$

On the one hand, $H^1(Y, \mathcal{O}_Y) = 0$ since Y is a complete intersection of dimension $d \geq 2$. On the other hand, the same group is $\mathrm{Ext}^1_Y(\mathcal{O}_Y, \mathcal{O}_Y)$ and hence is the tangent space to the Picard scheme at the identity; see [19, Lemma 08VW]. Thus $\mathrm{Pic}^0_{Y/k} = 0$ and so $\mathrm{Pic}^0_{X/k} = \mu_p$. In particular, the Picard scheme of the smooth proper variety X is nonreduced. □

Example 5.8.3 (Pointless conic) We will give an example of a section of the Picard functor that cannot be represented by a line bundle. Let

$$X := \mathrm{Proj}(\mathbf{R}[x, y, z]/(x^2 + y^2 + z^2)) \subset \mathbf{P}^2_{\mathbf{R}}.$$

Then X is a smooth curve over $\mathbf{R}$ such that $X(\mathbf{R}) = \emptyset$ and $X_{\mathbf{C}} \cong \mathbf{P}^1_{\mathbf{C}}$. By [19, Lemma 0D27], there is an exact sequence

$$0 \to \mathrm{Pic}(\mathrm{Spec}(\mathbf{R})) \to \mathrm{Pic}(X) \to \mathrm{Pic}_{X/\mathbf{R}}(\mathbf{R}).$$

We will show that $\mathrm{Pic}(X) \to \mathrm{Pic}_{X/\mathbf{R}}(\mathbf{R})$ is not surjective.

First, observe that $\mathrm{Pic}_{X/\mathbf{R}}(\mathbf{R}) \to \mathrm{Pic}_{X/\mathbf{R}}(\mathbf{C})$ is an isomorphism. It is injective since $\mathrm{Spec}(\mathbf{C}) \to \mathrm{Spec}(\mathbf{R})$ is a covering for the fppf (and even étale) topology. For the surjectivity, note that, by [19, Lemmas 0D28 and 0BXJ],

$$\mathrm{Pic}_{X/\mathbf{R}}(\mathbf{C}) = \mathrm{Pic}_{X_{\mathbf{C}}/\mathbf{C}}(\mathbf{C}) = \mathrm{Pic}(\mathbf{P}^1_{\mathbf{C}}) = \mathbf{Z} \cdot \mathcal{O}_{\mathbf{P}^1_{\mathbf{C}}}(1) =: \mathbf{Z} \cdot \mathcal{O}(1).$$

The class of $\mathcal{O}(1)$ descends along the étale covering $\mathrm{Spec}(\mathbf{C}) \to \mathrm{Spec}(\mathbf{R})$. Indeed, its two pullbacks to $X_{\mathbf{C} \otimes_{\mathbf{R}} \mathbf{C}} \cong \mathbf{P}^1_{\mathbf{C}} \amalg \mathbf{P}^1_{\mathbf{C}}$ must give $\mathcal{O}(1)$ on each copy of $\mathbf{P}^1_{\mathbf{C}}$, since for an isomorphism $f: \mathbf{P}^1_{\mathbf{C}} \to \mathbf{P}^1_{\mathbf{C}}$ of schemes, not necessarily over $\mathbf{C}$, $f^*\mathcal{O}(1) \cong \mathcal{O}(1)$: the pullback must generate the Picard group and, among the two possibilities, only one of them has a nonzero global section.

Thus to show that $\mathrm{Pic}(X) \to \mathrm{Pic}_{X/\mathbf{R}}(\mathbf{R})$ is not surjective, we just have to show that $\mathcal{O}(1)$ is not the pullback of a line bundle on X. If it were, then X would have a degree-1 line bundle $\mathcal{L}$ with a nonzero section, as

$$H^0(X, \mathcal{L}) \otimes_{\mathbf{R}} \mathbf{C} \cong H^0(\mathbf{P}^1_{\mathbf{C}}, \mathcal{O}(1)) \neq 0.$$

But then the zero locus of any nonzero section of $\mathcal{L}$ is an $\mathbf{R}$-rational point, contradicting the fact that $X(\mathbf{R}) = \emptyset$. □

Example 5.8.4 (Nodal curves) Let X be a nodal curve over an algebraically closed field k; see [19, Section 0C47]. Let $p_1, \ldots, p_n \in X$ be the nodes of X and let $\nu\colon X^\nu \to X$ be its normalization. We will show that there is a short exact sequence of group schemes

$$0 \to \mathbf{G}_m^n \to \mathrm{Pic}_{X/k} \to \mathrm{Pic}_{X^\nu/k} \to 0.$$

This shows that if X is not smooth, then $\mathrm{Pic}^0_{X/k}$ is not proper, since it contains $\mathbf{G}_m^n$ as a closed subgroup scheme with $n \geq 1$.

The normalization morphism gives a short exact sequence

$$0 \to \mathcal{O}_X \xrightarrow{\nu^\#} \nu_*(\mathcal{O}_{X^\nu}) \to \bigoplus_{i=1}^n \mathcal{O}_{p_i} \to 0$$

of coherent sheaves on X. Indeed, the cokernel consists of skyscraper sheaves supported on the p_i since there are exactly two points in X^ν lying above each p_i; see [19, Lemma 0CBW].

For any flat morphism $Y \to X$, pullback yields a short exact sequence

$$0 \to \mathcal{O}_Y \to \nu_{Y,*}(\mathcal{O}_{X^\nu \times_X Y}) \to \bigoplus_{i=1}^n \mathcal{O}_{Y_{p_i}} \to 0,$$

which remains valid upon replacing Y by any object in the small étale site $Y_{\text{ét}}$ of Y, since such objects are flat over Y. Thus upon taking units we obtain a short exact sequence

$$0 \to \mathbf{G}_m \to \nu_{Y,*}\mathbf{G}_{m,X^\nu \times_X Y} \to \bigoplus_{i=1}^n (Y_{p_i} \to Y)_* \mathbf{G}_{m,Y_{p_i}} \to 0$$

of sheaves on $Y_{\text{ét}}$. Since $\nu_Y : X^\nu \times_X Y \to Y$ and $Y_{p_i} \to Y$ are finite, the long exact sequence of cohomology gives

$$0 \to H^0(Y, \mathcal{O}_Y)^\times \to H^0(X^\nu \times_X Y, \mathcal{O}_{X^\nu \times_X Y})^\times \to \bigoplus_{i=1}^n H^0(Y_{p_i}, \mathcal{O}_{Y_{p_i}})^\times$$

$$\to \mathrm{Pic}(Y) \to \mathrm{Pic}(X^\nu \times_X Y) \to \bigoplus_{i=1}^n \mathrm{Pic}(Y_{p_i}) \to \cdots.$$

Now take $Y = X_T$ for T an object in $(\mathrm{Sch}/k)_{\mathrm{fppf}}$. Since $H^0(X, \mathcal{O}_X) = H^0(X^\nu, \mathcal{O}_{X^\nu}) = k$,

$$0 \to H^0(T, \mathcal{O}_T)^\times = H^0(T, \mathcal{O}_T)^\times \xrightarrow{0} \bigoplus_{i=1}^n H^0(T, \mathcal{O}_T)^\times$$

$$\to \mathrm{Pic}(X_T) \to \mathrm{Pic}((X^\nu)_T) \to \bigoplus_{i=1}^n \mathrm{Pic}(T) \to \cdots.$$

Furthermore, elements of $\mathrm{Pic}(T)$ vanish Zariski-locally on T, and therefore elements of $\mathrm{Pic}((X^\nu)_T)$ Zariski-locally on T come from $\mathrm{Pic}(X_T)$. Since we obtain $\mathrm{Pic}_{X/k}$ by sheafifying the rule $T \mapsto \mathrm{Pic}(X_T)$ and similarly for X^ν, we see that we have obtained a short exact sequence

$$0 \to \mathbf{G}_m^n \to \mathrm{Pic}_{X/k} \to \mathrm{Pic}_{X^\nu/k} \to 0,$$

as promised.

Acknowledgements This chapter is the result of a project led by Martin Olsson during the **S**tacks **P**roject **ON**line **G**eometry **E**vent, held during the first week of August in 2020. Much thanks goes to Martin for setting the topic and for the hours of helpful conversation during the workshop. We also thank Remy van Dobben de Bruyn, who was a member of our group during the workshop, for his contributions. Finally, thanks to the organizers for putting together an excellent event in such strange times.

References

[1] *Théorie des intersections et théorème de Riemann–Roch*, volume 225 of *Lecture Notes in Mathematics*. Springer, 1971. Séminaire de Géométrie Algébrique du Bois-Marie 1966–1967 (SGA 6), Dirigé par P. Berthelot, A. Grothendieck et L. Illusie. Avec la collaboration de D. Ferrand, J. P. Jouanolou, O. Jussila, S. Kleiman, M. Raynaud et J. P. Serre.

[2] *Théorie des topos et cohomologie étale des schémas. Tome 3*, volume 305 of *Lecture Notes in Mathematics*. Springer, 1973. Séminaire de Géométrie Algébrique du Bois-Marie 1963–1964 (SGA 4), Dirigé par M. Artin, A. Grothendieck et J. L. Verdier. Avec la collaboration de P. Deligne et B. Saint-Donat.

[3] M. Artin. Algebraization of formal moduli. I. In *Global analysis (Papers in honor of K. Kodaira)*, pages 21–71. University of Tokyo Press, 1969.

[4] S. Bosch, W. Lütkebohmert, and M. Raynaud. *Néron models*, volume 21 of *Ergebnisse der Mathematik und ihrer Grenzgebiete. 3. Folge*. Springer, 1990. `doi:10.1007/978-3-642-51438-8`.

[5] A. J. de Jong. Smoothness, semi-stability and alterations. *Inst. Hautes Études Sci. Publ. Math.*, 83:51–93, 1996.

[6] B. Fantechi, L. Göttsche, L. Illusie, S. L. Kleiman, N. Nitsure, and A. Vistoli. *Fundamental algebraic geometry*, volume 123 of *Mathematical Surveys and Monographs*. American Mathematical Society, 2005. Grothendieck's FGA explained. `doi:10.1090/surv/123`.

[7] A. Grothendieck. Éléments de géométrie algébrique. III. Étude cohomologique des faisceaux cohérents. II. *Inst. Hautes Études Sci. Publ. Math.*, 17:91, 1963.

[8] S. T. Jensen. Picard schemes of quotients by finite commutative group schemes. *Math. Scand.*, 42(2):197–210, 1978. `doi:10.7146/math.scand.a-11748`.

[9] S. L. Kleiman. Toward a numerical theory of ampleness. *Ann. Math. (2)*, 84:293–344, 1966. `doi:10.2307/1970447`.

[10] S. Lang and A. Néron. Rational points of abelian varieties over function fields. *Amer. J. Math.*, 81:95–118, 1959. `doi:10.2307/2372851`.

[11] R. Lazarsfeld. *Positivity in algebraic geometry. I*, volume 48 of *Ergebnisse der Mathematik und ihrer Grenzgebiete. 3. Folge*. Springer, 2004. `doi:10.1007/978-3-642-18808-4`.

[12] T. Matsusaka. The criteria for algebraic equivalence and the torsion group. *Amer. J. Math.*, 79:53–66, 1957. doi:10.2307/2372383.

[13] D. Mumford. *Lectures on curves on an algebraic surface*, volume 59 of *Annals of Mathematics Studies*. Princeton University Press, 1966. With a section by G. M. Bergman.

[14] A. Néron. Problèmes arithmétiques et géométriques rattachés à la notion de rang d'une courbe algébrique dans un corps. *Bull. Soc. Math. France*, 80:101–166, 1952.

[15] E. Picard. Sur quelques questions se rattachant à la connexion linéaire dans la théorie des fonctions algébriques de deux variables indépendantes. *J. Reine Angew. Math.*, 129:275–286, 1905. doi:10.1515/crll.1905.129.275.

[16] J.-P. Serre. Sur la topologie des variétés algébriques en caractéristique p. In *Symposium internacional de topología algebraica International symposium on algebraic topology*, pages 24–53. Universidad Nacional Autónoma de México and UNESCO, Mexico City, 1958.

[17] F. Severi. Sulla totalità delle curve algebriche tracciate sopra una superficie algebrica. *Math. Ann.*, 62(2):194–225, 1906. doi:10.1007/BF01449978.

[18] F. Severi. La base per le varietà algebriche di dimensione qualunque contenute in una data e la teoria generale delle corrispondenze fra punti di due superficie algebriche. *Mem. Accad. Ital.*, 5(6), 1934.

[19] The Stacks Project Authors. *The Stacks project*. https://stacks.math.columbia.edu.

6

Weil restriction for schemes and beyond

Lena Ji
University of Michigan

Shizhang Li
University of Michigan

Patrick McFaddin
Fordham University

Drew Moore
University of Chicago

Matthew Stevenson
Google

Abstract The aim of this note is to discuss the Weil restriction of schemes and algebraic spaces, highlighting pathological phenomena that appear in the theory and are not widely known. It is shown that the Weil restriction of a locally finite algebraic space along a finite flat morphism is an algebraic space.

6.1 Weil restriction of schemes

Briefly, the Weil restriction is the algebro-geometric analogue of viewing a complex-analytic manifold as a real-analytic manifold of twice the dimension. Often referred to as "restriction of scalars," this construction is adjoint to base extension or "extension of scalars," and may be phrased quite generally as the pushforward of a functor along a morphism.

The Weil restriction has proven to be an extremely interesting construction, utilized by a variety of researchers in algebra, algebraic geometry, and number theory. Specific analyses may be found in, e.g., representation theory [6, 14], Galois theory and arithmetic geometry [2, 4, 9, 19], arithmetic dynamics [10], (noncommutative) motives [11, 17], as well as generalizations to Quot schemes [15] and stacks [8]. Let us also highlight the influential work of Grothendieck, who used the Hilbert scheme to give a construction of the Weil restriction [7, 4c]. Namely, he observed that if $f\colon Z \to X$ is a morphism of S-schemes with $Z \to S$ quasi-projective and $X \to S$ flat and proper, then the pushforward map f_* to the Hilbert scheme of X is defined on an open subset of the Hilbert scheme of Z, and the Weil restrictions are realized by the fibers of f_*.

In this chapter, we give some background on the Weil restriction of schemes before moving on to algebraic spaces. We show that the Weil restriction of a locally finite algebraic space along a finite flat morphism is an algebraic space.

Notation 6.1.1 Throughout, we let S be a scheme. Given two S-schemes X and T, we use $X(T)$ to denote the collection $\mathrm{Hom}_S(T, X)$ of S-morphisms $T \to X$.

Let us begin with the definition of the Weil restriction of schemes before introducing the general definition of the restriction of a functor along a morphism.

Definition 6.1.2 Let $S' \to S$ be a morphism of schemes. Given an S'-scheme X', we denote by $\mathrm{R}_{S'/S}(X')\colon (\mathrm{Sch}/S)^{\mathrm{opp}} \to (\mathrm{Sets})$ the contravariant functor given by

$$T \mapsto X'(T \times_S S').$$

If the functor $\mathrm{R}_{S'/S}(X')$ is representable by an S-scheme X, then we say that X is the *Weil restriction of X' along $S' \to S$.*

Remark 6.1.3 Given any morphism of schemes $f\colon S' \to S$ and any contravariant functor $F\colon (\mathrm{Sch}/S')^{\mathrm{opp}} \to (\mathrm{Sets})$, one can define the pushforward functor $f_*F\colon (\mathrm{Sch}/S)^{\mathrm{opp}} \to (\mathrm{Sets})$ given on objects by $T \mapsto F(T \times_S S')$. If F is representable by an S'-scheme X', then $f_*F = \mathrm{R}_{S'/S}(X')$ is given by the Weil restriction. This approach is later used to define the Weil restriction of algebraic spaces.

Theorem 6.1.4 *Let $S' \to S$ be a finite flat morphism. Let X' be an S'-scheme such that, for any $s \in S$ and any finite set $P \subseteq X' \times_S \mathrm{Spec}(\kappa(s))$, there exists an affine open subscheme $U' \subseteq X'$ containing P. Then, the functor $\mathrm{R}_{S'/S}(X')$ is representable by an S-scheme. In particular, if X' is a quasi-projective S'-scheme, then the Weil restriction of X' exists.*

Proof Omitted. See [19, Theorem 7.6/4]. □

In Proposition 6.2.10, we will show that $\mathrm{R}_{S'/S}(X')$ is quasi-projective over S when X' is quasi-projective over S', where "quasi-projective" is defined as in [5, II, Définition 5.3.1]. In [18, Section 1.3], Weil introduced his version of the Weil restriction of a quasi-projective scheme over a field. His construction is presented below and it is shown to coincide with the modern definition in the case of quasi-projective schemes over a field.

Theorem 6.1.5 *Let k'/k be a finite separable extension, let X' be a quasi-projective k'-scheme, and let F/k be a finite Galois extension that splits k'/k. Let*

$$\overline{X} := \prod_{j\colon k' \hookrightarrow F} X' \times_{k',j} F,$$

where the product runs over all embeddings $k' \hookrightarrow F$ over k. Then, for $\sigma \in \mathrm{Gal}(F/k)$, there exists an isomorphism $\varphi_\sigma \colon \overline{X} \simeq \overline{X}$ over $\mathrm{Spec}(\sigma)$, such that $\left(\overline{X}, \{\varphi_\sigma\}_{\sigma \in \mathrm{Gal}(F/k)}\right)$ is an effective descent data, giving the k-scheme $\mathrm{R}_{k'/k}(X')$.

Proof Let J denote the set of embeddings $j\colon k' \hookrightarrow F$ over k and let $G = \mathrm{Gal}(F/k)$. If $j \in J$, let F_j denote the field F viewed as a k'-algebra via the embedding j, and let $X'_j = X' \times_{k'} F_j$. If $j \in J$ and $\sigma \in G$, then $\sigma \circ j \in J$, and this induces an isomorphism $X'_{\sigma \circ j} \simeq X'_j$ over $\mathrm{Spec}(\sigma)$. Taking the product over all $j \in J$ yields an isomorphism $\varphi_\sigma \colon \overline{X} \simeq \overline{X}$ over $\mathrm{Spec}(\sigma)$. For any $\sigma, \tau \in G$, it is clear that $\varphi_\sigma \circ \varphi_\tau = \varphi_{\sigma \circ \tau}$, so $\left(\overline{X}, \{\varphi_\sigma\}_{\sigma \in G}\right)$ is a descent datum.

Let $X := \mathrm{R}_{k'/k}(X')$. The F-scheme $X \times_k F$ has the canonical descent datum $\{\phi_\sigma\}_{\sigma \in G}$, where ϕ_σ is the automorphism of $X \times_k F$ over $\mathrm{Spec}(\sigma)$ given by $1 \times \mathrm{Spec}(\sigma)$. It suffices to construct an F-isomorphism $\psi \colon X \times_k F \simeq \overline{X}$ such that, for any $\sigma \in G$, $\psi \circ \phi_\sigma \circ \psi^{-1} = \varphi_\sigma$ (i.e. under the isomorphism ψ, the descent data $\{\phi_\sigma\}$ is sent to $\{\varphi_\sigma\}$).

To construct the isomorphism ψ, consider

$$\begin{aligned} X \times_k F &\simeq \mathrm{R}_{(k' \otimes_k F)/F} \left(X' \times_{k'} (k' \otimes_k F)\right) \\ &\simeq \mathrm{R}_{(\bigsqcup_{j \in J} F_j)/F} \left(\bigsqcup_{j \in J} X' \times_{k'} F_j\right) \\ &\simeq \prod_{j \in J} X' \times_{k'} F_j \\ &= \overline{X}, \end{aligned}$$

where the first isomorphism follows from Proposition 6.2.2 (ii) below, and the third isomorphism follows from Lemma 6.2.3. One can verify that the canonical descent datum $\{\phi_\sigma\}_{\sigma \in G}$ is carried to the descent datum $\{\varphi_\sigma\}_{\sigma \in G}$ under this isomorphism. □

6.2 Basic properties of Weil restriction

Proposition 6.2.1 *Let $S' \to S$ be a morphism of schemes, let X be an S-scheme, and let $X' = X \times_S S'$. Then we have*

$$\mathrm{R}_{S'/S}(X') = \underline{\mathrm{Hom}}_S(S', X).$$

Proof Recall that the right-hand side is defined as the functor which sends an S-scheme T to $\underline{\mathrm{Hom}}_S(S', X)(T) := \mathrm{Hom}_S(T \times_S S', X)$. The equality follows from the universal property of Weil restriction and the definition of the right-hand side. □

Proposition 6.2.2 *Let $S' \to S$ be a finite flat morphism of schemes and let X' be an S'-scheme such that the Weil restriction $\mathrm{R}_{S'/S}(X')$ exists as an S-scheme.*

(i) *If X', S', and S are affine with X' of finite type over S', then $\mathrm{R}_{S'/S}(X')$ is affine of finite type over S.*
(ii) *If T is an S-scheme and $T' = T \times_S S'$, then there is an isomorphism*

$$\mathrm{R}_{T'/T}(X' \times_{S'} T') \simeq \mathrm{R}_{S'/S}(X') \times_S T.$$

of functors on (Sch/T).
(iii) *If $X' \to Z'$ and $Y' \to Z'$ are morphisms of S'-schemes, then there is an isomorphism*

$$\mathrm{R}_{S'/S}(X' \times_{Z'} Y') \simeq \mathrm{R}_{S'/S}(X') \times_{\mathrm{R}_{S'/S}(Z')} \mathrm{R}_{S'/S}(Y')$$

of functors on (Sch/S).

Proof Omitted. See the proof of [3, Proposition A.5.2]. □

Lemma 6.2.3 *Let S be a scheme. Given a finite collection $\{S_i\}_{i \in I}$ of S-schemes, consider the S-scheme $S' = \bigsqcup_{i \in I} S_i \to S$. Given a collection, for each $i \in I$, of S_i-schemes X_i, consider the S'-scheme*

$$X' = \bigsqcup_{i \in I} X_i.$$

Then, there is an isomorphism

$$\mathrm{R}_{S'/S}(X') \simeq \prod_{i \in I} \mathrm{R}_{S_i/S}(X_i)$$

of functors on (Sch/S).

Proof For any S-scheme Y, observe that

$$\mathrm{Hom}_S(Y, \mathrm{R}_{S'/S}(X')) \simeq \mathrm{Hom}_{S'}(Y \times_S S', X')$$

$$\simeq \prod_{i\in I} \mathrm{Hom}_{S_i}(Y \times_S S_i, X_i)$$
$$\simeq \prod_{i\in I} \mathrm{Hom}_S(Y, \mathrm{R}_{S_i/S}(X_i))$$
$$\simeq \mathrm{Hom}_S\left(Y, \prod_{i\in I} \mathrm{R}_{S_i/S}(X_i)\right).$$

□

Proposition 6.2.4 *Let k be a field and let k' be a nonzero, finite, reduced k-algebra. If G' is a k'-group scheme of finite type, then $\mathrm{R}_{k'/k}(G')$ exists as a k-scheme and is a k-group scheme of finite type.*

Proof Omitted. See [3, Proposition A.5.1]. □

Proposition 6.2.5 *Let $S' \to S$ be a finite flat morphism of schemes. Assume that either S is locally Noetherian or $S' \to S$ is étale. Let X' be an S'-scheme such that the Weil restriction $\mathrm{R}_{S'/S}(X')$ exists as an S-scheme. If $X' \to S'$ is quasi-compact, then so is $\mathrm{R}_{S'/S}(X') \to S$.*

Proof Omitted. See [19, Proposition 7.6/5(a)]. □

Proposition 6.2.6 *Let $S' \to S$ be a finite flat morphism of schemes. Let* P *be one of the following properties of morphisms:*

(i) *monomorphism;*
(ii) *open immersion;*
(iii) *closed immersion;*
(iv) *separated.*

If $f' \colon X' \to Y'$ is a morphism of S'-schemes with the property P, *then the morphism $\mathrm{R}_{S'/S}(f') \colon \mathrm{R}_{S'/S}(X') \to \mathrm{R}_{S'/S}(Y')$ also has the property* P.

Moreover, if the Weil restrictions $\mathrm{R}_{S'/S}(X')$ and $\mathrm{R}_{S'/S}(Y')$ exist as S-schemes, then P *may be one of the following properties of morphisms:*

(v) *smooth;*
(vi) *étale;*
(vii) *locally of finite type;*
(viii) *locally of finite presentation;*
(ix) *finite presentation.*

Proof The condition (i) follows immediately from the definition of the Weil restriction, and (iv) follows immediately from (iii). For (ii) and (iii), see [19, Proposition 7.6/2]. For conditions (v)–(ix), we may assume that

S and S' are affine, by working Zariski-locally on S. For (v)–(vi), see the proof of [3, Proposition A.5.2(4)]. For (vii)–(ix), see [19, Proposition 7.6/5]. □

Proposition 6.2.7 *Let $S' \to S$ be a finite flat morphism of schemes. Let $f' \colon X' \to Y'$ be a smooth surjective S'-morphism between schemes locally of finite type over S', and assume that the Weil restrictions of X' and Y' exist as S-schemes. Then, $\mathrm{R}_{S'/S}(f') \colon \mathrm{R}_{S'/S}(X') \to \mathrm{R}_{S'/S}(Y')$ is smooth and surjective.*

Proof Omitted. See the proof of [3, Corollary A.5.4(1)]. □

Proposition 6.2.8 *Let $S' \to S$ be a finite flat morphism of Noetherian schemes that is surjective and radicial.*

(i) *If $X'_1, \ldots, X'_n$ are quasi-projective S'-schemes, then*

$$\mathrm{R}_{S'/S}\left(\bigsqcup_{i=1}^{n} X'_i\right) = \bigsqcup_{i=1}^{n} \mathrm{R}_{S'/S}(X'_i).$$

(ii) *If $\{U'_i\}$ is an open (resp. étale) cover of a quasi-projective S'-scheme X' then $\{\mathrm{R}_{S'/S}(U'_i)\}$ is an open (resp. étale) cover of $\mathrm{R}_{S'/S}(X')$.*

Proof For (i), denote $X' = \bigsqcup_{i=1}^{n} X'_i$, $X = \mathrm{R}_{S'/S}(X')$, and $X_i = \mathrm{R}_{S'/S}(X'_i)$ for each i. To show that the X_i are disjoint and that they cover X it suffices to check this on each fiber of $X \to S$, so by base changing to an algebraic closure of $\kappa(s)$ for each $s \in S$, we may assume that $S = \operatorname{Spec} k$ for an algebraically closed field k. Then $S' = \operatorname{Spec} B'$ where B' is a finite flat k-algebra and B' is local, by the radicial hypothesis. Each intersection is

$$X_i \cap X_j = X_i \times_X X_j \cong \mathrm{R}_{S'/S}(X'_i \times_{X'} X'_j) = \mathrm{R}_{S'/S}(\emptyset),$$

which is empty because $\mathrm{Hom}_k(Y, \mathrm{R}_{S'/S}(\emptyset)) = \mathrm{Hom}_{B'}(Y_{B'}, \emptyset)$ for any k-scheme Y (and since $B' \neq 0$ then $Y_{B'}$ is nonempty unless Y is the empty scheme). To show that the X_i cover X it suffices to show that the $X_i(k)$ cover $X(k)$. As B' is local, any morphism $\operatorname{Spec}(B') \to X' = \bigsqcup X'_i$ factors through some X'_i, so $X(k) = X'(B')$ is covered by the sets $X_i(k) = X'_i(B')$.

For (ii), let $\{U'_i\}_{i=1}^{n}$ be a finite subcover and let $U' := \bigsqcup_{i=1}^{n} U'_i$. Then, U' is quasi-projective over S' and it admits an étale surjection to X', and hence Proposition 6.2.7 implies, via (i), that $\mathrm{R}_{S'/S}(U') = \bigsqcup \mathrm{R}_{S'/S}(U'_i) \to \mathrm{R}_{S'/S}(X')$ is a smooth surjection.

Moreover, by Proposition 6.2.6, $\mathrm{R}_{S'/S}(U_i') \hookrightarrow \mathrm{R}_{S'/S}(X')$ is an open immersion. Thus, $\{\mathrm{R}_{S'/S}(U_i')\}$ is an open cover of $\mathrm{R}_{S'/S}(X')$.

□

Proposition 6.2.9 *Let $S' \to S$ be a finite étale morphism of schemes. If $f' : X' \to Y'$ is flat (resp. proper) and the Weil restrictions of X' and Y' exist as S-schemes, then $\mathrm{R}_{S'/S}(f') : \mathrm{R}_{S'/S}(X') \to \mathrm{R}_{S'/S}(Y')$ is flat (resp. proper).*

Proof Omitted. See [19, Proposition 7.6/5(f,g)]. □

Proposition 6.2.10 *Let $S' \to S$ be a finite flat morphism of Noetherian schemes. If X' is a quasi-projective S'-scheme, then $\mathrm{R}_{S'/S}(X')$ is a quasi-projective S-scheme.*

Proof Let $X = \mathrm{R}_{S'/S}(X')$. Let $g' : X' \to S'$ and $f : X \to S$ denote the structure morphisms of X' and X, respectively. By [16, Tag 01VW], it suffices to construct an f-relatively ample invertible $\mathcal{O}_X$-module. Let $\mathcal{N}'$ be a g'-relatively ample invertible $\mathcal{O}_{X'}$-module, and consider the adjunction morphism $q : X \times_S S' \to X'$ and the projection morphism $\pi : X \times_S S' \to X$. Note that π is finite and flat, and $\mathcal{L}' := q^*(\mathcal{N}')$ is an invertible sheaf on $X \times_S S'$. We claim that its norm $\mathcal{L} = \mathrm{Norm}_\pi(\mathcal{L}')$, as defined in [5, II, Section 6.5], is f-relatively ample on X.

Relative ampleness is local on the base, so in order to verify that $\mathcal{L}$ is f-relatively ample, we may assume that S is affine and $\mathcal{N}'$ is ample on X'. By [16, Tag 01PS], it suffices to show that, for every $x \in X$, there exist $n \geq 1$ and $s \in \Gamma(X, \mathcal{L}^{\otimes n})$ such that

$$X_s := \{z \in X : s(z) \notin \mathfrak{m}_z(\mathcal{L}^{\otimes n})_z\}$$

is an affine neighborhood of x.

Note that if $\{U_i'\}$ is an open cover of X' such that every finite subset of X' of bounded size is contained in some U_i', then the open sets $U_i := X \setminus \pi(q^{-1}(X' \setminus U_i'))$ form an open cover of X. Here, a subset of X' has bounded size if its image in S' lies in fibers of $S' \to S$ with maximal fiber degree. Indeed, for any $x \in X$, $E := \pi^{-1}(x)$ is such a finite subset of $X \times_S S'$, so $q(E) \subseteq U_i'$ for some index i. It follows that $E \cap q^{-1}(X' \setminus U_i') = \emptyset$, and hence $x \in U_i$.

For any $n \geq 1$, we have $\mathcal{L}^{\otimes n} = \mathrm{Norm}_\pi(\mathcal{L}'^{\otimes n})$. For any $s' \in \Gamma(X', \mathcal{N}'^{\otimes n})$, define $\mathrm{N}(s') \in \Gamma(X, \mathcal{L}^{\otimes n})$ to be the norm of the section $q^*(s') \in \Gamma(X_{S'}, \mathcal{L}'^{\otimes n})$. Since $\mathcal{N}'$ is ample, [16, Tag 09NV] asserts that there is a collection $\{s_j'\}$ of global sections of various powers $\mathcal{N}'^{\otimes n_j}$

of $\mathcal{N}'$, with $n_j \geq 1$, such that all loci $X'_{s'_j}$ are affine and cover X', and every finite subset of X' of bounded size is contained in some $X'_{s'_j}$. We claim that the subsets $\{X_{\mathrm{N}(s'_j)}\}$ form an open affine cover of X. By the above observation and Proposition 6.2.2, it suffices to show that

$$X_{\mathrm{N}(s')} = X \setminus \pi(q^{-1}(X' \setminus X'_{s'})) = \mathrm{R}_{S'/S}(X'_{s'}).$$

The first equality is an application of [5, II, Corollaire 6.5.7]. The second equality holds by functorial considerations. □

6.3 Geometric connectedness

Proposition 6.3.1 *Let k be a field and k' a finite local k-algebra with maximal ideal $\mathfrak{m}$ and residue field k. Assume that A is étale over $k[x_1, \ldots, x_n]$. Then, there is a noncanonical isomorphism*

$$\mathrm{R}_{k'/k}(\mathrm{Spec}(A \otimes_k k')) \simeq \mathrm{Spec}(A) \times \mathbb{A}^d_k$$

over $\mathrm{Spec}(A)$, *where* $d = (\dim_k(k') - 1) \cdot n$.

Remark 6.3.2 Let k, k' be as in Proposition 6.3.1, let X' be a scheme over k', and denote the special fiber by $X'_0 := X' \times_{\mathrm{Spec}(k')} \mathrm{Spec}(k)$. There is a natural morphism $q_{X'} \colon \mathrm{R}_{k'/k}(X') \to X'_0$ defined as follows: to a k-scheme T and a T-point $t \in \mathrm{R}_{k'/k}(X')(T) = X'(T \times_k k')$, we associate the reduction modulo $\mathfrak{m}$, i.e., the morphism

$$T = (T \times_k k') \times_{k'} (k'/\mathfrak{m}) \to X'_0.$$

Thus, $q_{X'}$ is surjecive when X' is k'-smooth, and, in the setting of Proposition 6.3.1, $\mathrm{R}_{k'/k}(\mathrm{Spec}(A \otimes_k k'))$ is *naturally* a $\mathrm{Spec}(A)$-scheme.

Proof By Proposition 6.2.1, it suffices to show that

$$\underline{\mathrm{Hom}}_{\mathrm{Spec}(k)}(\mathrm{Spec}(k'), \mathrm{Spec}(A)) = \mathrm{Spec}(A[x_1, \ldots, x_d]).$$

Consider the surjective multiplication map $A \otimes_k A \twoheadrightarrow A$ and denote the kernel by I. By our assumption, we have a noncanonical isomorphism

$$(A \otimes_k A)/I^m \simeq A[x_1, \ldots, x_n]/(x_i)^m$$

as A-algebras, for any $m \geq 1$. Let M denote the nilpotence order of the maximal ideal $\mathfrak{m} \subset k'$.

For every k-algebra B, an element

$$\phi \in \underline{\mathrm{Hom}}_{\mathrm{Spec}(k)}(\mathrm{Spec}(k'), \mathrm{Spec}(A))(B)$$

corresponds to a k-algebra homomorphism $\Phi\colon A \to k' \otimes_k B$.

Given Φ as above, we get an A-algebra structure on B via the composition

$$p_1^*\colon A \xrightarrow{\Phi} k' \otimes_k B \twoheadrightarrow B,$$

where the morphism $k' \otimes_k B \twoheadrightarrow B$ is induced from the surjection $k' \twoheadrightarrow k$. Similarly, we have the composition

$$p_1'^*\colon A \xrightarrow{p_1^*} B \to k' \otimes_k B,$$

where the second morphism is induced by the inclusion from $k \hookrightarrow k'$. Observe that $p_1'^*$ and Φ give rise to a k-algebra homomorphism

$$\Psi := (p_1'^* \otimes \Phi)\colon A \otimes_k A \to k' \otimes_k B.$$

This morphism has the property that the "first projection" (i.e., the precomposition of Ψ with the inclusion $A \xrightarrow{\mathrm{id}\otimes 1} A \otimes_k A$ into the first factor) agrees with $p_1'^*$, and the ideal $I \subseteq A \otimes_k A$ is mapped to $\mathfrak{m} \otimes B$ under Ψ. Given a pair $(p_1'^*, \Psi)$ as above, we can consider the "second projection," i.e., the composition

$$\Phi\colon A \xrightarrow{1\otimes \mathrm{id}} A \otimes_k A \xrightarrow{\Psi} k' \otimes_k B.$$

It is easy to check that these procedures invert one another, and they are functorial in B.

The discussion above shows that the datum of Φ is equivalent to the following data:

(i) a k-algebra homomorphism $p_1^*\colon A \to B$,
(ii) a k-algebra homomorphism $\Psi\colon A \otimes_k A \to k' \otimes_k B$

satisfying the two conditions

(i) the "first projection" $A \xrightarrow{\mathrm{id}\otimes 1} A \otimes_k A \xrightarrow{\Psi} k' \otimes_k B$ agrees with $p_1'^*$;
(ii) the ideal $I \subseteq A \otimes_k A$ is mapped to $\mathfrak{m} \otimes B$ under Ψ.

As $\mathfrak{m}$ has nilpotence order M, it follows that to give a k-algebra homomorphism Ψ is equivalent to giving a k-algebra homomorphism

$$\Psi'\colon (A \otimes_k A)/I^M \simeq A[x_1,\ldots,x_n]/(x_i)^M \to k' \otimes_k B$$

with the properties that

(i) the induced morphism $A \to A[x_1,\ldots,x_n] \to k' \otimes_k B$ agrees with $p_1'^*$;
(ii) the image of the $\overline{x_i}$ lands in $\mathfrak{m} \otimes B$.

Hence, the datum of Φ is functorially equivalent to a k-algebra homomorphism $A \to B$ along with n elements of $\mathfrak{m} \otimes_k B$. Therefore, by examining the coefficients of the n elements with respect to some chosen k-basis of $\mathfrak{m}$, we obtain an isomorphism

$$\underline{\mathrm{Hom}}_{\mathrm{Spec}(k)}(\mathrm{Spec}(k'), \mathrm{Spec}(A)) \to \mathrm{Spec}(A[x_1, \ldots, x_d]).$$

□

Lemma 6.3.3 *Let k be a field, let k' be a finite local k-algebra with maximal ideal $\mathfrak{m}$ and residue field k, and let X' be a smooth k'-scheme. Consider the surjective morphism $q_{X'}\colon \mathrm{R}_{k'/k}(X') \to X'_0$ as defined in Remark 6.3.2. The geometric fibers of $q_{X'}$ are affine spaces; in particular, the geometric fibers of $q_{X'}$ are connected.*

Proof The problem is local on the base, so we may assume that X' is affine and that $\Omega^1_{X'/k'}$ is free (the latter is equivalent to saying that $\Omega^1_{X'_0/k}$ is free, by Nakayama's lemma). By [1, Exposé III, Corollaire 6.8], there is a noncanonical k'-isomorphism $X' \simeq X'_0 \times_k k'$. The conclusion follows from Proposition 6.3.1. □

Lemma 6.3.4 *Let S be a scheme. Let $f\colon X \to Y$ be a morphism of algebraic spaces of finite type over S. Suppose that:*

(i) *the geometric fibers of f are connected;*
(ii) *étale-locally on Y, f admits a section;*
(iii) *Y is geometrically connected over S.*

Then, X is geometrically connected over S.

Remark 6.3.5 Below, we recall certain definitions on the (geometric) connectedness of algebraic spaces.

(i) Let X be an algebraic space locally of finite type over an algebraically closed field k, and denote by $|X|$ the topological space associated to X (as in [16, Tag 03BY]). We say X is *connected* if the associated topological space $|X|$ is connected. Note that, with our assumptions on X, the topological space $|X|$ is automatically locally connected.
(ii) A morphism of algebraic spaces is *geometrically connected* if all geometric fibers are nonempty and connected.

Lemma 6.3.6 *If X is a nonempty algebraic space locally of finite type over an algebraically closed field k, then X is connected if and only if, for any two points $x, x' \in |X|$, there is a finite chain $x_1, \ldots, x_n \in |X|$ of* *points and connected étale scheme neighborhoods U_{x_i} of x_i with $x_1 = x$ and $x_n = x'$ and such that $U_{x_i} \times_X U_{x_{i+1}} \neq \emptyset$ for all $i = 1, \dots, n-1$.*

Proof It is clear that the existence of such finite chains of points and neighborhoods as in the statement implies that $|X|$ is connected. Conversely, we can define an equivalence relation on $|X|$, where two points $x, x' \in |X|$ are equivalent if there is a finite chain of points and neighborhoods as in the statement. To each equivalence class $\widetilde{x}$, consider $W_{\widetilde{x}} = \bigcup_{x \in \widetilde{x}} W_x$, where W_x denotes the image of U_x in X. As subsets of $|X|$, the $W_{\widetilde{x}}$ are easily seen to be open and disjoint, hence closed. The connectedness of $|X|$ implies that any two points are equivalent, which completes the proof. □

Proof of Lemma 6.3.4 We may assume that S is the spectrum of an algebraically closed field, owing to the definition of geometric connectedness as in Remark 6.3.5 (ii). Given a point $x \in |X|$, there is a connected scheme $V_{f(x)}$, étale over Y, whose image contains $f(x)$ and for which the pullback morphism $U_x := X \times_Y V_{f(x)} \to V_{f(x)}$ admits a section (this is guaranteed by condition (ii) of Lemma 6.3.4). Now, condition (i) of Lemma 6.3.4 along with the existence of a section and the connectedness of $V_{f(x)}$ implies that U_x is connected by Lemma 6.3.6.

Furthermore, for any two points $x, x' \in |X|$, there is a finite collection

$$\{x_1, \dots, x_n\} \subseteq |X|$$

of points with $x_1 = x$, $x_n = x'$, and such that $V_{f(x_i)} \cap V_{f(x_{i+1})} \neq \emptyset$ for all $i = 1, \dots, n-1$. By condition (i), we must have that $U_{x_i} \cap U_{x_{i+1} \neq \emptyset}$. Therefore, X is connected. □

Proposition 6.3.7 *Let k be a field, k' a nonzero finite k-algebra, and $X' \to \operatorname{Spec} k'$ a smooth surjective quasi-projective morphism of schemes with geometrically connected fibers. Then, the smooth k-scheme $X = \mathrm{R}_{k'/k}(X')$ is nonempty and geometrically connected.*

An alternative proof is given in [3, Proposition A.5.9]. If X' is not smooth, then the result fails: in Example 6.4.6, we describe an example due to Gabber of a quasi-projective scheme X' over a field k' that is geometrically connected and smooth away from one point, and whose Weil restriction $\mathrm{R}_{k'/k}(X')$ is not geometrically connected (in fact, it has a nowhere-reduced connected component!).

Proof By Proposition 6.2.2 (ii), we may assume that k is algebraically closed. In particular, k' is a finite algebra over an algebraically closed field, and thus it is a semi-local ring over k. Applying Lemma 6.2.3, we can assume that k' is a local ring over k with nilpotent maximal ideal $\mathfrak{m}$. Furthermore, since k is algebraically closed, the map $k \to k'/\mathfrak{m}'$ is an isomorphism. In the sequel, we identify k with the residue field $k'/\mathfrak{m}'$ via this isomorphism.

Let X'_0 be the fiber of X' over $\operatorname{Spec} k = \operatorname{Spec}(k'/\mathfrak{m}') \to \operatorname{Spec} k'$. As in Lemma 6.3.3, there is a natural morphism $q_{X'}\colon \mathrm{R}_{k'/k}(X') \to X'_0$ of k-schemes. Let V' be a nonempty open affine subset of X', and V'_0 the special fiber over $\operatorname{Spec} k$. Since V' is k'-smooth and affine, it is necessarily a trivial deformation of V'_0, i.e., $V \cong V'_0 \times_k k'$. In particular, the adjunction map $V'_0 \to \mathrm{R}_{k'/k}(V')$ gives a section of $\mathrm{R}_{k'/k}(X') \to X'_0$ over V'_0. Thus, $\mathrm{R}_{k'/k}(V')$ is nonempty.

The geometric fibers of $\mathrm{R}_{k'/k}(X') \to X'_0$ are connected by Lemma 6.3.3. The preceding discussion shows that, Zariski-locally on X'_0, the morphism $q_{X'}\colon \mathrm{R}_{k'/k}(X') \to X'_0$ admits a section. By assumption, X'_0 is connected, and hence $\mathrm{R}_{k'/k}(X')$ is geometrically connected by Lemma 6.3.4. □

The methods of this section can also be applied to show the following result on the dimension of the Weil restriction.

Corollary 6.3.8 *Let k'/k be a finite field extension of degree d and let X' be a smooth quasi-projective k'-scheme of pure dimension d'. Then, the smooth k-scheme $X = \mathrm{R}_{k'/k}(X')$ is of pure dimension $d \cdot d'$.*

Proof By Proposition 6.2.2 (ii) and Lemma 6.2.3, we may assume that k is algebraically closed and k' is a finite local k-algebra. In particular, if the original field extension k'/k is finite *separable*, then we are done. As dimension is an (étale) local property, we may assume by Proposition 6.2.8 (ii) that X' is as in Proposition 6.3.1, in which case the result immediately follows from Proposition 6.3.1. □

If k'/k is a finite separable extension, then we may drop the smoothness assumption on X' appearing in Corollary 6.3.8. On the other hand, if k'/k is non-separable, then the smoothness hypothesis in Corollary 6.3.8 is necessary: see Example 6.4.7.

6.4 Examples

Example 6.4.1 Let $S' \to S$ be a finite flat morphism of schemes, let U'_i be an open cover of an S'-scheme X', and assume that the Weil restrictions $\mathrm{R}_{S'/S}(U'_i), \mathrm{R}_{S'/S}(X')$ exist for each i. By Proposition 6.2.6

(ii) we know that $R_{S'/S}(U'_i)$ are open subschemes of $R_{S'/S}(X')$, but it may be the case that they do not jointly cover $R_{S'/S}(X')$! Let us provide such an example.

Let k'/k be a finite separable extension of degree $d > 1$ and let $X' = \mathbb{A}^1_{k'}$. Consider the open cover $\{U'_0, U'_1\}$ of X', where $U'_i = \{t \neq i\}$. Then $X = R_{k'/k}(\mathbb{A}^1_{k'})$ contains the Weil restrictions $U_i = R_{k'/k}(U'_i)$ as open subschemes by Proposition 6.2.6 (ii). We claim that $\{U_0, U_1\}$ does not cover X. Indeed, if k_s denotes a separable closure of k, there is a canonical k_s-algebra isomorphism $k_s \otimes_k k' \simeq k_s^d$ and an identification

$$X_{k_s} \simeq \mathbb{A}^d_{k_s}.$$

If $(t_1, \ldots, t_d)$ are the coordinates on $\mathbb{A}^d_{k_s}$ induced by the isomorphism $k_s \otimes_k k' \simeq k_s^d$, then

$$(U_j)_{k_s} = \{(t_1, \ldots, t_d) \in \mathbb{A}^d_{k_s} \mid t_i \neq j \text{ for } i = 1, \ldots, d\}$$

for $j = 0, 1$. Hence, $(U_0)_{k_s} \cup (U_1)_{k_s}$ does not contain points such as $(1, 0, \ldots, 0)$. In particular, $\{U_0, U_1\}$ does not cover X.

In fact, given a Zariski-open cover, it is possible that the (set of points of the) Weil restriction of each member of this cover is empty! Let k be a field and let $k' = k^d$, with $d > 1$, so that $\mathrm{Spec}(k')$ is just d copies of $\mathrm{Spec}(k)$. For any quasi-projective k'-scheme X' with fibers X_i, we have $R_{k'/k}(X') = \prod_{i=1}^d X_i$ by Lemma 6.2.3. Thus, $R_{k'/k}(X')$ is nonempty if and only if all fibers X_i are nonempty. Let us fix an X' with all fibers X_i nonempty. We may view each X_i as an open subset U'_i of X', so it has fiber X_i over the ith point of $\mathrm{Spec}\, k'$ and empty fibers over the other points. Then clearly $\{U'_i\}$ is an open cover of X', but $R_{k'/k}(X'_i) = \emptyset$ for all i!

Note that this phenomenon of failure to preserve Zariski-open covers cannot occur when k'/k is purely inseparable, by Proposition 6.2.8 (ii).

Example 6.4.2 Let k be a field and let $k[\epsilon] := k[x]/(x^2)$. If X is a smooth quasi-projective k-scheme, then the functor of points of $R_{k[\epsilon]/k}(X_{k[\epsilon]})$ is given by

$$U \mapsto \mathrm{Hom}_{k[\epsilon]}\left(U \times_k k[\epsilon], X_{k[\epsilon]}\right) = \mathrm{Hom}_k\left(U \times_k k[\epsilon], X\right).$$

In particular, $R_{k[\epsilon]/k}(X_{k[\epsilon]})$ is isomorphic to the tangent bundle T_X of X as a k-scheme. Moreover, the structure maps $R_{k[\epsilon]/k}(X_{k[\epsilon]}) \to X$ and $T_X \to X$ are the adjunction morphisms obtained from the respective universal properties, so $R_{k[\epsilon]/k}(X_{k[\epsilon]})$ and T_X are in fact isomorphic as X-schemes.

Example 6.4.3 Let k be a field and $k[\epsilon] := k[x]/(x^2)$ be the ring of dual numbers. We consider the Weil restriction of a smooth scheme X' over $k[\epsilon]$ along the structure morphism $\mathrm{Spec}(k[\epsilon]) \to \mathrm{Spec}(k)$. If $X_0 := X' \times_{k[\epsilon]} k$ is the special fiber of X', then we claim that $\mathrm{R}_{k[\epsilon]/k}(X')$ is a principal homogeneous space over the tangent bundle of X_0.

Let $X := X_0 \times_k k[\epsilon]$ be the trivial deformation of X_0 over $k[\epsilon]$. We claim that there is a canonical action

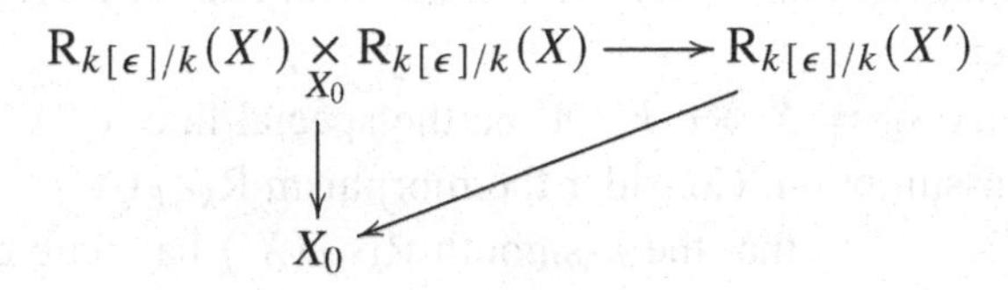

where $\mathrm{R}_{k[\epsilon]/k}(X) = T_{X_0}$ via Example 6.4.2 (an X_0-group), and the morphism $\mathrm{R}_{k[\epsilon]/k}(X') \to X_0$ is as in Remark 6.3.2. To see this, it suffices to construct the *canonical* action for affine schemes X' by Proposition 6.2.8 (ii), so we may assume $X' = \mathrm{Spec}(B)$ for some smooth $k[\epsilon]$-algebra B. Then, we have $X_0 = \mathrm{Spec}(B_0)$ and $X = \mathrm{Spec}(B_0[\epsilon])$, where $B_0 = B/(\epsilon)$. For any k-algebra A, we seek an action

$$\mathrm{Hom}_{k[\epsilon]}(B, A[\epsilon]) \times_{\mathrm{Hom}_k(B_0, A)} \mathrm{Hom}_{k[\epsilon]}(B_0[\epsilon], A[\epsilon]) \longrightarrow \mathrm{Hom}_{k[\epsilon]}(B, A[\epsilon])$$

$$\downarrow \qquad \swarrow$$

$$\mathrm{Hom}_k(B_0, A)$$

that is functorial in A and B.

Let $\phi' \in \mathrm{Hom}_{k[\epsilon]}(B, A[\epsilon])$ and $\widetilde{\phi} \in \mathrm{Hom}_{k[\epsilon]}(B_0[\epsilon], A[\epsilon])$ be such that both have the same image $\phi \in \mathrm{Hom}_k(B_0, A)$. Given the data above, we must produce a $\psi \in \mathrm{Hom}_{k[\epsilon]}(B, A[\epsilon])$ whose reduction is again ϕ. We have

$$\widetilde{\phi}(b_0) = \phi(b_0) + \epsilon \cdot D(b_0), \quad \forall b_0 \in B_0,$$

where $D \colon B_0 \to A$ is k-linear and satisfies the condition

$$D(b_0 \cdot b_0') = \phi(b_0) \cdot D(b_0') + \phi(b_0') \cdot D(b_0). \tag{1}$$

Define

$$\psi(b) := \phi'(b) + \epsilon \cdot D(\overline{b}), \quad \forall b \in B,$$

where $\overline{b} \in B_0$ is the image of b under the canonical reduction $B \twoheadrightarrow B_0$. This map ψ is a $k[\epsilon]$-algebra homomorphism precisely because of (1) and the fact that $\epsilon^2 = 0$. The set of ψ's is clearly in bijection with the set of $\widetilde{\phi}$'s, and $(\phi', \widetilde{\phi}) \mapsto \psi$ is easily seen to be an action with respect to addition in D's and it has the desired functoriality in A and B. Thus, $\mathrm{R}_{k[\epsilon]/k}(X')$ is a principal homogeneous space of T_{X_0}, the tangent bundle of X_0.

Example 6.4.4 This is a continuation of the discussion of Example 6.4.3. Let k' be a local finite k-algebra with nonzero maximal ideal $\mathfrak{m}'$, where k is an algebraically closed field, and let X' be a smooth proper k'-scheme of positive dimension. We claim that the Weil restriction $\mathrm{R}_{k'/k}(X')$ is never proper over k.

Indeed, let $X_0 := X' \times_{k'} k'/\mathfrak{m}'$ be the special fiber of X', which is proper by assumption. Consider the morphism $\mathrm{R}_{k'/k}(X') \to X_0$ as in Remark 6.3.2. Note that the k-smooth $\mathrm{R}_{k'/k}(X')$ has pure dimension $\dim_k(k') \cdot \dim(X_0')$ by Corollary 6.3.8. The geometric fiber at any $x_0 \in X_0(k)$ is a principal homogeneous space for the nonzero vector space $T_{x_0}(X_0)$ by Example 6.4.3. Thus, it cannot be proper.

Example 6.4.5 If k is an imperfect field, k'/k is a nontrivial finite inseparable extension, and A' is a nonzero abelian variety over k', then we claim that $\mathrm{R}_{k'/k}(A')$ violates the conclusion of Chevalley's structure theorem over perfect fields. Recall that Chevalley's structure theorem asserts that for a smooth connected group G over a perfect field k, there is a unique short exact sequence of smooth connected k-groups

$$1 \to H \to G \to A \to 1,$$

where H is affine and A is an abelian variety. See [3, Theorem A.3.7] for a reference for its proof.

Note that $\mathrm{R}_{k'/k}(A')$ exists as a k-group scheme by Proposition 6.2.4, and it is smooth and connected by Proposition 6.2.6 (v). Suppose there exists a smooth, connected, affine k-group H, an abelian variety A over k, and an exact sequence of k-group homomorphisms

$$1 \to H \to \mathrm{R}_{k'/k}(A') \to A \to 1.$$

By the universal property of the Weil restriction, the k-group homomorphism $H \to \mathrm{R}_{k'/k}(A')$ corresponds to a k'-group homomorphism $H_{k'} \to A'$. If $K \subseteq H_{k'}$ is the kernel of $H_{k'} \to A'$, then $H_{k'}/K$ exists as an affine k'-group scheme of finite type, and the induced k'-group homomorphism $H_{k'}/K \to A'$ is a closed immersion, by [3, Proposition A.2.1]. Thus, the image of $H_{k'}/K \to A'$ is an affine abelian subvariety of A', i.e., it is the inclusion of 0. Thus, $H \to \mathrm{R}_{k'/k}(A')$ must be the constant map given by 0. It follows that $\mathrm{R}_{k'/k}(A') \simeq A$; however, $\mathrm{R}_{k'/k}(A')$ cannot be proper owing to Example 6.4.4 (applied over $\overline{k}$) and Lemma 6.2.3 (applied to $S' = \mathrm{Spec}(k' \otimes_k \overline{k})$), a contradiction.

Example 6.4.6 (Gabber) If k is a field, k' is a nonzero finite k-algebra, and X' is a quasi-projective k'-scheme with geometrically connected fibers, then it is not necessarily the case that $\mathrm{R}_{k'/k}(X')$ is geometrically connected (unless we also assume that X' is k'-smooth and the structure morphism $X' \to \mathrm{Spec}(k')$ is surjective).

Let k be an imperfect field of characteristic $p > 0$, and let k' be a nontrivial purely inseparable finite extension of k of degree $d > 1$. Let $q = p^e$ be a prime power of p such that $(k')^q \subseteq k$, and let $m > 2q$ be an integer not divisible by p. Lastly, let us fix an $a' \in k' \setminus k$. Let X' be the geometrically integral curve over k' given by $y^{qp} = a'x^q + x^m$. The open subscheme $U' = X' \setminus \{(0,0)\} = X' \cap \{x \neq 0\}$ is k'-smooth. Therefore, $U := \mathrm{R}_{k'/k}(U')$ is smooth and geometrically connected, by Proposition 6.3.7, and it is an open subscheme of $X := \mathrm{R}_{k'/k}(X')$ by Proposition 6.2.6 (ii). We claim that X is the disjoint union of U and a nonempty open subscheme V which is nowhere reduced.

Choose a k-basis $\{a'_i\}_{i=0}^{d-1}$ of k' with $a'_0 = 1$ and $a'_1 = a'$, and substitute $\sum_{i=0}^{d-1} a'_i x_i$ for x and $\sum_{i=0}^{d-1} a'_i y_i$ for y in the equation defining X'. If $c_i := a'^q_i \in k^\times$, we have that X is cut out, as a closed subscheme of $\mathrm{R}_{k'/k}(\mathbb{A}^2_{k'}) \simeq \mathbb{A}^{2d}_k$, by the equation

$$\left(\sum_{i=0}^{d-1} c_i y_i^q\right)^p = a'_1 \left(\sum_{i=0}^{d-1} c_i x_i^q\right) + \left(\sum_{i=0}^{d-1} a'_i x_i\right)^{m-2q} \left(\sum_{i=0}^{d-1} c_i x_i^q\right)^2.$$

Expanding the $(m-2q)$th power, we may write

$$\left(\sum_{i=0}^{d-1} a'_i x_i\right)^{m-2q} = \sum_{j \geq 0} a'_j f_j(x)$$

for some $f_j(x) \in k[x_0, x_1, \ldots, x_{d-1}]$. By our choice of m, the polynomial $f_1(x)$ in the above expansion is nonzero. Comparing the coefficients of the a'_i, we observe that X is defined by the following system of equations on $\mathbb{A}^{2d}_k$:

$$\begin{cases} \left(\sum_{i=0}^{d-1} c_i y_i^q\right)^p = f_0(x) \left(\sum_{i=0}^{d-1} c_i x_i^q\right)^2, \\ \sum_{i=0}^{d-1} c_i x_i^q + f_1(x) \left(\sum_{i=0}^{d-1} c_i x_i^q\right)^2 = 0, \\ f_j(x) \left(\sum_{i=0}^{d-1} c_i x_i^q\right)^2 = 0 \quad (j \geq 2). \end{cases} \tag{2}$$

If $h := \sum_{i=0}^{d-1} c_i x_i^q$, then the relation $h(1 + f_1 h) = 0$ guarantees that the loci $\{h = 0\}$ and $\{h \text{ is a unit}\}$ define a separation of X. For $P \in X(\overline{k})$, the associated point $P' \in X'(k' \otimes_k \overline{k})$ corresponds to a pair $(x, y) \in (k' \otimes_k \overline{k})^2$ satisfying the equation $y^{qp} = a' x^q + x^m$, with $x = \sum_{i=0}^{d-1} a_i' \otimes x_i$ and $y = \sum_{i=0}^{d-1} a_i' \otimes y_i$. Since $\overline{h} = \overline{x}^q$ in the residue field of the Artinian local ring $k' \otimes_k \overline{k}$, we see that the locus $\{h \text{ is a unit}\}$ coincides with U, as defined above.

Moreover, if $V := X \backslash U$ denotes the open subscheme of X defined by $h = 0$, then setting $h = 0$ in the equations (2) shows that V is cut out by h and $\left(\sum_{i=0}^{d-1} c_i y_i^q\right)^p$; in particular, V is nowhere reduced.

Example 6.4.7 This is a continuation of the discussion of Example 6.4.6. In the absence of a smoothness hypothesis, Corollary 6.3.8 fails: the dimension of the nonsmooth plane curve X' is 1 and the degree of the extension k'/k is $d > 1$, but the dimension of the component V of X is $2d - 2$. As long as $d > 2$, the dimension of X' times the degree of k'/k will be less than the dimension of X on V.

Example 6.4.8 The Weil restriction of a smooth affine group scheme with positive dimensional fibers along a non-étale finite flat morphism of Noetherian rings is never reductive. More precisely, by considering the geometric fibers at the non-étale points of the base, one can show the following: if k is an algebraically closed field, k' is a nonreduced, local, finite k-algebra, and G is a nontrivial connected smooth affine k'-group scheme, then the smooth, connected, affine k-group scheme $\mathrm{R}_{k'/k}(G)$ is not reductive.

In the special case when $G = \mathrm{GL}_{n,k[\epsilon]}$, we have that $\mathrm{R}_{k[\epsilon]/k}(G) = \mathrm{GL}_n \ltimes \mathfrak{gl}_n$ where the semi-direct product is taken using the adjoint action of GL_n on $\mathfrak{gl}_n$. More generally, if G is a smooth affine k-group scheme with Lie algebra $\mathfrak{g}$, then one sees that $\mathrm{R}_{k[\epsilon]/k}(G_{k[\epsilon]}) = G \ltimes \mathfrak{g}$, and $G \ltimes \mathfrak{g}$ is clearly not reductive. For the general case, see [12, Proposition A.3.5].

Example 6.4.9 Proposition 6.2.7 (and Proposition 6.5.6) assert that the Weil restriction of a smooth and surjective morphism between schemes (or algebraic spaces) is again smooth and surjective; however, it is not true that the Weil restriction of a surjective morphism between quasi-compact smooth schemes (or algebraic spaces) is necessarily surjective. More precisely, if $k \subseteq k'$ is a finite field extension and $X' \to Y'$ is a surjective morphism of smooth quasi-projective k'-schemes, then it is not necessarily true that the induced map $\mathrm{R}_{k'/k}(X') \to \mathrm{R}_{k'/k}(Y')$ is surjective. Let k be an imperfect field of characteristic $p > 0$. Pick

$a \in k \setminus k^p$ and consider the finite degree-p extension $k' = k(a^{1/p})$ of k. Consider the exact sequence

$$1 \to \mu_p \to \mathrm{SL}_p \to \mathrm{PGL}_p \to 1$$

of k'-group homomorphisms. As Weil restriction is left exact on the category of k'-groups, there is an exact sequence

$$1 \to \mathrm{R}_{k'/k}(\mu_p) \to \mathrm{R}_{k'/k}(\mathrm{SL}_p) \to \mathrm{R}_{k'/k}(\mathrm{PGL}_p)$$

of k-groups. However, $\mathrm{R}_{k'/k}(\mathrm{SL}_p)$ and $\mathrm{R}_{k'/k}(\mathrm{GL}_p)$ are smooth with the same dimension $[k'\colon k] \cdot (p^2 - 1)$ and $\mathrm{R}_{k'/k}(\mu_p)$ is positive dimensional (see [3, Example 1.3.2]); hence, $\mathrm{R}_{k'/k}(\mathrm{SL}_p) \to \mathrm{R}_{k'/k}(\mathrm{PGL}_p)$ cannot be surjective.

Example 6.4.10 Let $S' \to S$ be a finite étale morphism between connected schemes, and let π (resp. π') denote the étale fundamental group of S (resp. S'). Let X' be a finite étale cover of S', corresponding to the finite, discrete π'-set A'. Then, $\mathrm{R}_{S'/S}(X') \to S$ is a finite étale cover, by Proposition 6.2.6 (vi) and Proposition 6.2.9; it corresponds to a finite, discrete π-set A. We claim that there is a canonical identification

$$A = \mathrm{Ind}_{\pi'}^{\pi}(A'),$$

where $\mathrm{Ind}_{\pi'}^{\pi}(A')$ denotes the induced representation. The above assertion can be verified by combining the universal properties of the induced representation and of the Weil restriction: if B is a finite, discrete π-set, corresponding to the finite étale map $Y \to S$, then

$$\begin{aligned}\mathrm{Hom}_\pi(B, \mathrm{Ind}_{\pi'}^{\pi}(A)) &= \mathrm{Hom}_\pi(B, A')\\ &= \mathrm{Hom}_{S'}(Y \times_S S', X')\\ &= \mathrm{Hom}_S(Y, \mathrm{R}_{S'/S}(X'))\\ &= \mathrm{Hom}_\pi(B, A).\end{aligned}$$

Example 6.4.11 Let k'/k be a finite separable extension of fields. Let A' be an abelian variety over k'. Then, $A := \mathrm{R}_{k'/k}(A')$ is an abelian variety over k by Proposition 6.2.9, Proposition 6.2.6 (v), and Proposition 6.2.2 (iii). If ℓ is a prime that is different from the characteristic of k, then there is a canonical isomorphism

$$\mathrm{T}_\ell(A) = \mathrm{Ind}_{\mathrm{Gal}_{k'}}^{\mathrm{Gal}_k}(\mathrm{T}_\ell(A'))$$

of Galois representations of Gal_k. Here, Gal_k and $\mathrm{Gal}_{k'}$ denote the absolute Galois groups of k and k', respectively. This follows from the corresponding statements on ℓ-power torsion points, which are a consequence of Example 6.4.10.

6.5 Weil restriction of algebraic spaces

Definition 6.5.1 Let $S' \to S$ be a morphism of schemes and let X' be an algebraic space over S'. Consider the contravariant functor

$$\mathrm{R}_{S'/S}(X')\colon (\mathrm{Sch}/S)^{\mathrm{opp}} \to (\mathrm{Sets})$$

given by

$$T \mapsto X'(T \times_S S').$$

If the functor $\mathrm{R}_{S'/S}(X')$ is an algebraic space X over S, then we say that X is the *Weil restriction of X' along $S' \to S$*, and we denote it by $\mathrm{R}_{S'/S}(X')$.

Thinking of the algebraic space X' as a sheaf on the big étale site of (Sch/S'), the Weil restriction $\mathrm{R}_{S'/S}(X')$ is the pushforward sheaf along $S' \to S$.

Theorem 6.5.2 *Let $S' \to S$ be a finite flat morphism of schemes and let X' be an algebraic space of finite presentation over S'. Then, the Weil restriction $\mathrm{R}_{S'/S}(X')$ exists as an algebraic space over S.*

Proof *Step 1 (Construction of a candidate).* In order to verify that $\mathrm{R}_{S'/S}(X')$ is an algebraic space over S, we may assume that S is affine, by [16, Tag 04SK]; say, $S = \mathrm{Spec}(B)$. It follows that S' is affine as well, say $S' = \mathrm{Spec}(B')$, because $S' \to S$ is finite (in particular, affine).

As $X' \to S'$ is quasi-compact, there exists an affine étale chart $U' \to X'$ (for example, the disjoint union of a finite open affine cover of some étale chart of X'). If $R' = U' \times_{X'} U'$, then $R' \rightrightarrows U'$ is an étale equivalence relation such that $X' = U'/R'$. There is a Cartesian square

$$\begin{array}{ccc} R' & \xrightarrow{\ \delta\ } & U' \times_{S'} U' \\ \downarrow & & \downarrow \\ X' & \xrightarrow{\ \Delta_{X'}\ } & X' \times_{S'} X' \end{array}$$

of algebraic spaces over S'. As $X' \to S'$ is finitely presented, the diagonal $\Delta_{X'}$ is quasi-compact and hence the base change δ is quasi-compact and locally of finite type. Similarly, since the diagonal $\Delta_{X'}$ is a monomorphism, we have that δ is also a monomorphism. In particular δ has finite fibers and is separated. Therefore δ, being a quasi-compact morphism of two finitely presented S'-schemes having finite fibers, is quasi-finite. By [5, IV$_4$, 18.12.12], this implies that R' is quasi-affine; in particular, R' is quasi-projective, and so one has that the Weil restiction $\mathrm{R}_{S'/S}(R')$ is a scheme.

By Proposition 6.2.6 (i) and (vi), the induced map

$$\mathrm{R}_{S'/S}(R') \to \mathrm{R}_{S'/S}(U' \times_{S'} U') \simeq \mathrm{R}_{S'/S}(U') \times_S \mathrm{R}_{S'/S}(U')$$

is a monomorphism such that postcomposing with either projection gives an étale S-morphism $\mathrm{R}_{S'/S}(R') \to \mathrm{R}_{S'/S}(U')$. Implicitly, we have used that $\mathrm{R}_{S'/S}(S') = S$. Thus, $\mathrm{R}_{S'/S}(R') \rightrightarrows \mathrm{R}_{S'/S}(U')$ is an étale equivalence relation and we can form the algebraic space $X := \mathrm{R}_{S'/S}(U')/\mathrm{R}_{S'/S}(R')$ over S. By construction, there is a coequalizer diagram

$$\mathrm{R}_{S'/S}(R') \rightrightarrows \mathrm{R}_{S'/S}(U') \to X,$$

so the universal property of the coequalizer gives a morphism $X \to \mathrm{R}_{S'/S}(X')$ of étale sheaves on (Sch/S).

To see that $X \to \mathrm{R}_{S'/S}(X')$ is monic, it suffices to consider the T-valued points of X that lift to $\mathrm{R}_{S'/S}(U')$, by working étale-locally on T. It then suffices to show that the commutative diagram

$$\begin{array}{ccc} \mathrm{R}_{S'/S}(R') & \longrightarrow & \mathrm{R}_{S'/S}(U') \\ \downarrow & & \downarrow \\ \mathrm{R}_{S'/S}(U') & \longrightarrow & \mathrm{R}_{S'/S}(X') \end{array}$$

is Cartesian. This holds since

$$\mathrm{R}_{S'/S}(U') \times_{\mathrm{R}_{S'/S}(X')} \mathrm{R}_{S'/S}(U') = \mathrm{R}_{S'/S}(U' \times_{X'} U') = \mathrm{R}_{S'/S}(R').$$

Step 2 (Reduction to strictly henselian points). It remains to show that the map $X \hookrightarrow \mathrm{R}_{S'/S}(X')$ is a surjective morphism of étale sheaves. Recall that $S = \mathrm{Spec}(B)$ and $S' = \mathrm{Spec}(B')$. We claim that it suffices to show that, for any strictly henselian local B-algebra A, the induced map

$$U'(A \otimes_B B') \longrightarrow X'(A \otimes_B B') \qquad (*)$$

is surjective. Take an S-scheme T and $\xi \in \mathrm{R}_{S'/S}(X')(T) = X'(T \times_{\mathrm{Spec}\, B} \mathrm{Spec}\, B')$. We want to lift ξ to U' étale-locally on T. We may assume T is affine, and then that T is of finite type over S because X' is assumed to be of finite presentation over B' by [16, Tag 04AK].

For every $t \in T$, fix a separable closure $k(t)^{\mathrm{sep}}$ of the residue field $k(t)$ of t. If $A = \mathcal{O}_{T,t}^{\mathrm{sh}}$ is the strict henselization of $\mathcal{O}_{T,t}$, then the surjectivity of (∗) implies that we can find a lift $\xi_t \in U'(A \otimes_B B')$ of $\xi|_{\mathrm{Spec}\, A} \in X'(A \otimes_B B')$. By a standard direct-limit argument, we can find an étale neighborhood V_t of $t \in T$ and a "spreading-out" $\widetilde{\xi_t} \in U'(V_t \times_{\mathrm{Spec}(B)} \mathrm{Spec}(B'))$ of ξ_t. As T is quasi-compact, the disjoint union of finitely many of the V_t gives an étale cover V of T and a lift

$$\widetilde{\xi} \in U'(V' \times_{\mathrm{Spec}(B)} \mathrm{Spec}(B'))$$

of ξ. Thus, the morphism $X \to \mathrm{R}_{S'/S}(X')$ is surjective as étale sheaves, granting the claim.

Step 3 (Surjectivity for strictly henselian points). It remains to show that if A is a strictly henselian B-algebra, then

$$U'(A \otimes_B B') \to X'(A \otimes_B B')$$

is surjective. As $B \to B'$ is (module) finite, the base change $A \otimes_B B'$ is a (module) finite A-algebra. By [5, IV$_4$, Proposition 18.8.10], there are finitely many strictly henselian local rings $C_1, \ldots, C_n$ such that

$$A \otimes_B B' = C_1 \times \cdots \times C_n.$$

The map $U'(A \otimes_B B') \to X'(A \otimes_B B')$ decomposes as the product of the maps $U'(C_i) \to X'(C_i)$, hence it suffices to show that, for a strictly henselian local ring C, the map $U'(C) \to X'(C)$ is surjective. Given a point $\gamma \in X'(C)$, consider the Cartesian square

$$\begin{array}{ccc} Z & \longrightarrow & U' \\ \downarrow & & \downarrow \\ \mathrm{Spec}(C) & \xrightarrow{\gamma} & X' \end{array}$$

where Z is a scheme and $Z \to \mathrm{Spec}(C)$ is étale and surjective since $U' \to X'$ is étale and surjective. By [5, IV$_4$, Théorème 18.5.11(b)], the étale cover $Z \to \mathrm{Spec}(C)$ has a section $\sigma\colon \mathrm{Spec}(C) \to Z$, and hence $\mathrm{Spec}(C) \xrightarrow{\sigma} Z \to U'$ lifts γ, as required. □

Remark 6.5.3 If $S' \to S$ is a finite flat morphism of schemes and X' is a finitely presented algebraic space over S', then $\mathrm{R}_{S'/S}(X')$ is finitely presented over S by construction (in particular, it is quasi-compact over S).

Proposition 6.5.4 *Let $S' \to S$ be a morphism of Noetherian schemes, and let X' be an algebraic space over S'.*

(i) *If T is an S-scheme and $T' = T \times_S S'$, then there is an isomorphism*

$$\mathrm{R}_{T'/T}(X' \times_{S'} T') \simeq \mathrm{R}_{S'/S}(X') \times_S T$$

of functors on (Sch/T).

(ii) *If $X' \to Z'$ and $Y' \to Z'$ are morphisms of algebraic spaces over S', then there is an isomorphism*

$$\mathrm{R}_{S'/S}(X' \times_{Z'} Y') \simeq \mathrm{R}_{S'/S}(X') \times_{\mathrm{R}_{S'/S}(Z')} \mathrm{R}_{S'/S}(Y')$$

of functors on (Sch/S).

Proof The proof is identical to that of Proposition 6.2.2 (ii), (iii). □

Proposition 6.5.5 *Let $S' \to S$ be a finite flat morphism of Noetherian schemes. If $f' \colon X' \to Y'$ is a smooth (resp. étale) morphism between finitely presented algebraic spaces over S', then $\mathrm{R}_{S'/S}(f') \colon \mathrm{R}_{S'/S}(X') \to \mathrm{R}_{S'/S}(Y')$ is smooth (resp. étale).*

Proof The proof is identical to the case of schemes in Proposition 6.2.6 (v), (vi), since the smoothness of algebraic spaces can be checked using the infinitesimal lifting criterion, by [16, Tag 04AM]. □

Proposition 6.5.6 *Let $S' \to S$ be a finite flat morphism of Noetherian schemes. If $f' \colon X' \to Y'$ is a smooth surjective morphism between finitely presented algebraic spaces over S', then $\mathrm{R}_{S'/S}(f') \colon \mathrm{R}_{S'/S}(X') \to \mathrm{R}_{S'/S}(Y')$ is smooth and surjective.*

Proof Write $X = \mathrm{R}_{S'/S}(X')$, $Y = \mathrm{R}_{S'/S}(Y')$, and $f = \mathrm{R}_{S'/S}(f')$. By Proposition 6.5.5, f is smooth. It suffices to show that, for any geometric point $s \colon \mathrm{Spec}(k) \to S$, the k-morphism $f_s \colon X_s \to Y_s$ is surjective. Thus, we may assume that $S = \mathrm{Spec}(k)$ for an algebraically closed field k. As $S' \to S$ is finite, it follows that S' is affine, say $S' = \mathrm{Spec}(k')$; k' is a finite k-algebra, hence it decomposes as a finite product of finite local k-algebras $k' = \prod_{i=1}^{m} A_i'$.

The morphism $X(k) \to Y(k)$ is, by the universal property of the Weil restriction, exactly the map $X'(k') \to Y'(k')$. Given a point $y' \in Y'(k')$, it suffices to construct a point $x' \in X'(k')$ such that the following diagram commutes:

$$\begin{array}{ccccc} \mathrm{Spec}(k') & \xrightarrow{x'} & X'_{y'} & \longrightarrow & X' \\ & \searrow^{=} & \downarrow & & \downarrow f' \\ & & \mathrm{Spec}(k') & \xrightarrow{y'} & Y'. \end{array}$$

Since $X'(k') = \prod_i X'(A'_i)$ and $Y'(k') = \prod_i Y'(A'_i)$, then y' corresponds to an m-tuple of points $y'_i \in Y'(A'_i)$ and it suffices to show that each y'_i lifts to $X'(A'_i)$. So we may assume that k' is a finite local k-algebra (with residue field k).

As k is algebraically closed, there is a point $z' \in X'_{y'}(k)$. The morphism $X'_{y'} \to \mathrm{Spec}(k')$ is smooth, because it is the base change of f', and so the infinitesimal lifting criterion gives a lift $x' \in X'_{y'}(k')$ of z', as required. □

Proposition 6.5.7 *Let $S' \to S$ be a finite flat morphism of Noetherian schemes. Let* P *be one of the following properties of morphisms:*

(i) *monomorphism;*
(ii) *open immersion;*
(iii) *closed immersion;*
(iv) *separated.*

If $f' : X' \to Y'$ is a morphism of finitely presented algebraic spaces over S' with the property P*, then the morphism $\mathrm{R}_{S'/S}(f') : \mathrm{R}_{S'/S}(X') \to \mathrm{R}_{S'/S}(Y')$ also has the property* P.

Proof As in the proof of Proposition 6.2.6, the real content is in (ii) and (iii), so we focus on those assertions. Without loss of generality, we may assume that S and S' are affine. If $V' \to Y'$ is an étale cover of Y' by an affine scheme (in particular V' is quasi-projective over S'), we can form the fiber product

$$\begin{array}{ccc} U' & \xrightarrow{f} & V' \\ \downarrow g & & \downarrow h \\ X & \xrightarrow{f'} & Y' \end{array}$$

as algebraic spaces over S'. Then, U' is an S'-scheme and g is étale and surjective. The property P is stable under base change, so f has P. By Proposition 6.5.4 (ii), there is a Cartesian diagram

$$\begin{array}{ccc} \mathrm{R}_{S'/S}(U') & \xrightarrow{\mathrm{R}_{S'/S}(f)} & \mathrm{R}_{S'/S}(V') \\ \downarrow \mathrm{R}_{S'/S}(g) & & \downarrow \mathrm{R}_{S'/S}(h) \\ \mathrm{R}_{S'/S}(X) & \xrightarrow{\mathrm{R}_{S'/S}(f')} & \mathrm{R}_{S'/S}(Y'). \end{array}$$

The morphism $\mathrm{R}_{S'/S}(f)$ has property P by Proposition 6.2.6, and the morphisms $\mathrm{R}_{S'/S}(g)$ and $\mathrm{R}_{S'/S}(h)$ are étale and surjective, by Proposition 6.5.6. It follows that $\mathrm{R}_{S'/S}(f')$ has property P by étale descent. □

Proposition 6.5.8 *Let $S' \to S$ be a finite flat morphism of Noetherian schemes that is surjective and radicial.*

(i) *If $X'_1, \ldots, X'_n$ are finitely presented algebraic spaces over S', then*

$$\mathrm{R}_{S'/S}\left(\coprod_{i=1}^{n} X'_i\right) = \coprod_{i=1}^{n} \mathrm{R}_{S'/S}(X'_i).$$

(ii) *If $\{U'_i\}_{i\in I}$ is an open (resp. étale) cover of a smooth finitely presented algebraic space X' over S', then $\{\mathrm{R}_{S'/S}(U'_i)\}_{i\in I}$ is an open (resp. étale) cover of $\mathrm{R}_{S'/S}(X')$.*

Proof The proof is identical to the case of schemes, as in Proposition 6.2.8. Note that, since the $\mathrm{R}_{S'/S}(X'_i)$ and $\mathrm{R}_{S'/S}(U'_i)$ are algebraic spaces, the need for the quasi-projectivity hypotheses (that appear in Proposition 6.2.8) disappears. □

Proposition 6.5.9 *Let k be a field and let k' a nonzero finite k-algebra. Let X' be a smooth finitely presented algebraic space over* $\mathrm{Spec}(k')$ *such that $X' \to$* $\mathrm{Spec}(k')$ *is surjective with geometrically connected fibers. Then, $X = \mathrm{R}_{S'/S}(X)$ is nonempty and geometrically connected.*

Proof The proof is identical to that of Proposition 6.3.7. Note that a crucial ingredient, Lemma 6.3.4, was stated for algebraic spaces. □

6.6 Olsson's theorem

The aim of this section is to discuss Olsson's result [13, Theorem 1.5] on the Weil restriction of certain Artin stacks along a *proper* flat morphism of schemes. The notation for this section is fixed below.

Setup 6.6.1 Let S be a Noetherian affine scheme, let S' be a proper flat algebraic S-space, and let $\mathcal{X}' \to S'$ be one of the following:

(i) a separated Artin stack of finite type over S', with finite diagonal;
(ii) a Deligne–Mumford stack of finite type over S', with finite diagonal;
(iii) an algebraic space, separated and of finite type over S'.

Definition 6.6.2 The *Weil restriction* $\mathrm{R}_{S'/S}(\mathcal{X}')$ *of $\mathcal{X}'$ along $S' \to S$* is the fibered category over S which, to any S-scheme T, associates the groupoid $\mathcal{X}'(T \times_S S')$.

As in the case of schemes (cf. Proposition 6.2.1), if the stack/space $\mathcal{X}'$ arises via base change from a stack/space over S, then the Weil restriction recaptures the $\underline{\text{Hom}}$-stack/space of $\mathcal{X}'$:

Proposition 6.6.3 *If $\mathcal{X}$ is as in Setup 6.6.1, $S' \to S$ is a morphism of schemes, and $\mathcal{X}' = \mathcal{X} \times_S S'$, then there is an isomorphism*

$$\mathrm{R}_{S'/S}(\mathcal{X}') = \underline{\mathrm{Hom}}_S(S', \mathcal{X})$$

of functors on (Sch/S).

If S is a scheme and $\mathcal{Y}$, $\mathcal{Z}$ are Artin stacks over S, recall that the $\underline{\text{Hom}}$-stack $\underline{\mathrm{Hom}}_S(\mathcal{Y}, \mathcal{Z})$ is the fibered category over (Sch/S) that, to an S-scheme T, assigns the groupoid of T-morphisms $\mathcal{Y} \times_S T \to \mathcal{Z} \times_S T$.

Proof This is immediate from the universal property of Weil restriction. □

The following theorem is due to Olsson; see [13, Theorem 1.5].

Theorem 6.6.4 *With notation as in Setup 6.6.1, the Weil restriction $\mathrm{R}_{S'/S}(\mathcal{X}')$ is respectively the following:*

(i) *an Artin stack, locally of finite type over S, with quasi-compact and separated diagonal;*
(ii) *a Deligne–Mumford stack, locally of finite type over S, with quasi-compact and separated diagonal;*
(iii) *an algebraic space, locally of finite type over S, with quasi-compact diagonal.*

Olsson applied Theorem 6.6.4 to show [13, Theorem 1.1], which states that if S is a Noetherian scheme, $\mathcal{X} \to S$ is a proper flat algebraic space of finite type, and $\mathcal{Y} \to S$ is as in Setup 6.6.1 (i), then the associated $\underline{\text{Hom}}$-stack $\underline{\mathrm{Hom}}_S(\mathcal{X}, \mathcal{Y})$ is an Artin stack, locally of finite type over S, with quasi-compact and separated diagonal (and moreover, if $\mathcal{Y}$ is a Deligne–Mumford stack or an algebraic space, then so too is $\underline{\mathrm{Hom}}_S(\mathcal{X}, \mathcal{Y})$).

Example 6.6.5 Although the Weil restriction of $\mathcal{X}' \to S'$ through a proper flat morphism $S' \to S$ preserves the property of being locally of finite presentation over the base, it may not preserve quasi-compactness. In fact, it is possible that there are connected components of $\mathrm{R}_{S'/S}(\mathcal{X}')$ that are *not* quasi-compact!

To explain this, let $\mathcal{X} \to S$ be as in Setup 6.6.1, and let $\mathcal{X}' = \mathcal{X} \times_S S'$. Then, by Proposition 6.6.3, we have

$$\mathrm{R}_{S'/S}(\mathcal{X}') = \underline{\mathrm{Hom}}_S(S', \mathcal{X}).$$

When $\mathcal{X}$ is a proper algebraic space over S, the Weil restriction $\mathrm{R}_{S'/S}(\mathcal{X}')$ (as a $\underline{\mathrm{Hom}}$-space) will be an algebraic space *locally* of finite type. We will show that quasi-compactness fails *very* badly, in the sense that *most* connected components are not quasi-compact.

Let k be an algebraically closed field, let $S = \operatorname{Spec} k$, and let $T = \mathbb{P}^1_k$. Consider a line L and a conic C in $\mathbb{P}^3$ that do not intersect one another. Take two copies of $\mathbb{P}^3$ named P_1 and P_2 containing lines L_i and conics C_i respectively, then take two copies of $\mathbb{P}^1$ named Z_1 and Z_2 respectively. Now consider closed embeddings from $Z := Z_1 \sqcup Z_2$ to P_1 and P_2: the closed embedding i_1 sends Z_1 to L_1 and Z_2 to C_1 on P_1, whereas the closed embedding i_2 sends Z_1 to C_2 and Z_2 to L_2 on P_2. Let X be the gluing of P_i along Z; below we collect some facts concerning the gluing from [16, Tag 0ECH].

(i) The existence of X, as a scheme, is guaranteed by [16, Tag 0E25];
(ii) In the course of construction one knows that the map $P_1 \sqcup P_2 \to X$ is surjective and universally closed, hence X is separated over $\operatorname{Spec}(k)$ and is quasi-compact;
(iii) By [16, Tag 0E27] we know that X is locally of finite type over $\operatorname{Spec}(k)$. Combining with the previous statement, we know that X is a variety over $\operatorname{Spec}(k)$;
(iv) Combining the last two statements, we furthermore know that X is a proper variety over $\operatorname{Spec}(k)$;
(v) Both P_1 and P_2 embed as closed subvarieties inside X.

In particular, by Proposition 6.6.3 and Theorem 6.6.4, we know that $\mathrm{R}_{T/S}(X_T) = \underline{\mathrm{Hom}}_k(\mathbb{P}^1, X)$ is an algebraic space, locally of finite type over $\operatorname{Spec}(k)$, with quasi-compact diagonal.

We claim that all components of $\mathrm{R}_{T/S}(X_T) = \underline{\mathrm{Hom}}_k(\mathbb{P}^1, X)$ are not quasi-compact, except for the components corresponding to constant maps. Suppose otherwise, so that every component is quasi-compact and of finite type over $\operatorname{Spec}(k)$; then we shall arrive at a contradiction by exhibiting monomorphisms from increasingly higher-dimensional varieties to every component corresponding to nonconstant maps.

To that end, let us take a k-point of $\mathrm{R}_{T/S}(X_T)$ belonging to a connected component consisting of nonconstant maps; it corresponds to a nonconstant map $\varphi\colon \mathbb{P}^1 \to X$. Without loss of generality, let us assume φ

has image contained in P_1. First, degenerate φ into a degree-d covering of a line on P_1. Then, we can move the image to L_1, which is identified with $C_2 \subset P_2$. Next, we may move it outside C_2 and degenerate it again into a double line on P_2, and by the same process we may move the image back to P_1 and double the degree one more time. The above argument tells us that given a degree-d map from $\mathbb{P}^1$ to $P_1 \subset X$, the corresponding connected component of $\underline{\mathrm{Hom}}_k(\mathbb{P}^1, X)$ receives monomorphisms from all degree-$4^n d$ components of $\underline{\mathrm{Hom}}_k(\mathbb{P}^1, P_1)$ which have dimensions at least a positive constant multiple of $(4^n d)^3$. In particular these spaces have dimensions tending to positive infinity. This gives the desired contradiction we alluded to above.

Acknowledgments The authors would like to express their sincere gratitude to Brian Conrad for suggesting this topic and for his guidance throughout the drafting of this chapter. We would also like to thank the anonymous referee for a careful reading and many insightful suggestions.

References

[1] *Revêtements étales et groupe fondamental*, volume 224 of *Lecture Notes in Mathematics*. Springer, 1971. Séminaire de Géométrie Algébrique du Bois Marie 1960–1961 (SGA 1), Dirigé par Alexandre Grothendieck. Augmenté de deux exposés de M. Raynaud.

[2] Alessandra Bertapelle and Cristian D. González-Avilés. Galois sets of connected components and Weil restriction. *Math. Z.*, 285(1-2):607–612, 2017. doi:10.1007/s00209-016-1723-9.

[3] Brian Conrad, Ofer Gabber, and Gopal Prasad. *Pseudo-reductive groups*, volume 26 of *New Mathematical Monographs*. Cambridge University Press, second edition, 2015. doi:10.1017/CBO9781316092439.

[4] Sławomir Cynk and Matthias Schütt. Generalised Kummer constructions and Weil restrictions. *J. Number Theory*, 129(8):1965–1975, 2009. doi:10.1016/j.jnt.2008.09.010.

[5] Jean Dieudonné and Alexander Grothendieck. Éléments de géométrie algébrique. *Inst. Hautes Études Sci. Publ. Math.*, 4, 8, 11, 17, 20, 24, 28, 32, 1961–1967.

[6] Eric M. Friedlander. Weil restriction and support varieties. *J. Reine Angew. Math.*, 648:183–200, 2010. doi:10.1515/CRELLE.2010.083.

[7] Alexander Grothendieck. Techniques de construction et théorèmes d'existence en géométrie algébrique. IV. Les schémas de Hilbert. In *Séminaire Bourbaki*, volume 6, pages 249–276. Soc. Math. France, 1995.

[8] Jack Hall and David Rydh. General Hilbert stacks and Quot schemes. *Michigan Math. J.*, 64(2):335–347, 2015. doi:10.1307/mmj/1434731927.

[9] Masanari Kida. Descent Kummer theory via Weil restriction of multiplicative groups. *J. Number Theory*, 130(3):639–659, 2010. doi:10.1016/j.jnt.2009.10.001.

[10] Daniel Loughran. Rational points of bounded height and the Weil restriction. *Israel J. Math.*, 210(1):47–79, 2015. doi:10.1007/s11856-015-1245-x.

[11] Johannes Nicaise and Julien Sebag. Motivic Serre invariants and Weil restriction. *J. Algebra*, 319(4):1585–1610, 2008. doi:10.1016/j.jalgebra.2007.11.006.

[12] Joseph Oesterlé. Nombres de Tamagawa et groupes unipotents en caractéristique p. *Invent. Math.*, 78(1):13–88, 1984. doi:10.1007/BF01388714.

[13] Martin C. Olsson. $\underline{\mathrm{Hom}}$-stacks and restriction of scalars. *Duke Math. J.*, 134(1):139–164, 2006. doi:10.1215/S0012-7094-06-13414-2.

[14] Richard Pink. On Weil restriction of reductive groups and a theorem of Prasad. *Math. Z.*, 248(3):449–457, 2004. doi:10.1007/s002090100339.

[15] Roy Mikael Skjelnes. Weil restriction and the Quot scheme. *Algebr. Geom.*, 2(4):514–534, 2015. doi:10.14231/AG-2015-023.

[16] The Stacks Project Authors. *The Stacks project*. https://stacks.math.columbia.edu.

[17] Gonçalo Tabuada. Weil restriction of noncommutative motives. *J. Algebra*, 430:119–152, 2015. doi:10.1016/j.jalgebra.2015.02.013.

[18] André Weil. *Adeles and algebraic groups*, volume 23 of *Progress in Mathematics*. Birkhäuser, 1982. With appendices by M. Demazure and Takashi Ono.

[19] Tim Wouters. The elementary obstruction and the Weil restriction. *Manuscripta Math.*, 128(2):137–146, 2009. doi:10.1007/s00229-008-0219-2.

7

Heights over finitely generated fields

Stephen McKean
Harvard University

Soumya Sankar
The Ohio State University

Abstract This is an expository account of height functions and Arakelov theory in arithmetic geometry. We recall Conrad's description of generalized global fields in order to describe heights over function fields of higher transcendence degree. We then give a brief overview of Arakelov theory and arithmetic intersection theory. Our exposition culminates in a description of Moriwaki's Arakelov-theoretic formulation of heights, as well as a comparison of Moriwaki's construction to various versions of heights.

7.1 Introduction

A central goal in arithmetic geometry is to measure and compare the arithmetic complexity of points on an algebraic variety. For example, $[0 : 1]$ and $[49 : 54]$ are both rational points of the projective line, but the latter point is "more complicated" in a tractable way. The theory of heights provides such measures of complexity in the form of real-valued functions. Studying points of bounded height is of great interest from the point of view of arithmetic statistics and arithmetic geometry. Some such areas of extensive work include Manin's conjecture [11, 25, 2], Vojta's conjecture [31], and Bogomolov's conjecture (proved by Ullmo [30] and Zhang [35]). The theory of heights has also proved to be a powerful tool in arithmetic and algebraic geometry – many classical finiteness theorems, such as the Mordell–Weil theorem [33] and Faltings's theorem [9, 32], rely heavily on heights. See also Chambert-Loir's surveys on the subject [5, 6].

The formulation of geometric versions of the above conjectures and results (for instance the Lang–Néron theorem [7] or the geometric

Bogomolov conjecture [23]) requires making sense of heights over fields of arbitrary transcendence degree. At a first pass, one usually constructs height functions on projective varieties over global fields. The set of valuations on a global field gives convenient real-valued functions, and the product formula enables one to fit these valuations together to obtain a well-defined height function. In this chapter, we discuss some heights over fields of higher transcendence degree. There are several possible approaches to heights in this case (see for instance [8] or recent work of Yuan and Zhang [34], which generalizes Moriwaki's height, discussed in this chapter). We choose two perspectives. First, following Conrad [7, Section 8], we give an introduction to *generalized global fields* in Section 7.2. The structure of these fields includes a set of valuations satisfying an appropriate generalization of the product formula, thus making it possible to construct a naïve height over these fields in a similar manner to the case of global fields. We give a brief survey of naïve and geometric height functions in Section 7.3. See [19] for a classical discussion of these topics.

Second, we discuss Moriwaki's height ([23]) over finitely generated extensions of **Q**. Moriwaki's definition of height is based on Arakelov theory and arithmetic intersection theory. A desirable property of height functions is that they reflect the geometry of the underlying variety in some sense. This leads to a "geometric" definition of height in terms of the degree of a line bundle. Classically, the geometric approach to height functions was used for varieties over function fields of curves. This was generalized to number fields using the work of Arakelov in [1]. In Part B of [17], Hindry and Silverman gave a detailed exposition of the number-field–function-field analogy in the context of heights. In [23], Moriwaki generalized geometric heights to projective varieties over finitely generated extensions of **Q**, using arithmetic intersection theory as developed by Gillet and Soulé ([14]). The second half of this chapter is an exposition of this height function defined by Moriwaki. After reviewing the necessary ideas from Arakelov theory and arithmetic intersection theory in Section 7.5, we discuss Moriwaki's height in Section 7.6. We also discuss how Moriwaki's height recovers more familiar height functions over global fields [23]. In Theorem 7.6.6, we show that Moriwaki's height is induced by a generalized global field structure in certain cases.

We will assume that the reader is comfortable with line bundles on algebraic varieties. Some basic familiarity with valuation theory (see [22, 24]) and intersection theory (see [13]) is also desirable.

Definition 7.1.1 A *valuation* on a field K is a map $v\colon K \to \Gamma \cup \{\infty\}$, where Γ is a totally ordered abelian group (we will take $\Gamma = \mathbf{R}$ or $\mathbf{Z}$) such that (i) $v^{-1}(\infty) = 0$; (ii) $v(ab) = v(a) + v(b)$; and (iii) $v(a + b) \geq v(a) + v(b)$. Two valuations are *equivalent* if they differ by an order-preserving group isomorphism on the target. A *place* on K is an equivalence class of valuations on K. A valuation v is called *non-Archimedean* if $v(a+b) \geq \min\{v(a), v(b)\}$ for all $a, b \in K$, with equality if $a \neq b$. A valuation that is not non-Archimedean is called *Archimedean.* To any valuation v, one can associate an absolute value $|\cdot|_v : K \to \mathbf{R}$ by setting $|a|_v := q_v^{-v(a)}$ for some real number $q_v > 1$. An absolute value defines a metric on K and hence a topology. Equivalent valuations define the same topology on K. Any absolute value satisfies the triangular inequality $|a + b|_v \leq |a|_v + |b|_v$. The non-Archimedean absolute values (coming from non-Archimedean valuations) are exactly those that satisfy a stronger inequality, namely, $|a + b|_v \leq \max\{|a|_v, |b|_v\}$. For a more detailed discussion on valuations, we refer the reader to [24].

Notation 7.1.2 A *number field* is a finite extension of $\mathbf{Q}$. Given a number field K, we denote its *ring of integers* by $\mathcal{O}_K$. Similarly, we denote the *structure sheaf* of a scheme X by $\mathcal{O}_X$. A *global field* is either a number field or the function field of a curve over a finite field. Given a global field K, the *set of places* of K will be denoted M_K.

7.2 Generalized global fields

In Section 7.3, we will discuss various ways to define height functions on algebraic varieties over a field K. If K is a global field, we can use the theory of valuations to define heights on varieties over K. The key property of global fields that allows us to define a height in terms of valuations on K is the *product formula.* It turns out that if a field K satisfies a more general version of the product formula, we can still define heights on varieties over K in terms of valuations on K. This leads us to the notion of *generalized global fields.* We follow [7, Section 8] for our discussion of generalized global fields.

Notation 7.2.1 Given a field K with a valuation v, denote the completion of $(K, |\cdot|_v)$ by K_v. If v is Archimedean and $K_v \cong \mathbf{C}$, let $e_v := 2$. Let $e_v := 1$ in all other cases.

Definition 7.2.2 A *generalized global field* is a field K with infinitely many nontrivial places v and a choice of absolute value $|\cdot|_v$ for each v such that

(i) all but finitely many v are non-Archimedean,
(ii) each non-Archimedean v is discretely valued,
(iii) K_v/K is a separable extension for all non-Archimedean v,
(iv) for each $x \in K^\times$, we have $v(x) = 0$ for all but finitely many v, and
(v) for each $x \in K^\times$, the *generalized product formula* holds:

$$\prod_{\text{places } v} |x|_v^{e_v} = 1. \tag{1}$$

In order to show that the term "generalized" is justified, we need to check that global fields are examples of generalized global fields.

Proposition 7.2.3 *Every number field is a generalized global field.*

Proof Let K be a number field. Let M_K be the set of places of K. Given $v \in M_K$, let κ_v be the residue field of v if v is finite. The global field structure for K is given by the absolute values $\|\cdot\|_v$, where

$$\|\cdot\|_v = \begin{cases} |\kappa_v|^{-\operatorname{ord}_v(\cdot)} & v \text{ finite,} \\ |\cdot| & v \text{ infinite.} \end{cases}$$

Galois theory tells us that $[K : \mathbf{Q}]$ is the number of $\mathbf{Q}$-linear embeddings of K into $\overline{\mathbf{Q}} \subset \mathbf{C}$. Each of the infinitely many prime ideals $\mathfrak{p} \subset \mathcal{O}_K$ defines a distinct, discrete, non-Archimedean valuation. This verifies conditions (i) and (ii) in Definition 7.2.2. Since char $K = 0$, any extension of K is separable. Further, any element of K is a product of finitely many prime ideals in $\mathcal{O}_K$, so its $\mathfrak{p}$-adic valuation is 0 for all but finitely many finite places.

Finally, we need to show that the generalized product formula holds. Since the usual product formula holds with respect to the absolute values $\{\|\cdot\|_v\}$, the generalized product formula holds with respect to the absolute values $\{\|\cdot\|_v^{1/e_v}\}$. Thus the set $\{\|\cdot\|_v^{1/e_v}\}$ gives K the structure of a generalized global field. □

Global fields in positive characteristic do not have Archimedean places, so we do not need to check any of the Archimedean criteria for generalized global fields. Instead, we need to check that the completion of a global field K is a separable extension of K, which was not a concern in characteristic 0.

Proposition 7.2.4 *Every global field of positive characteristic is a generalized global field.*

Proof Let K be a finite extension of $\mathbf{F}_p(t)$ for some prime p. Then K is the function field of some complete irreducible curve C over an extension $\mathbf{F}_q$ of $\mathbf{F}_p$. The *finite* places of K are in bijection with the closed points of C. Let $\mathcal{O}$ denote the integral closure in K of $\mathbf{F}_p[t]$. Then each maximal ideal of $\mathcal{O}$ defines a closed point of C, and distinct maximal ideals define distinct closed points. In particular, since there are infinitely many irreducible polynomials over $\mathbf{F}_p$ and K is a finite extension of $\mathbf{F}_p(t)$, there are infinitely many places of K. For a closed point $\mathfrak{m} \in C$, the corresponding absolute value is $\|\cdot\|_{\mathfrak{m}} = |\kappa_{\mathfrak{m}}|^{-\operatorname{ord}_{\mathfrak{m}}(\cdot)}$, where $\kappa_{\mathfrak{m}}$ denotes the size of the residue field. Since, for any $x \in K^*$, $\operatorname{ord}_{\mathfrak{m}}(x)$ is the order of vanishing of the function x along $\mathfrak{m}$, criteria (ii) and (iv) in Definition 7.2.2 are satisfied by these absolute values. Criterion (v) follows from the fact that a rational function on a complete, irreducible curve has degree 0, since

$$\prod_{\mathfrak{m}} \|x\|_{\mathfrak{m}} = \prod_{\mathfrak{m}} |\kappa_{\mathfrak{m}}|^{-\operatorname{ord}_{\mathfrak{m}}(x)} = \prod_{\mathfrak{m}} p^{-[\kappa_{\mathfrak{m}}:\mathbf{F}_p]\cdot\operatorname{ord}_{\mathfrak{m}}(x)} = p^{-\deg(x)}.$$

It remains to show that K_v/K is a separable extension for all places of K. Since K is a finite extension of $\mathbf{F}_p(t)$, [3, Theorem 1] implies that $K \cong \mathbf{F}_p(t, \alpha)$, where $\alpha \in K$ is a root of some irreducible polynomial in $\mathbf{F}_p(t)[x]$. At any place v, [26, II.4, Theorem 2] implies that $K_v \cong \mathbf{F}_{p^m}((u))$, where m is a positive integer and $u \in \mathcal{O}_v$ is a uniformizer. Since $\mathcal{O}_v \subseteq \mathbf{F}_p[t^{\pm 1}, \alpha^{\pm 1}]$, we can express u as a rational function in t and α over $\mathbf{F}_p$. In particular, K is an intermediate field of the extension $\mathbf{F}_{p^m}((u))/\mathbf{F}_p(u)$.

It thus suffices to prove that $\mathbf{F}_{p^m}((u))/\mathbf{F}_p(u)$ is separable. Since $\mathbf{F}_{p^m}/\mathbf{F}_p$ is separable, we just need to prove that $\mathbf{F}_s((u))/\mathbf{F}_s(u)$ is separable, where $s = p^m$. By [12, Lemma 2.6.1 (b)], it suffices to prove that if $f_1, \ldots, f_r \in \mathbf{F}_s((u))$ are linearly independent over $\mathbf{F}_s(u)$, then $f_1^p, \ldots, f_r^p$ are as well.

Suppose $\sum_{i=1}^r g_i f_i^p = 0$ for some $g_1, \ldots, g_r \in \mathbf{F}_s(u)$. By clearing denominators, we may assume that $f_1^p, \ldots, f_r^p \in \mathbf{F}_s[[u]]$ and $g_1, \ldots, g_r \in \mathbf{F}_s[u]$. We then write $g_i = \sum_{j=0}^{p-1} g_{ij}(u^p)u^j$, where each $g_{ij} \in \mathbf{F}_s[u]$. Since $x \mapsto x^p$ is an automorphism of $\mathbf{F}_s$, it follows that we may write $g_{ij}(u^p) = h_{ij}^p$ for some $h_{ij} \in \mathbf{F}_s[u]$. We now have

$$0 = \sum_{i=1}^r g_i f_i^p = \sum_{i=1}^r \left(\sum_{j=0}^{p-1} h_{ij}^p u^j \right) f_i^p = \sum_{j=0}^{p-1} \left(\sum_{i=1}^r h_{ij}^p f_i^p \right) u^j.$$

Note that $\sum_{i=1}^r h_{ij}^p f_i^p \in \mathbf{F}_s[[u^p]]$ (by the freshman's dream) and $j < p$

for all j, so the terms of $\sum_{i=1}^{r} g_i f_i^p$ of degree-(j mod p) all belong to the $(\sum_{i=1}^{r} h_{ij}^p f_i^p)u^j$ summand. Since the degree j mod p terms of $\sum_{i=1}^{r} g_i f_i^p$ must sum to 0, we conclude that $\sum_{i=1}^{r} h_{ij}^p f_i^p = 0$ for all j. Thus $\sum_{i=1}^{r} h_{ij}^p f_i^p = (\sum_{i=1}^{r} h_{ij} f_i)^p = 0$ for all j, so $\sum_{i=1}^{r} h_{ij} f_i = 0$ for all j. By the $\mathbf{F}_s(u)$-linear independence of $f_1, \ldots, f_r$, it follows that $h_{ij} = 0$ for all i, j. Thus $g_i = 0$ for all i, so $f_1^p, \ldots, f_r^p$ are linearly independent over $\mathbf{F}_s(u)$. □

7.2.1 *General function fields*

While function fields of transcendence degree 1 are global fields, we would also like to describe heights associated to function fields of higher-dimensional varieties. This is the main motivation behind generalized global fields: function fields of transcendence degree at least 2 are generalized global fields that are not global fields.

Example 7.2.5 (See also [27]) Let K/k be a finitely generated field extension with k algebraically closed in K. Assume that $\operatorname{trdeg}(K/k) > 0$. We will describe a generalized global field structure on K. For a concrete example, one can take $K = k(t_1, \ldots, t_n)$.

Let V be a normal, integral, projective k-scheme such that $k(V) = K$. If $\operatorname{trdeg}(K/k) = 1$, then there is a unique such V. For each codimension-1 point $v \in V$, the order of vanishing of a rational function along v induces a valuation $\operatorname{ord}_v \colon K \to \mathbf{Z} \cup \{\infty\}$ with valuation ring $\mathcal{O}_{V,v}$. The valuation ord_v can be recovered from the valuation ring $\mathcal{O}_{V,v}$ (see e.g., [29, Tag 00I8]). The valuation rings $\mathcal{O}_{V,v}$, and hence the valuations ord_v, depend on the choice of model V (which is not unique when $\operatorname{trdeg}(K/k) > 1$), so we consider the model V of the extension K/k to be part of the generalized global field structure in this case.

We now check that K/k satisfies the criteria listed in Definition 7.2.2. Each ord_v is non-Archimedean and nontrivial. Moreover, ord_v and ord_w induce different topologies on K if $v \neq w$. Since there are infinitely many codimension-1 $\overline{k}$-points of V, we thus have infinitely many nontrivial, non-Archimedean places of K. By construction, $\operatorname{ord}_v(f) = 0$ if and only if f and $1/f$ do not vanish along v. A nonzero function vanishes or has poles at only finitely many v, so $\operatorname{ord}_v(f) = 0$ for all but finitely many v. Moreover, $\operatorname{ord}_v(f) = 0$ for all v if and only if f is a nonzero constant.

The separability criterion for K_v/K follows from the fact that finite type schemes over a field are excellent [16, Scholie 7.8.3(iii) and Proposition 7.8.6(i)]. For a hands-on approach, we can modify Proposition 7.2.4 in the case of $K = \mathbf{F}_{p^m}(t_1,\ldots,t_n)$. The completion K_v is the fraction field of the completion $\widehat{\mathscr{O}}_{V,v}$, so K_v is an intermediate field in the extension $\mathbf{F}_{p^m}((u_1,\ldots,u_n))/K$ for some local coordinates $u_1,\ldots,u_n$. As in Proposition 7.2.4, it suffices to show that $\mathbf{F}_s((u_1,\ldots,u_n))/\mathbf{F}_s(u_1,\ldots,u_n)$ is a separable extension with s a power of a prime p. Again by [12, Lemma 2.6.1(b)], it suffices to prove the following proposition.

Proposition 7.2.6 *If $f_1,\ldots,f_r \in \mathbf{F}_s((u_1,\ldots,u_n))$ are linearly independent over $\mathbf{F}_s(u_1,\ldots,u_n)$, then $f_1^p,\ldots,f_r^p$ are linearly independent over $\mathbf{F}_s(u_1,\ldots,u_n)$.*

Proof Suppose $\sum_{i=1}^r g_i f_i^p = 0$ for some $g_1,\ldots,g_r \in \mathbf{F}_s(u_1,\ldots,u_n)$. By clearing denominators, we may assume $f_1^p,\ldots,f_r^p \in \mathbf{F}_s[[u_1,\ldots,u_n]]$ and $g_1,\ldots,g_r \in \mathbf{F}_s[u_1,\ldots,u_n]$. Given $\mathbf{d} = (d_1,\ldots,d_n) \in \mathbf{Z}^n_{\geq 0}$, let $\mathbf{u}^{\mathbf{d}} := u_1^{d_1}\cdots u_n^{d_n}$. Let $P = \{\mathbf{d} \in \mathbf{Z}^n_{\geq 0} : d_i < p \text{ for all } i\}$. Then there exist $\{g_{1,\mathbf{d}},\ldots,g_{r,\mathbf{d}}\}_{\mathbf{d}\in P} \subset \mathbf{F}_s[u_1,\ldots,u_n]$ such that

$$g_i = \sum_{\mathbf{d}\in P} g_{i,\mathbf{d}}(u_1^p,\ldots,u_n^p)\mathbf{u}^{\mathbf{d}}.$$

We write $g_{i,\mathbf{d}}(u_1^p,\ldots,u_n^p) = h_{i,\mathbf{d}}^p$ for some $h_{i,\mathbf{d}} \in \mathbf{F}_s[u_1,\ldots,u_n]$, so that

$$0 = \sum_{i=1}^r g_i f_i^p = \sum_{\mathbf{d}\in P}\left(\sum_{i=1}^r h_{i,\mathbf{d}}^p f_i^p\right)\mathbf{u}^{\mathbf{d}}.$$

Note that $\sum_{i=1}^r h_{i,\mathbf{d}}^p f_i^p \in \mathbf{F}_s[u_1^p,\ldots,u_n^p]$ and $d_1,\ldots,d_n < p$ for all $\mathbf{d}$. Following the proof of Proposition 7.2.4, we can thus conclude that $(\sum_{i=1}^r h_{i,\mathbf{d}} f_i)^p = 0$ for all $\mathbf{d} \in P$ by considering terms of multidegree $\mathbf{d}$ mod p. As in Proposition 7.2.4, it follows that $f_1^p,\ldots,f_r^p$ are linearly independent over $\mathbf{F}_s(u_1,\ldots,u_n)$. □

It remains to address the generalized product formula for K. For each $v \in V$ of codimension-1, we will construct a constant $0 < c_v < 1$ such that the absolute values $\|\cdot\|_v := c_v^{\operatorname{ord}_v(\cdot)}$ satisfy the generalized product formula. Since V is a projective k-variety, there is a closed embedding $i\colon V \hookrightarrow \mathbf{P}^n_k$ over k. Let $\overline{i(v)}$ be the closure of $i(v) \in \mathbf{P}^n_k$, so that $\overline{i(v)}$ is an integral closed subscheme of $\mathbf{P}^n_k$. Let $\deg_{k,i}(v)$ be the degree of $\overline{i(v)}$, and set

$$c_v := \begin{cases} |k|^{-\deg_{k,i}(v)} & |k| < \infty, \\ e^{-\deg_{k,i}(v)} & \text{otherwise.} \end{cases} \tag{2}$$

This choice of c_v allows us to deduce the generalized product formula geometrically. In particular, given a rational function $f \in K^\times$, the principal Weil divisor $\mathrm{div}(f) = \sum_{v \in V} \mathrm{ord}_v(f) \cdot v$ has degree 0. That is,

$$0 = \deg_{k,i}(\mathrm{div}(f)) = \sum_{v \in V} \mathrm{ord}_v(f) \deg_{k,i}(v),$$

so $\prod_{v \in V} \|f\|_v = c^0 = 1$, where $c = |k|$ if $|k| < \infty$ and $c = e$ otherwise.

Remark 7.2.7 If we are given a very ample line bundle L on V instead of a specified projective embedding $i \colon V \hookrightarrow \mathbf{P}^n_k$, we can still define a generalized global field structure on K. We simply replace $\deg_{k,i}(v)$ in equation (2) with $\deg_{k,L}(v) := \deg_{k,\overline{v}}(c_1(L|_{\overline{v}})^{\dim \overline{v}})$.

7.2.2 Extensions of generalized global fields

We now discuss a generalized global field structure on finite extensions of generalized global fields. Let F be a finite extension of a generalized global field K. Since F/K is finite, each place v on K lifts to finitely many places w on F. Since at most finitely many places of K are Archimedean, these lift to the finitely many Archimedean places of F. Since each non-Archimedean place v is discretely valued, the same holds for each non-Archimedean lift w.

If $w(x) \neq 0$ for some $x \in F^\times$, then x (or $1/x$) is non-integral at w. This implies that one of the coefficients of the minimal polynomial of x (or $1/x$) over K is non-integral in the valuation ring of v. Since for each $y \in K^\times$ we have $v(y) = 0$ for all but finitely many v, it follows that $w(x) = 0$ for all but finitely many w.

By assumption, K_v/K is a separable extension for all non-Archimedean v. Given a non-Archimedean lift w of v, we need to check that F_w/F is a separable extension. Since F/K is finite, there are generators $\alpha_1, \ldots, \alpha_n \in F$ such that $F = K(\alpha_1, \ldots, \alpha_n)$. We will show that $F_w = K_v(\alpha_1, \ldots, \alpha_n)$. The separability of F_w/F will then follow from the separability of K_v/K.

Proposition 7.2.8 *Let v be a valuation on a field K. Let w be a valuation on the field $F = K(\alpha_1, \ldots, \alpha_n)$ that is an extension of v. Then $(K(\alpha_1, \ldots, \alpha_n))_w = K_v(\alpha_1, \ldots, \alpha_n)$.*

Proof First, we note that $\alpha_1, \ldots, \alpha_n \in F \hookrightarrow F_w$. We also have that F_w is an extension of K_v, so it follows that $K_v(\alpha_1, \ldots, \alpha_n) \subseteq F_w$. Since $K \subseteq K_v$, we have $F \subseteq K_v(\alpha_1, \ldots, \alpha_n) \subseteq F_w$, so $(K_v(\alpha_1, \ldots, \alpha_n))_w = F_w$. Finally, $K_v(\alpha_1, \ldots, \alpha_n)$ is complete with respect to w, since any finite extension of a complete valued field is complete with respect to the corresponding extension of the valuation. □

We now choose a unique representative of each $\|\cdot\|_w$ by specifying

$$\|\cdot\|_w|_K = \|\cdot\|_v^{[F_w:K_v]e_v/e_w}. \tag{3}$$

For example, if v is Archimedean, then we are requiring $\|\cdot\|_w|_K = \|\cdot\|_v$. Indeed, if v is complex, then $F_w \cong K_v \cong \mathbf{C}$ and $e_v = e_w = 2$. If v and w are both real, then $F_w \cong K_v \cong \mathbf{R}$ and $e_v = e_w = 1$. If v is real and w is complex, then $[F_w : K_v] = [\mathbf{C} : \mathbf{R}] = 2$, while $e_v = 1$ and $e_w = 2$.

We need to check that our choices of $\|\cdot\|_w$ satisfy the generalized product formula. The trick here is to reduce to the generalized product formula over K using field norms. Since K_v/K is separable for all v, the ring $K_v \otimes_K F$ is reduced for all v. This induces an isomorphism

$$K_v \otimes_K F \to \prod_{w|v} F_w \quad \text{given by} \quad a \otimes b \mapsto (ab, \dots, ab)$$

for all v. Also note that any basis of F as a K-vector space is also a basis of $K_v \otimes_K F$ as a K_v-vector space. Given $x \in F^\times$, we thus have

$$N_{F/K}(x) = N_{(K_v \otimes_K F)/K_v}(x) = \prod_{w|v} N_{F_w/K_v}(x),$$

where $N_{E'/E}$ is the norm of the extension E'/E. In particular,

$$\prod_{w|v} \|N_{F_w/K_v}(x)\|_w = \|N_{F/K}(x)\|_v$$

for all v. Since $\|x\|_w = \|N_{F_w/K_v}(x)\|_w^{1/[F_w:K_v]}$, it follows that

$$\begin{aligned}
\prod_{w|v} \|x\|_w^{e_w} &= \prod_{w|v} \|N_{F_w/K_v}(x)\|_w^{e_w/[F_w:K_v]} \\
&= \prod_{w|v} \left(\|N_{F_w/K_v}(x)\|_v^{e_w/[F_w:K_v]}\right)^{[F_w:K_v]e_v/e_w} \\
&= \prod_{w|v} \|N_{F_w/K_v}(x)\|_v^{e_v} = \|N_{F/K}(x)\|_v^{e_v}.
\end{aligned}$$

The generalized product formula for $\{\|\cdot\|_w\}_w$ on F thus follows from the generalized product formula for $\{\|N_{F/K}(\cdot)\|_v\}_v$ on K.

7.3 Heights

Given an algebraic variety X over a field k, a *height* is a function $h\colon X(\overline{k}) \to \mathbf{R}_{\geq 0}$, with $h(x)$ a measure of the complexity of x. Using the ordering on $\mathbf{R}$, we can filter $X(\overline{k})$ by height, which allows us to study rational points using limits and induction. Ideally, one would like points of bounded height and bounded degree to be finite sets. This property (known as the *Northcott property*) holds for many, but not all, of the height functions that we will describe.

Following [17, Part B] and [7, Section 9], we now discuss a few classical height functions. A central theme is that valuations and the product formula are useful in constructing heights, both for global and generalized global fields. To conclude this section, we briefly describe a geometric approach to heights over finitely generated fields of transcendence degree 1 over $\mathbf{Q}$. These geometric heights will serve as an analogy for Moriwaki's Arakelov-theoretic heights discussed in Section 7.6.

7.3.1 Naïve and logarithmic heights

Any nonzero element of $\mathbf{Q}$ can be written uniquely as a fraction $\frac{a}{b}$, where $a, b \in \mathbf{Z}$, $b > 0$, and $\gcd(a, b) = 1$. We define the *naïve height* of $\frac{a}{b}$ to be $h(\frac{a}{b}) = \max\{|a|, |b|\}$. For scaling reasons and to ensure that the minimum value attained by the height is 0, one defines the *logarithmic height* $\log h(\frac{a}{b}) = \log \max\{|a|, |b|\}$. We can mimic these definitions to obtain our first height functions on $\mathbf{P}^n(\mathbf{Q})$.

Definition 7.3.1 Any rational point $x \in \mathbf{P}^n(\mathbf{Q})$ can be written uniquely (up to scaling the coordinates by ± 1) as $x = [x_0 : \cdots : x_n]$, where $x_0, \ldots, x_n \in \mathbf{Z}$ and $\gcd(x_0, \ldots, x_n) = 1$. The *naïve (multiplicative) height* and *naïve logarithmic height* of x are defined to be $h(x) := \max\{|x_0|, \ldots, |x_n|\}$ and $\log h(x)$, respectively. Note that

$$\max\{|x_0|, \ldots, |x_n|\} = \max\{|-x_0|, \ldots, |-x_n|\},$$

so these heights are well defined.

Remark 7.3.2 Since $\{n \in \mathbf{Z} : |n| \leq H\}$ is a finite set for any positive bound H, it follows that sets of bounded naïve or logarithmic height are finite. As mentioned earlier, this is known as the Northcott property. Later, we will discuss how to define height functions on varieties over number fields. A height function h on a variety V over a number field K is said to satisfy the *Northcott property* if, for any real numbers $D, H > 0$, the set

$$\{x \in V(\overline{\mathbf{Q}}) : [\mathbf{Q}(x) : \mathbf{Q}] < D \text{ and } h(x) < H\}$$

is finite. This is a desirable property for many applications, and plays a central role in results such as the Mordell–Weil or Lang–Néron theorems.

We now define naïve and logarithmic heights on $\mathbf{P}^n(K)$ for any number field K using the global field structure on K, as in Proposition 7.2.3. Let M_K denote the set of places on K.

Definition 7.3.3 Let K be a number field, and let $x = [x_0 : \cdots : x_n] \in \mathbf{P}^n(K)$. The *naïve (multiplicative) height* and *naïve logarithmic height* of x with respect to K are defined to be

$$h_K(x) := \prod_{v \in M_K} \max\{\|x_0\|_v, \ldots, \|x_n\|_v\}$$

and $\log h_K(x)$, respectively.

Remark 7.3.4 The naïve multiplicative and logarithmic heights with respect to K are well defined by the product formula. Indeed, for any $c \neq 0$,

$$\begin{aligned}\prod_{v \in M_K} \max_{0 \le i \le n} \{\|cx_i\|_v\} &= \left(\prod_{v \in M_K} \|c\|_v\right)\left(\prod_{v \in M_K} \max_{0 \le i \le n} \{\|x_i\|_v\}\right) \\ &= \prod_{v \in M_K} \max_{0 \le i \le n} \{\|x_i\|_v\}.\end{aligned}$$

Going beyond number fields, we would like to define a notion of height on $\mathbf{P}^n(\overline{\mathbf{Q}})$. To do this, we will have to keep track of the field of definition of a given $\overline{\mathbf{Q}}$-rational point of $\mathbf{P}^n$. Given a finite extension F/K, we can naturally view $\mathbf{P}^n(K)$ as a subset of $\mathbf{P}^n(F)$. For any $x \in \mathbf{P}^n(K)$, one can show that $h_F(x) = h_K(x)^{[F:K]}$ [17, Lemma B.2.1(c)].

Definition 7.3.5 Let $x \in \mathbf{P}^n(\overline{\mathbf{Q}})$. The *absolute (multiplicative) height* and *absolute logarithmic height* of x are defined to be

$$h_{\text{abs}}(x) = h_K(x)^{1/[K:\mathbf{Q}]},$$

$$\log h_{\text{abs}}(x) = \frac{1}{[K:\mathbf{Q}]} \log h_K(x),$$

respectively, where K is any number field over which x is defined.

As one would hope, the absolute height satisfies a Northcott property, albeit in a slightly different form than for the naïve height on $\mathbf{P}^n(\mathbf{Q})$. We will see that we need to bound both the height and degree of the field of definition to get a finite set of points.

Theorem 7.3.6 *[17, Theorem B.2.3] For any $H, D \ge 0$, the set*

$$\{x \in \mathbf{P}^n(\overline{\mathbf{Q}}) : h_{\text{abs}}(x) \le H \text{ and } [\mathbf{Q}(x) : \mathbf{Q}] \le D\}$$

is finite.

It follows that, for any fixed number field K, h_K and $\log h_K$ satisfy the Northcott property. The absolute height is also invariant under Galois action [17, Proposition B.2.2].

7.3.2 Weil heights

Given a projective variety X over a number field K with a very ample line bundle L, we get an embedding $\phi\colon X \to \mathbf{P}^n_K$. This enables us to define a height $\log h_{L,K}\colon X(K) \to \mathbf{R}_{\geq 0}$ by setting $\log h_{L,K}(x) := \log h_K(\phi(x))$ (and similarly for $\log h_{L,\mathrm{abs}}$). Of course, one needs to ask how this depends on the embedding ϕ; it turns out that $\log h_{L,K}$ is well-defined *up to a bounded function* [17, Theorem B.3.1]. This leads us to the notion of *Weil heights*.

Notation 7.3.7 Given any set S, let $O(1)$ be the set of bounded functions $S \to \mathbf{R}$. Given a function $f\colon S \to \mathbf{R}$, we denote the set of functions $\{g\colon S \to \mathbf{R}$ such that $g - f$ is bounded on $S\}$ by $f + O(1)$.

Definition 7.3.8 Let X be a projective variety over a number field K with a line bundle L. Then there exist very ample line bundles L_1, L_2 such that $L \cong L_1 \otimes L_2^{-1}$. The *Weil height* with respect to L is the difference

$$\log h_{K,L} := \log h_{K,L_1} - \log h_{K,L_2} + O(1)\colon X(K) \to \mathbf{R}.$$

Similarly, we define the *absolute Weil height* to be

$$\log h_{\mathrm{abs},L} := \log h_{\mathrm{abs},L_1} - \log h_{\mathrm{abs},L_2} + O(1)\colon X(\overline{\mathbf{Q}}) \to \mathbf{R}.$$

The differences $\log h_{K,L_1} - \log h_{K,L_2}$ and $\log_{h_{\mathrm{abs},L_1}} - \log h_{\mathrm{abs},L_2}$ depend on the choice of L_1, L_2 only up to $O(1)$, so the Weil height and absolute Weil height are well defined.

As with absolute naïve heights, the absolute Weil height is invariant under Galois actions.

Remark 7.3.9 Weil heights are additive in L. That is, given two line bundles L, L', we have $\log h_{K,L\otimes L'} = \log h_{K,L} + \log h_{K,L'}$ (and likewise for absolute Weil heights).

Remark 7.3.10 In some circumstances, there is a particular representative of a Weil height in its $O(1)$-equivalence class that satisfies nice properties. The *canonical* or *Néron–Tate height* is an important example of such a height function [17, Section B.4].

7.3.3 Heights over generalized global fields

When constructing heights on projective varieties over a number field K, we saw that the global field structure of K played an essential role. The defining characteristics of generalized global fields encapsulate the properties of a global field that allow one to construct height functions. Following [7, Section 9], we can construct height functions on projective varieties over generalized global fields in a manner analogous to the height functions discussed thus far. We start with a generalization of absolute (logarithmic) heights on $\mathbf{P}^n(\overline{\mathbf{Q}})$.

Definition 7.3.11 Let $(K, \{\|\cdot\|_v\}_v)$ be a generalized global field. The *standard K-height* and *logarithmic K-height* are functions $\mathbf{P}^n(\overline{K}) \to \mathbf{R}_{\geq 0}$ defined by

$$H_K(x) = \prod_w \max\{\|x_0\|_w^{e_w/[F:K]}, \ldots, \|x_n\|_w^{e_w/[F:K]}\},$$

$$\log H_K(x) = \frac{1}{[F:K]} \sum_w \log \max\{\|x_0\|_w^{e_w}, \ldots, \|x_n\|_w^{e_w}\},$$

respectively, where F is any finite extension of K over which x is defined, endowed with a generalized global field structure as described in Section 7.2.2.

Remark 7.3.12 Let F' be a finite extension of F. Since $[F' : K] = \sum_{w'|w} [F'_{w'} : F_w]$ for all w on F, (3) in Section 7.2 implies that H_K and $\log H_K$ do not depend on the choice of field of definition F. Moreover, the generalized product formula implies that $H_K(x)$ and $\log H_K(x)$ do not depend on the choice of projective coordinates of x (compare Remark 7.3.4). One can also prove $\mathrm{Aut}(\overline{K}/K)$-invariance, so that the K-height does not depend on the choice of algebraic closure $\overline{K}$.

Remark 7.3.13 Note that $\log H_{\mathbf{Q}} = h_{abs}$ on $\mathbf{P}^n(\overline{K})$.

We now extend the definition of absolute Weil heights to generalized global fields. Given a projective variety X over a field K with a very ample line bundle L, let $H_{K,L} = H_K \circ \phi$, where $\phi\colon X \to \mathbf{P}^n_K$ is any projective embedding determined by L.

Definition 7.3.14 Let K be a generalized global field. Let X be a projective variety over K with a line bundle L. Let L_1, L_2 be very ample line bundles on X such that $L \cong L_1 \otimes L_2^{-1}$. The *generalized Weil height* is defined as

$$\log H_{K,L} := \log H_{K,L_1} - \log H_{K,L_2} + O(1)\colon X(\overline{K}) \to \mathbf{R}.$$

Generalized Weil heights satisfy many nice properties. For example, generalized Weil heights are additive in L (see Remark 7.3.9). Moreover, generalized Weil heights are functorial: given a map $f\colon X \to Y$ of projective K-varieties and a line bundle L on Y, we have

$$\log H_{K,f^*L} = \log H_{K,L} \circ f + O(1)$$

as functions on $X(\overline{K})$.

7.3.4 Geometric heights

We now discuss a method for defining heights in terms of the degree of a line bundle. This approach will be mirrored by Moriwaki's height function, which we describe in Section 7.6. Let K be a finitely generated field of transcendence degree 1 over a prime field k. Let C be a curve over k such that $k(C) = K$. A point $x \in \mathbf{P}^n(K)$ determines a map $\phi_x\colon C \to \mathbf{P}^n$. The pullback $\phi_x^*\mathcal{O}_{\mathbf{P}^n}(1)$ is a line bundle on C.

Definition 7.3.15 The *geometric height* of $x \in \mathbf{P}^n(K)$ is $h_{\mathrm{geom}}(x) := \deg(\phi_x^*\mathcal{O}_{\mathbf{P}^n}(1))$.

Remark 7.3.16 Let M_K' be the places on K that are trivial on k (which correspond to the codimension-1 points of C). Given coordinates $x = [x_0 : \cdots : x_n]$, we have $h_{\mathrm{geom}}(x) = -\sum_{v\in M_K'} \min\{\mathrm{ord}_v(x_0), \dots, \mathrm{ord}_v(x_n)\}$.

To define the geometric height of points in $\mathbf{P}^n(\overline{K})$, we must keep track of the field of definition as we did for absolute heights. Any point $x \in \mathbf{P}^n(\overline{K})$ is defined over $k(C')$ for some finite cover $C' \to C$ of degree $[k(C') : k(C)]$. This defines a map $\phi_x\colon C' \to \mathbf{P}^n$, and we again get a line bundle $\phi_x^*\mathcal{O}_{\mathbf{P}^n}(1)$ on C'.

Definition 7.3.17 The *geometric height* of $x \in \mathbf{P}^n(\overline{K})$ is

$$h_{\mathrm{geom}}(x) := \frac{\deg(\phi_x^*\mathcal{O}_{\mathbf{P}^n}(1))}{[k(C') : k(C)]},$$

where C' is any finite cover of C such that x is defined over $k(C')$.

Finally, let X be a projective variety over K. Let L be an ample line bundle on X. A point $x \in X(\overline{K})$ determines a map $\phi_x\colon C' \to X$, where $C' \to C$ is a finite cover of degree $[k(C') : k(C)]$. As before, ϕ_x^*L is a line bundle on C'.

Definition 7.3.18 The *geometric height* of $x \in X(\overline{K})$ is

$$h_{\mathrm{geom},L}(x) := \frac{\deg(\phi_x^* L)}{[k(C') : k(C)]},$$

where C' is any finite cover of C such that x is defined over $k(C')$.

Remark 7.3.19 The geometric height $h_{\mathrm{geom},L} \colon X(\overline{K}) \to \mathbf{R}$ depends on the choice of ample line bundle L.

Remark 7.3.20 The assumption that L is ample is a *positivity assumption*. Because L is ample and ϕ_x is finite, $\phi_x^* L$ is an ample bundle on the curve C'. By the Riemann–Roch theorem, a line bundle on a curve is ample if and only if its degree is positive, so the assumption that L is ample guarantees that $h_{\mathrm{geom},L} \colon X(\overline{K}) \to \mathbf{R}$ takes only nonnegative values. We will see analogous positivity assumptions on the line bundles used to define Moriwaki heights in Section 7.6.

7.4 Analytic background

In order to construct geometric heights over more general fields of characteristic 0, one must make sense of the degree of a line bundle at the infinite place. In particular, one needs an intersection theory on the finite and infinite fibers of maps of the form $\mathcal{X} \to \operatorname{Spec} \mathbf{Z}$. Arakelov laid the groundwork for understanding intersection theory on surfaces in an arithmetic sense (even at the infinite place, which a priori has only a complex structure) [1]. This theory was developed further by Faltings in [10] and later was generalized to higher-dimensional varieties by Gillet and Soulé (see e.g., [14]). We will provide some analytic background for arithmetic intersection theory, following [14] and [18]. In this section, we restrict our attention to complex manifolds. All complex manifolds arising in our context will be algebraic varieties.

7.4.1 Differential forms

Let X be a complex manifold of dimension n. Let U be an open subset of X isomorphic to $\mathbf{C}^n$. Pick a system of local coordinates $z_1, z_2, \ldots, z_n$ on U and write $z_j = x_j + iy_j$. A function $f \colon U \to \mathbf{C}$ is said to be *holomorphic* if it satisfies the Cauchy–Riemann equations with respect to each pair (x_j, y_j). A function is holomorphic on X if it is holomorphic on each chart. Holomorphic functions are infinitely ($\mathbf{R}$-)differentiable, and we denote by $C^\infty(X)$ the class of infinitely **R**-differentiable functions on X. The structure sheaf $\mathscr{O}_X$ of X, is the sheaf of holomorphic functions on X. A holomorphic vector bundle on X is a vector bundle $p\colon E \longrightarrow X$ such that (i) p is holomorphic, and (ii) the local trivializations $p^{-1}(U) \cong U \times \mathbf{C}^{\operatorname{rank}(E)}$ are biholomorphic maps.

Definition 7.4.1 (Complexified tangent bundle) Let X be a complex manifold and let TX denote the tangent bundle on the underlying real manifold. The complexified tangent bundle is $T_{\mathbf{C}}X := TX \otimes \mathbf{C}$.

The complexified tangent bundle admits a decomposition

$$\mathrm{T}_{\mathbf{C}}X = \mathrm{T}^{1,0}X \oplus \mathrm{T}^{0,1}X$$

of complex vector bundles on X. The bundle $\mathrm{T}^{1,0}X$ is naturally isomorphic to the holomorphic tangent bundle of X, while the antiholomorphic tangent bundle $\mathrm{T}^{0,1}X$ of X is the complex conjugate of $\mathrm{T}^{1,0}X$.

Remark 7.4.2 In contrast with the analytic nature of Definition 7.4.1, one can define the holomorphic tangent bundle in a more algebraic manner. Define the holomorphic cotangent bundle Ω^1_X by setting $\Omega^1_X(U)$ to be the $\mathscr{O}_X(U)$-algebra generated by

$$\{\mathrm{d}f : f \in \mathscr{O}_X(U) \text{ and } \mathrm{d}f \text{ satisfies the Leibniz rule}\}.$$

Then $\mathrm{T}^{1,0}X$ is the dual of Ω^1_X.

For any integer k, let $A^k(X) := \bigwedge^k (T_{\mathbf{C}}X)^*$, where $(-)^*$ denotes the dual. This sheaf is often called the *space of k-forms* on X.

Definition 7.4.3 (Differential (p,q)-forms) Let $p, q \in \mathbf{Z}_{\geq 0}$. Define the *sheaf of (p,q)-forms* as $A^{p,q}(X) := \left(\bigwedge^p (\mathrm{T}^{1,0}X)^*\right) \otimes \left(\bigwedge^q (\mathrm{T}^{0,1}X)^*\right)$.

The sheaf $A^{p,q}(X)$ has an explicit description in local coordinates. Let $U \subset X$ be an open set with local coordinates $z_1, z_2, \ldots, z_n$. A differential (p,q)-form on U is a $\mathbf{C}(U)$-linear combination of the form

$$\sum a_{i_1 i_2 \ldots j_q} \mathrm{d}z_{i_1} \wedge \mathrm{d}z_{i_2} \wedge \cdots \wedge \mathrm{d}z_{i_p} \wedge \mathrm{d}\bar{z}_{j_1} \wedge \mathrm{d}\bar{z}_{j_2} \wedge \cdots \wedge \mathrm{d}\bar{z}_{j_q},$$

where $\bar{z}$ denotes complex conjugation, and the sum is over all tuples of size p and q.

In subsequent sections, we will use the maps $\partial\colon A^{p,q}(X) \longrightarrow A^{p+1,q}(X)$ and $\bar{\partial}\colon A^{p,q}(X) \longrightarrow A^{p,q+1}(X)$, which are given on coordinate charts by $\partial(f\,\omega) = \sum_{k=1}^n \frac{\partial f}{\partial z_k}\,\mathrm{d}z_k \wedge \omega$ and $\bar{\partial}(f\,\omega) = \sum_{k=1}^n \frac{\partial f}{\partial \bar{z}_k}\,\mathrm{d}\bar{z}_k \wedge \omega$.

Remark 7.4.4 The maps ∂ and $\overline{\partial}$ are closely related to the exterior derivative. For a local function f, the exterior derivative is defined as $\mathrm{d}(f) = \sum(\partial f/\partial z_i)dz_i + \sum(\partial f/\partial \overline{z}_j)\mathrm{d}z_j$. This can be extended to a map $\mathrm{d}\colon A^k(X) \to A^{k+1}(X)$ using the Leibniz rule: $\mathrm{d}(u \wedge v) = \mathrm{d}u \wedge v + (-1)^{\deg u} u \wedge \mathrm{d}v$. Using the decomposition $A^k(X) = \bigoplus_{p+q=k} A^{p,q}(X)$, we have that $\mathrm{d} = \partial + \overline{\partial}$. For more details, see [18].

Hermitian metrics

Before we proceed, we recall a few definitions about Hermitian vector bundles on a manifold.

Definition 7.4.5 A *Hermitian form* on a complex vector space V is a pairing $H\colon V \times V \to \mathbf{C}$ such that (i) $H(u,v)$ is $\mathbf{C}$-linear in the first variable, and (ii) $H(u,v) = \overline{H(v,u)}$ for all $u, v \in V$. Further, H is *positive definite* if $H(u,u) > 0$ for all $u \neq 0$. In this case, one can associate a metric to a Hermitian form by defining $\|u\|_H := \sqrt{H(u,u)}$. In what follows, we will suppress the subscript H whenever it is clear from context.

Definition 7.4.6 A *Hermitian metric H* on a holomorphic vector bundle $E \to X$ on a complex manifold X is a smoothly varying positive definite Hermitian form on each fiber. A *(Hermitian) metrized vector bundle* on X is a pair (E, H) consisting of a vector bundle E equipped with a (Hermitian) metric H.

Example 7.4.7 Let $(X, L) = (\mathbf{P}^n(\mathbf{C}), \mathcal{O}(1))$. For any point $x = [x_0 : \cdots : x_n]$ and section $s \in L$ that does not vanish in a neighborhood of x, define

$$\|s(x)\|_\infty = \frac{|s(x)|}{\max\{|x_0|, |x_1|, \ldots, |x_n|\}}.$$

Then $(L, \|\cdot\|_\infty)$ is a metrized line bundle on X.

Every complex vector bundle admits a Hermitian metric by gluing together the standard Hermitian metric on $\mathbf{C}^n$ (see [18, Proposition 4.1.4]).

7.4.2 Currents

A *current* is an element of the dual space of the space of differential forms that satisfies some additional completeness properties. In this chapter, we will not define currents in full generality. Instead, we will give some key examples that are sufficient for our purposes here. We denote by $D_{p,q}(X) := (A^{p,q}(X))^*$ and $D_d(X) := (A^d(X))^*$ the space of currents of bidimension (p, q) and the space of currents of dimension d, respectively.

Example 7.4.8 (Current associated to a subspace) Let $\iota\colon Y \to X$ be an analytic subspace of X of dimension k, and let $\alpha \in A^{2k}(X)$ be a differential form on X. We define a current $\delta_Y \in D_{2k}(X)$ by

$$\delta_Y(\alpha) = \int_Y \iota^*\alpha.$$

Note that this definition can be extended to any analytic cycle, i.e., any $\mathbf{Z}$-linear combination of analytic subspaces. Also note that if $\beta \in A^{p,q}(X)$ with $p + q = 2k$, then $\iota^*(\beta) = 0$ unless $p = q = k$. It follows that $\delta_Y \in D_{k,k}(X)$.

Example 7.4.9 (Current associated to a differential form) Let α denote a (p, q)-form. The current associated to α is the map

$$[\alpha]\colon A^{n-p,n-q}(X) \to \mathbf{C} \quad \text{given by} \quad \beta \mapsto \int_X \alpha \wedge \beta.$$

This defines a map $A^{p,q}(X) \to D_{n-p,n-q}(X)$ sending α to $[\alpha]$. Alternatively, one may think of this as a pairing:

$$A^{p,q}(X) \times A^{n-p,n-q}(X) \to \mathbf{C} \quad \text{given by} \quad (\alpha, \beta) \mapsto \int_X \alpha \wedge \beta.$$

Example 7.4.10 (Logarithmic current associated to a line bundle) Let Y be a divisor on X, and let L be the line bundle corresponding to Y. Let s be a section of L. Choose a smooth Hermitian norm $\|\cdot\|$ on L. Then $\log\|s\|^2$ is a $(0, 0)$-form on X, which has an associated current $[-\log\|s\|^2]$. Further, $[-\log\|s\|^2]$ is a *Green current* for Y; that is, there exists a smooth closed (1,1)-form β on X such that

$$\frac{i}{2\pi}\partial\bar{\partial}\log\|s\|^2 = \delta_Y - \beta.$$

The form $\beta \in A^{1,1}(X)$ is known as the *Chern form* and will be discussed in the following section.

7.4.3 Chern classes

Given a line bundle L on X, one can define its first Chern class $c_1(L)$ in a variety of ways. In arithmetic intersection theory, one needs both the algebraic and the analytic descriptions of the Chern class. In this subsection, we give a brief analytic description of $c_1(L)$. Let us first recall the definition of $c_1(L)$.

Definition 7.4.11 (First Chern class) Let $L \in H^1(X, \mathcal{O}_X^*)$ be a line bundle. Then $c_1(L)$ is the image of L in $H^2(X, \mathbf{Z})$ under the boundary map of the long exact sequence induced by the exponential exact sequence $0 \to \mathbf{Z} \to \mathcal{O}_X \to \mathcal{O}_X^* \to 0$.

Definition 7.4.12 Let X be a manifold with a metrized Hermitian holomorphic line bundle L on X, and let s be a section of L. The *first Chern class*, which by an abuse of notation we also denote by $c_1(L)$, is the de Rham cohomology class of the differential form whose associated current is given by

$$\delta_{\mathrm{div}(s)} - \frac{i}{2\pi}\partial\bar{\partial}[\log \|s\|^2]. \tag{1}$$

This is independent of the choice of s, by the Poincaré–Lelong formula ([15, Chapter 3, Section 2]), which states that $\delta_{\mathrm{div}(f)} + \frac{i}{2\pi}\partial\bar{\partial}[\log \|f\|^2] = 0$ for any meromorphic function f on X. Since $\log \|s\| \in A^{0,0}(X)$ and $\delta_{\mathrm{div}(s)} \in D_{n-1,n-1}(X)$, we have $c_1(L) \in A^{1,1}(X)$.

Remark 7.4.13 We now give a brief justification for the abuse of notation in Definition 7.4.12.

(i) Recall that the de Rham cohomology of X is defined as the cohomology of the complex $A^\bullet(X)$. Further, $c_1(L)$ is closed and invariant under complex conjugation and thus defines a cohomology class in $H^2_{dR}(X, \mathbf{R}) \subset H^2_{dR}(X, \mathbf{C})$. The divisor $\mathrm{div}(s)$ also defines a class in $H^2(X, \mathbf{R})$ via the map $H^2(X, \mathbf{Z}) \to H^2_{dR}(X, \mathbf{R})$. The Poincaré–Lelong formula can be used to show that these two classes in $H^2_{dR}(X, \mathbf{R})$ are the same (see e.g., [18, Proposition 4.4.123]).

(ii) The "analytic" Chern class is usually not defined as in equation (1), but rather as the failure of a certain complex to be exact. This approach gives an explicit way to calculate $c_1(L)$. We omit the details here for brevity and refer the reader to [18, Chapter 4] for details.

7.4.4 Arakelov–Green currents

In this subsection, we briefly define Arakelov–Green functions on a Riemann surface X. These functions were used by Arakelov in [1] to define an Archimedean version of the local intersection number. In essence, they play the same role as a uniformizer in the non-Archimedean case. This comes up in Section 7.5. See [20] for more details.

Definition 7.4.14 Let X be a Riemann surface and let μ be a Hermitian metric on X with volume element $d\mu$. The *Arakelov–Green function* $G\colon X \times X \to \mathbf{R}_{\geq 0}$ for μ is the unique function satisfying all the following properties:

(i) $G(P,Q)^2 \in C^\infty(X \times X)$ and vanishes only on the diagonal Δ_X. For a fixed $P \in X$, an open neighborhood U of P, and a local coordinate z on U such that $z(P) = 0$, there exists $f \in C^\infty(X)$ such that $\log G(P,Q) = \log |z(Q)| + f(Q)$ for all $Q \in U \backslash \{P\}$.

(ii) For all $P \in X$, we have $\partial_Q \overline{\partial}_Q \log(G(P,Q))dxdy = 2\pi\imath d\mu(Q)$ for any $Q \neq P$.

(iii) G is symmetric, i.e., $G(P,Q) = G(Q,P)$.

(iv) For all $P \in X$, we have $\int_X \log G(P,Q) d\mu(Q) = 0$.

Condition (i) allows us to think of $G(P,-)$ as a uniformizer around P. The rest of the conditions uniquely determine G among the class of possible uniformizers.

Example 7.4.15 Let $X = \mathbf{P}^1_{\mathbf{C}}$ with the metric given by $d\mu = \frac{1}{2\pi}\frac{|dz|^2}{(1+|z|^2)^2}$. (This is a normalized version of the Fubini–Study metric.) Then the corresponding Green function is given by

$$G^2(w,z) = e\frac{|w-z|^2}{(1+|w|^2)(1+|z|^2)}.$$

Remark 7.4.16 A Green function defines a Hermitian metric on the line bundle $\mathscr{O}_{X\times X}(\Delta_X)$ via $\|s_\Delta\| = G(P,Q)$, where s_Δ is the image of the unit section of $\mathscr{O}_{X\times X}$.

7.5 Arithmetic intersection theory and Arakelov theory

We now describe intersection theory on arithmetic varieties. Roughly speaking, arithmetic varieties are varieties over rings which have both finite and infinite places (e.g., the ring of integers over a number field). This differs from intersection theory in more classical settings (e.g., as in [13]) in that it takes into account the sizes of the residue fields at the finite places as well as making sense of what it means for divisors to "intersect at infinity."

We begin by giving some definitions in Section 7.5.1. In Section 7.5.2, we describe intersection theory on surfaces, following [1]. In the remaining subsections, we give a description of intersections on higher-dimensional arithmetic varieties, following [14] and [23].

7.5.1 Arithmetic varieties

The main reference for this subsection is [14]. Arithmetic varieties in Gillet and Soulé are defined over *arithmetic rings*, which are essentially rings equipped with embeddings into $\mathbf{C}$ and a notion of complex conjugation. For the purpose of this subsection we will consider arithmetic varieties over the ring of integers in number field K. We will let $B = \operatorname{Spec} \mathcal{O}_K$, where $\mathcal{O}_K$ denotes the ring of integers of K.

Definition 7.5.1 An *arithmetic variety* $\mathcal{X}$ over B is a flat, finite type scheme over B. We write $\mathcal{X}_K$ for the generic fiber of $\mathcal{X}$. For any point $s \in B$, we denote by $\mathcal{X}_s$ the fiber over s. If $\sigma\colon \mathcal{O}_K \to \mathbf{C}$ is an embedding of $\mathcal{O}_K$, we write $\mathcal{X}_\sigma := \mathcal{X} \otimes_{\sigma,\mathcal{O}_K} \mathbf{C}$. If Σ denotes the set of embeddings of $\mathcal{O}_K$ into $\mathbf{C}$, we write $\mathcal{X}_\Sigma := \coprod_{\sigma\in\Sigma} \mathcal{X}_\sigma$. The analytic subspace $\mathcal{X}_\Sigma(\mathbf{C})$ comes equipped with an involution, which we call F_∞. An arithmetic surface is an arithmetic variety $\mathcal{X} \to B$ such that the generic fiber $\mathcal{X}_K$ is a geometrically connected curve over K.

From now on, we will assume that $\mathcal{X}_K$ is smooth for convenience.

Example 7.5.2 $\mathbf{P}^n$ is an arithmetic variety over $\mathbf{Z}$, where the infinite fiber is the complex manifold $\mathbf{P}^n(\mathbf{C})$. In this case, F_∞ is just complex conjugation on the coordinates.

Example 7.5.3 (Néron model) The Néron model of a rational elliptic curve is an arithmetic variety over $\mathbf{Z}$. More precisely, given an elliptic curve $E_{/\mathbf{Q}}$, there exists a smooth (commutative group) scheme $\mathcal{E}$ over $\mathbf{Z}$ whose generic fiber $\mathcal{E}_\mathbf{Q}$ is isomorphic to E. Among such schemes, there is one (unique up to unique isomorphism) that is universal with respect to the Néron mapping property, called the *Néron model* of E. Its existence is a deep theorem. For more details, we refer the reader to [28] for the elliptic curve case and to [4] for the case of general abelian varieties. In the elliptic curve case, the infinite fiber is a torus and F_∞ is the complex conjugation induced from $\mathbf{C}$.

We will write $A^{p,q}(\mathcal{X})$ for $\bigoplus_{\sigma\in\Sigma} A^{p,q}(\mathcal{X}_\sigma)$. Any integral subscheme $\mathcal{Y}$ of $\mathcal{X}$ is an arithmetic variety if and only if $\mathcal{Y} \to B$ is surjective. In this case, $\mathcal{Y}_\Sigma(\mathbf{C})$ is also a (disjoint union of) complex manifold(s).

Definition 7.5.4 An *Arakelov divisor* on $\mathcal{X}$ is the sum of a Weil divisor on $\mathcal{X}$ and an infinite contribution $\sum_\sigma \alpha_\sigma \mathcal{X}_\sigma$, where the sum is over all embeddings $\sigma\colon K \hookrightarrow \mathbf{C}$.

Let $\mathcal{X}$ be an arithmetic variety, with a choice of volume form μ_σ on each infinite fiber $\mathcal{X}_\sigma$. Let D be a Weil divisor on $\mathcal{X}$. Then, the choice of Green currents (see Section 7.4.2) $\{[g_\sigma]\}_{\sigma\in\Sigma}$ for D_σ on $\mathcal{X}_\Sigma$ turns D into an Arakelov divisor. In this case, $\alpha_\sigma = \int g_\sigma \cdot d\mu_\sigma$.

Definition 7.5.5 A *principal* Arakelov divisor is of the form

$$\operatorname{div}(f) + \sum_{\sigma\in\Sigma} \nu_\sigma(f)\mathcal{X}_\sigma$$

for a rational function f on $\mathcal{X}$, where $\nu_\sigma(f) := -\int_{\mathcal{X}_\sigma} \log |f|_\sigma \cdot d\mu_\sigma$.

Let $\pi\colon \mathcal{X} \to B$ be an arithmetic surface. For any point $s \in B$ (or infinite place σ) the fiber $\mathcal{X}_s$ (resp. $\mathcal{X}_\sigma$) is a *vertical* divisor. In general, a *vertical* divisor is a linear combination of irreducible components of closed fibers. A prime divisor D is called *horizontal* if $\pi(D) = B$. A divisor is called *horizontal* if it is a finite sum of horizontal prime divisors. In particular, if a prime divisor D is horizontal, then there is a finite extension F/K and a map $\varepsilon\colon \operatorname{Spec}\mathcal{O}_F \to \mathcal{X}$ over B such that $D = \varepsilon(\operatorname{Spec}\mathcal{O}_F)$.

7.5.2 Intersections on an arithmetic surface

We now discuss intersections on an arithmetic surface (i.e., a two-dimensional scheme whose generic fiber is a smooth curve) as motivation for intersections on higher-dimensional varieties. We primarily follow [1]. A discussion on intersections on arithmetic surfaces can also be found in [21, Section 9.1]. For a detailed exposition on Arakelov theory for arithmetic surfaces, we refer the reader to [20]. We will assume that $\pi\colon \mathcal{X} \to B = \operatorname{Spec}\mathcal{O}_K$ is an arithmetic surface that is regular, with smooth generic fiber $\mathcal{X}_K$.

Definition 7.5.6 Let D_1, D_2 be distinct irreducible divisors on $\mathcal{X}$ and let $x \in \mathcal{X}$ be a closed point. Let f and g be two functions that cut out D_1 and D_2 locally around x. We define

$$\langle D_1, D_2\rangle_x := \operatorname{length}_{\mathcal{O}_{\mathcal{X},x}}(\mathcal{O}_{\mathcal{X},x}/(f,g)) \log |k(x)|,$$

where $k(x)$ denotes the residue field of x. The first part of this intersection number is "geometric" in nature, in that it looks like the intersection number in the algebraically closed case [13]. The second part, $\log |k(x)|$, keeps track of arithmetic information about the points of intersection. This intersection pairing can be extended by linearity to any pair of divisors on $\mathcal{X}$.

Any point $x \in \mathcal{X}$ is in $\mathcal{X}_b$ for some $b \in B$. Further, the residue field

$k(x)$ of x is a finite extension of the residue field $k(b)$ of b. Let D_1 and D_2 be two divisors on $\mathcal{X}$ with no common components. Define the *total intersection* over b as

$$\langle D_1, D_2 \rangle_b := \sum_{x \in |D_1 \cap D_2|} \langle D_1, D_2 \rangle_x.$$

Lemma 7.5.7 ([1], Section 1) *Let D_1 be a horizontal prime divisor on $\mathcal{X}$, i.e., there is some finite extension F/K with ring of integers $\mathcal{O}_F$ such that D_1 is the image of a map ε: $\operatorname{Spec} \mathcal{O}_F \to \mathcal{X}$. Let D_2 be any other divisor on $\mathcal{X}$. Let $x \in D_1 \cap D_2$ be a closed point of $\mathcal{X}$ and suppose D_2 is defined locally around x by a function f. Let $\mathfrak{p}_1, \ldots, \mathfrak{p}_r$ be the primes of $\mathcal{O}_F$ such that $\varepsilon(\mathfrak{p}_i) = x$. Then*

$$\langle D_1, D_2 \rangle_x = \sum_{i=1}^{r} - \log \|\varepsilon^* f\|_{\mathfrak{p}_i}.$$

Proof Let $f|_{D_1}$ denote the restriction of f to D_1. Then, by definition of the intersection number, we have $\langle D_1, D_2 \rangle_x = \operatorname{ord}_x(f|_{D_1}) \log |k(x)|$. Thus we have

$$\operatorname{ord}_x(f|_{D_1}) = \sum_{i=1}^{r} \operatorname{ord}_{\mathfrak{p}_i}(\varepsilon^* f|_{D_1})[k(\mathfrak{p}_i) : k(x)],$$

where $k(\mathfrak{p}_i)$ denotes the residue field of $\mathfrak{p}_i$. By definition of the non-Archimedean absolute value, $\|\alpha\|_{\mathfrak{p}_i} = |k(\mathfrak{p}_i)|^{-\operatorname{ord}_{\mathfrak{p}_i}(\alpha)}$. Thus

$$\begin{aligned} -\sum_{i=1}^{r} \log \|\varepsilon^* f_2|_{D_1}\|_{\mathfrak{p}_i} &= \sum_{i=1}^{r} \operatorname{ord}_{\mathfrak{p}_i}(\varepsilon^* f|_{D_1}) \log |k(\mathfrak{p}_i)| \\ &= \sum_{i=1}^{r} \operatorname{ord}_{\mathfrak{p}_i}(\varepsilon^* f|_{D_1})[k(\mathfrak{p}_i) : k(x)] \log |k(x)|. \quad \square \end{aligned}$$

Given two Arakelov divisors D_1 and D_2 that are not contained in distinct fibers of $\mathcal{X} \to B$, their intersection (which we denote by $\langle D_1, D_2 \rangle$) has a finite and an infinite component. In [1], Arakelov defines the infinite part of the intersection by first defining an "intersection number" for two points P and Q on the Riemann surface $\mathcal{X}_\sigma$ using Green functions (see Section 7.4.4). Let $\mathcal{X}_\sigma$ be any Riemann surface with a Hermitian metric μ, and let $P, Q \in \mathcal{X}_\sigma$. Motivated by Lemma 7.5.7, one might want to define $\langle P, Q \rangle = -\log \phi_P(Q)$, where $\phi_P(z)$ is a function that vanishes to degree 1 at P (like a uniformizer). However, there are too many functions that satisfy this, so one insists on additional conditions. For example, one requires that ϕ is a nonnegative function with a unique zero at P, with a first-order zero at P. Imposing further conditions, such as symmetry of $\langle\cdot,\cdot\rangle$ and the normalization $\int \log \phi_P d\mu = 0$, leads to the concept of an *Arakelov–Green function*, as defined in Section 7.4.4. For more details on the significance of the properties of such functions, we refer the reader to [1]. We now define the total intersection product of two Arakelov divisors.

Definition 7.5.8 (Intersection of Arakelov divisors) Let D_1 and D_2 be two irreducible Arakelov divisors on an arithmetic surface $\mathcal{X}$. Then the intersection product $\langle D_1, D_2\rangle$ is defined as the symmetric $\mathbf{R}$-bilinear form satisfying the following conditions:

(i) If D_1 is a vertical divisor and D_2 has no components in common with D_1, then $\langle D_1, D_2\rangle := \sum_{b\in B}\langle D_1, D_2\rangle_b$ where the sum is over the closed points of B. This implies that if either D_1 or D_2 is a fiber of $\mathcal{X} \to B$, then there is no infinite component of the intersection.

(ii) Let D_1 be a horizontal divisor and $D_2 = \mathcal{X}_\sigma$ for some σ. Suppose D_1 is the image of a point $\varepsilon\colon B_F \to \mathcal{X}_K$ for a finite extension F/K. Then $\langle D_1, D_2\rangle$ is defined as the degree $[F : K]$. Equivalently, this is the degree of the residue field of D_1 over K.

(iii) If σ, σ' are two distinct embeddings $K \hookrightarrow \mathbf{C}$, then $\langle \mathcal{X}_\sigma, \mathcal{X}_{\sigma'}\rangle = 0$.

(iv) Suppose D_1 and D_2 are two horizontal sections of $X \to B$, defined over number fields F_1 and F_2 respectively. Fix a $\sigma\colon K \hookrightarrow \mathbf{C}$. Let $\tau_1\colon F_1 \hookrightarrow \mathbf{C}$ and $\tau_2\colon F_2 \hookrightarrow \mathbf{C}$ be embeddings that extend σ. These correspond to points P^{τ_1} and P^{τ_2} on the Riemann surface $\mathcal{X}_\sigma$. Define

$$\langle D_1, D_2\rangle_\sigma = \sum_{\tau_1,\tau_2} -\log G_\sigma(P^{\tau_1}, P^{\tau_2}),$$

where G_σ is the Arakelov–Green function attached to $\mathcal{X}_\sigma$.

Finally, the *total intersection* is defined as

$$\langle D_1, D_2\rangle := \sum_{b\in B}\langle D_1, D_2\rangle_b + \sum_{\sigma\colon K\hookrightarrow\mathbf{C}} \langle D_1, D_2\rangle_\sigma, \tag{1}$$

Proposition 7.5.9 ([1], Proposition 1.2) *Let D_1 be a principal Arakelov divisor on $\mathcal{X}$. Then $\langle D_1, D\rangle = 0$ for any divisor D.*

Using Lemma 7.5.7 and the properties of Arakelov–Green functions, this proposition reduces to the product formula on K.

7.5.3 Intersections on higher-dimensional varieties

We now take a different approach to intersection theory on higher-dimensional arithmetic varieties, following Fulton [13, Section 2.3], Moriwaki [23], and Gillet–Soulé [14].

Definition 7.5.10 Let X be any variety and let D be a Cartier divisor on X. Let $j\colon V \hookrightarrow X$ be a codimension-p irreducible subvariety of X. If V is not contained in the support of D, then define

$$D \cdot [V] = [j^*(D)],$$

where $[-]$ is the cycle corresponding to the respective subvariety. If V is contained in the support of D, then $j^*(D)$ is no longer a Cartier divisor on V. In this case, consider the line bundle $j^*\mathcal{O}_X(D)$. Let $[C]$ denote the Weil divisor on V corresponding to the line bundle $\mathcal{O}_V(C)$ that is isomorphic to $j^*\mathcal{O}_X(D)$. Define $D \cdot [V] = [C]$.

Let $Z^p(X)$ denote the set of codimension-p cycles on X. Then, extending the above definition linearly, we can define a homomorphism $Z^p(X) \to Z^{p+1}(X)$. Further, this map respects rational equivalence and thus descends to a homomorphism $\mathrm{CH}^p(X) \to \mathrm{CH}^{p+1}(X)$. When $L = \mathcal{O}_X(D)$, this is precisely the algebraic first Chern class. We now proceed to define a similar homomorphism in the arithmetic world.

Arithmetic Chow groups

Here we let $\mathcal{X}$ be an arithmetic variety over $\operatorname{Spec} \mathbf{Z}$ that is regular, with smooth generic fiber. By a metrized line bundle $\overline{\mathcal{L}}$ on $\mathcal{X}$, we will mean a line bundle $\mathcal{L}$ on $\mathcal{X}$ with a metric $\|\cdot\|$ on $\mathcal{L} \otimes \mathbf{C}$ on $\mathcal{X}_\infty(\mathbf{C})$.

Definition 7.5.11 An *arithmetic cycle* of codimension p is a pair (Z, g), where Z is a codimension-p algebraic cycle on $\mathcal{X}$ and g is a Green current for $Z(\mathbf{C})$. An *arithmetic D-cycle* of codimension p is a pair (Z, g) where Z is a codimension-p algebraic cycle on $\mathcal{X}$, and g is a current of types $(p-1, p-1)$ on $\mathcal{X}(\mathbf{C})$.

The set of all arithmetic cycles (resp. D-cycles) of codimension p is denoted $\widehat{Z}^p(\mathcal{X})$ (resp. $\widehat{Z}^p_D(\mathcal{X})$). Let $\widehat{R}^p(\mathcal{X})$ denote the subgroup of $\widehat{Z}^p(\mathcal{X})$ generated by:

(i) $(\operatorname{div}(f), [-\log|f|^2])$, where f is a rational function on some subvariety Y of codimension $p-1$, and $[\log|f|^2]$ is the current defined by $\phi \mapsto \int_{Y(\mathbf{C})} \log|f|^2 \wedge \phi$.

(ii) $(0, \partial\alpha + \bar{\partial}\beta)$ where α and β are forms of types $(p-2, p-1)$ and $(p-1, p-2)$ respectively.

Note that $\widehat{R}^p(\mathcal{X})$ can also be considered as subgroup of $\widehat{Z}^p_D(\mathcal{X})$. This allows us to make the following definitions.

Definition 7.5.12 Define the *arithmetic Chow group* and *arithmetic D-Chow group* of codimension p as $\widehat{\mathrm{CH}}^p(\mathcal{X}) = \widehat{Z}^p(\mathcal{X})/\widehat{R}^p(\mathcal{X})$ and $\widehat{\mathrm{CH}}^p_D(\mathcal{X}) = \widehat{Z}^p_D(\mathcal{X})/\widehat{R}^p(\mathcal{X})$, respectively.

Definition 7.5.13 Let $\overline{\mathcal{L}} = (\mathcal{L}, \|\cdot\|)$ be a metrized line bundle on $\mathcal{X}$. Let $(Z, g) \in \widehat{Z}^p_D(\mathcal{X})$ and suppose Z is integral. Further, let s be a rational section of $\mathcal{L}|_Z$. Define the map $\tilde{\phi}\colon \widehat{Z}^p_D(\mathcal{X}) \to \widehat{Z}^{p+1}_D(\mathcal{X})$ by

$$\tilde{\phi}\colon (Z, g) \mapsto (\mathrm{div}(s), [-\log \|s\|_Z^2] + c_1(\overline{\mathcal{L}}) \wedge g),$$

where $c_1(\overline{\mathcal{L}})$ is as in Section 7.4.3. Note that $c_1(\overline{\mathcal{L}})$ is a $(1,1)$-form, so $c_1(\overline{\mathcal{L}}) \wedge g$ is (dual to) a (p, p)-form. Remark 7.4.13 implies that the form $c_1(\overline{\mathcal{L}})$ vanishes on $\widehat{R}^p(\mathcal{X})$, so the map $\tilde{\phi}$ descends to a homomorphism

$$\widehat{c}_1(\overline{\mathcal{L}}) \cdot (-)\colon \widehat{\mathrm{CH}}^p_D(\mathcal{X}) \to \widehat{\mathrm{CH}}^{p+1}_D(\mathcal{X}).$$

We call $\widehat{c}_1(\overline{\mathcal{L}})$ the *first arithmetic Chern class*. Just as in the classical case, the first Chern class also admits a description as a cycle. This is given by $\widehat{c}_1(\overline{\mathcal{L}}) = (\mathrm{div}(s), -\log \|s\|^2)$ for some section s of $\mathcal{L}$.

The first Chern class satisfies the following projection formula.

Proposition 7.5.14 ([23], Proposition 1.2) *Let $f\colon \mathcal{X} \to \mathcal{Y}$ be a projective morphism of generically smooth arithmetic varieties. Let $\overline{\mathcal{L}}$ be a C^∞-Hermitian line bundle on $\mathcal{Y}$, and let $z \in \widehat{\mathrm{CH}}^p_D(\mathcal{X})$. Then $f_*(\widehat{c}_1(f^*\overline{\mathcal{L}}) \cdot z) = \widehat{c}_1(\overline{\mathcal{L}}) \cdot f_*(z)$ (see Definition 7.5.13).*

Definition 7.5.15 (Arakelov degree) Let d be the dimension of the generic fiber of $\mathcal{X} \to B$. Define the *arithmetic intersection number* $\widehat{\deg}\colon \widehat{\mathrm{CH}}^{d+1}_D(\mathcal{X}) \to \mathbf{R}$ by

$$\widehat{\deg}\left(\sum_P n_P P, T\right) = \sum_P n_P \log |k(P)| + \frac{1}{2} \int_{X(\mathbf{C})} T. \tag{2}$$

An inductive argument using the product formula for number fields implies that $\widehat{\deg}$ is 0 on $\widehat{R}^{d+1}(\mathcal{X})$, so this is well defined on $\widehat{\mathrm{CH}}^{d+1}_D(\mathcal{X})$. The projection formula for $\widehat{c}_1$ implies a similar projection formula for $\widehat{\deg}$ [23, Proposition 1.3].

7.6 Moriwaki heights

We now discuss the height function defined by Moriwaki [23]. Moriwaki heights are fairly general and specialize to some of the heights defined in Section 7.3. Roughly speaking, Moriwaki heights are the higher-transcendence-degree analogues of geometric heights over number fields (see Definition 7.3.18). They bridge the gap between geometric heights, which have a pleasant definition but are often poorly behaved, and naïve heights, which are generally well behaved.

Definition 7.6.1 Let $\mathcal{B}$ be an arithmetic variety. Let $\mathcal{X}$ be a normal projective arithmetic variety over $\mathcal{B}$. A metrized line bundle $\overline{\mathcal{L}}$ on $\mathcal{X}$ is *nef* if $c_1(\mathcal{L})$ is semipositive and $\widehat{\deg}(\overline{\mathcal{L}}|_\Gamma) \geq 0$ for all curves $\Gamma \subset \mathcal{B}$.

Definition 7.6.2 Let K be a finitely generated field over $\mathbf{Q}$, and let $\operatorname{trdeg}_{\mathbf{Q}}(K) = d$. A *polarization* of K is a collection $\overline{\mathcal{B}} = (\mathcal{B}; \overline{\mathcal{H}}_1, \ldots, \overline{\mathcal{H}}_d)$, where $\mathcal{B}$ is a normal, projective, arithmetic variety with fraction field K, and each $\overline{\mathcal{H}}_i$ is a nef C^∞-Hermitian line bundle on $\mathcal{B}$.

Now let $X \to \operatorname{Spec} K$ be a variety with an integral model $\pi\colon \mathcal{X} \to \mathcal{B}$, that is to say, $\pi\colon \mathcal{X} \to \mathcal{B}$ is an arithmetic variety whose generic fiber is isomorphic to X. Let L be a metrized $\mathbf{Q}$-line bundle on X that extends to a line bundle $\mathcal{L}$ on $\mathcal{X}$. The pair $(\mathcal{X}, \mathcal{L})$ is called a *model* for (X, L).

Finally, given $P \in X(\overline{K})$, let $\Delta_P \in \mathcal{X}$ be the Zariski closure of the image of P under $X \hookrightarrow \mathcal{X}$. The *Moriwaki height* corresponding to the polarization $\overline{\mathcal{B}}$ is

$$h^{\overline{\mathcal{B}}}_{(\mathcal{X},\mathcal{L})}(P) := \frac{1}{[K(P):K]}\widehat{\deg}\left(\widehat{c}_1(\mathcal{L}|_{\Delta_P}) \cdot \prod_{i=1}^{d} \widehat{c}_1(\pi^*\overline{\mathcal{H}}_i|_{\Delta_P})\right).$$

Up to bounded functions, the Moriwaki height is independent of the choice of model $(\mathcal{X}, \mathcal{L})$ [23, Corollary 3.3.5]. One may thus write $h^{\overline{\mathcal{B}}}_L := h^{\overline{\mathcal{B}}}_{(\mathcal{X},\mathcal{L})} + O(1)$. By making particular choices for our polarization $\overline{\mathcal{B}}$, we can recover some of the height functions introduced in Section 7.3.

Example 7.6.3 (Geometric height) Let K be a number field, so that $\operatorname{trdeg}_{\mathbf{Q}}(K) = 0$. Then

$$h^{\overline{\mathcal{B}}}_{(\mathcal{X},\mathcal{L})}(P) = \frac{\widehat{\deg}(\widehat{c}_1(\mathcal{L}|_{\Delta_P}))}{[K(P):K]},$$

which is the Arakelov-theoretic analogue of the geometric height (see Definition 7.3.18).

Definition 7.6.4 Let K be a finitely generated extension of $\mathbf{Q}$. Fix a polarization $\overline{\mathcal{B}}$ of K. Let $x := (x_0, \ldots, x_n) \in K^{n+1}\backslash\{0\}$. Let Γ be the set of all prime divisors in $\mathcal{B}$. Define the *naïve height* with respect to $\overline{\mathcal{B}}$ as

$$\begin{aligned} h_{nv}^{\overline{\mathcal{B}}}(x) := \sum_{\gamma\in\Gamma} \max_i\{-\operatorname{ord}_\gamma(x_i)\}\, \widehat{\deg}\left(\prod_{i=1}^{d} \widehat{c}_1(\overline{\mathcal{H}}_i)|_\gamma\right) \\ + \int_{P\in\mathcal{B}(\mathbf{C})} \log\max_i\{|x_i(P)|\} \bigwedge_{i=1}^{d} c_1(\overline{\mathcal{H}}_i). \end{aligned} \tag{1}$$

By [23, Section 3.2], $h_{nv}^{\overline{\mathcal{B}}}$ is well defined on $\mathbf{P}^n(K)$ and compatible with finite extensions K'/K. The motivation for calling $h_{nv}^{\overline{\mathcal{B}}}$ the naïve height is the following observation.

Remark 7.6.5 If K is a number field (i.e., $d = 0$), then $\prod_{i=1}^{d} \widehat{c}_1(\overline{\mathcal{H}}_i)|_\gamma = ([\gamma], 0)$ and $\bigwedge_{i=1}^{d} c_1(\overline{\mathcal{H}}_i) = 1$. Setting $\mathcal{B} = \operatorname{Spec} \mathcal{O}_K$, we find that $h_{nv}^{\overline{\mathcal{B}}}$ recovers the naïve logarithmic height (Definition 7.3.3):

$$\begin{aligned} h_{nv}^{\overline{\mathcal{B}}}(x) &= \sum_{\mathfrak{p}\nmid\infty} \log|k(\mathfrak{p})| \cdot \max_i\{-\operatorname{ord}_\mathfrak{p}(x_i)\} + \sum_{\mathfrak{p}\mid\infty} \log\max_i\{|x_i|_\mathfrak{p}\} \\ &= \log h_K(x). \end{aligned}$$

Similarly, $h_{nv}^{\overline{\mathcal{B}}}$ is induced by a generalized global field structure on $\mathbf{Q}(t_1, \ldots, t_d)$.

Theorem 7.6.6 *Let $K = \mathbf{Q}(\mathbf{P}^d)$. Let $\mathcal{B} = \mathbf{P}^d_\mathbf{Z}$ and $\overline{\mathcal{H}}_i = (\mathcal{O}_\mathcal{B}(1), \|\cdot\|_\infty)$ for $1 \le i \le d$. Then there exists a generalized global field structure on K such that $h_{nv}^{\overline{\mathcal{B}}}$ is the logarithmic standard height.*

Proof Let Γ be the set of Weil divisors $\gamma \subset \mathcal{B}$, and let $\Gamma_\infty = \Gamma \cup \{\infty\}$. For each $\gamma \in \Gamma$, set $\|x\|_\gamma = e^{-\lambda_\gamma \operatorname{ord}_\gamma(x)}$, where $\lambda_\gamma = \widehat{\deg}(\widehat{c}_1(\overline{\mathcal{O}_\mathcal{B}(1)}|_\gamma)^d)$. Define the absolute value at the Archimedean place to be $\|x\|_\infty$ (see Example 7.4.7). Then $(K, \{\|\cdot\|_v\}_{v\in\Gamma_\infty})$ is a generalized global field, and $h_{nv}^{\overline{\mathcal{B}}}$ is the corresponding logarithmic standard height:

$$\begin{aligned} h_{nv}^{\overline{\mathcal{B}}}(x) &= \sum_{\gamma\in\Gamma} \log\max_i\{\|x_i\|_\gamma\} + \int_{\mathcal{B}(\mathbf{C})} \log\max_i\{\|x_i\|_\infty\} \\ &= \sum_{v\in\Gamma_\infty} e_v \log\max_i\{\|x_i\|_v\}. \end{aligned}$$

(Note that the Archimedean place is real, so $e_v = 1$ for all $v \in \Gamma_\infty$.) To see that $(K, \{\|\cdot\|_v\}_{v\in\Gamma_\infty})$ is indeed a generalized global field, it suffices to verify the generalized product formula for $\{\|\cdot\|_v\}$. Given $x \in K^\times$, this is equivalent to computing $\sum_{v\in\Gamma_\infty} e_v \log \|x\|_v = 0$. Indeed, note that

$$\sum_{v\in\Gamma_\infty} e_v \log \|x\|_v = \sum_{\gamma\in\Gamma} (-\operatorname{ord}_\gamma(x))\lambda_\gamma + \int_{\mathcal{B}(\mathbf{C})} \log \|x\|_\infty \bigwedge_{i=1}^{d} c_1(\overline{\mathcal{O}_{\mathcal{B}}(1)})$$
$$= \widehat{\deg}(\widehat{c}_1(\overline{\mathcal{O}_{\mathcal{B}}(1)})^d \cdot \widehat{(x^{-1})}),$$

which is equal to 0 since $\widehat{(x^{-1})}$ is a principal divisor. □

By making a particular choice of (X, L), we recover $h_{nv}^{\overline{\mathcal{B}}}$ from $h_{(\mathcal{X},\mathcal{L})}^{\overline{\mathcal{B}}}$ [23, Proposition 3.3.2].

Proposition 7.6.7 *Let $X = \mathbf{P}_K^n$ and $L = (\mathcal{O}_X(1), \|\cdot\|_\infty)$. Then $h_{(\mathcal{X},\mathcal{L})}^{\overline{\mathcal{B}}} = h_{nv}^{\overline{\mathcal{B}}}$.*

Proof Let $(\mathcal{X}, \mathcal{L}) = (\mathbf{P}_{\mathcal{B}}^n, \mathcal{O}_{\mathbf{P}_{\mathcal{B}}^n}(1))$. Here, $\mathbf{P}_{\mathcal{B}}^n = \mathbf{P}_{\mathbf{Z}}^n \times_{\mathbf{Z}} \mathcal{B}$ has a projection to $\mathbf{P}_{\mathbf{Z}}^n$, and $(\mathcal{O}_{\mathbf{P}_B^n}(1), \|\cdot\|_\infty)$ is defined to be the pullback of $(\mathcal{O}_{\mathbf{P}_{\mathbf{Z}}^n}(1), \|\cdot\|_\infty)$. We prove the case $d = 1$ for simplicity. The case of general d can be proved similarly. Further, for simplicity of notation, suppose that P is defined over the field K. Thus

$$h_{(\mathcal{X},\mathcal{L})}^{\overline{\mathcal{B}}}(P) = \widehat{\deg}(\widehat{c}_1(\pi^*\overline{\mathcal{H}}|_{\Delta_P}) \cdot \widehat{c}_1(\mathcal{L}|_{\Delta_P})),$$

where $\pi : \mathbf{P}_{\mathcal{B}}^n \to \mathcal{B}$ is the structure map and $\overline{\mathcal{B}} = (\mathcal{B}, \overline{\mathcal{H}})$ is a polarization of $\mathcal{B}$. Since Δ_P is the closure of a map $\operatorname{Spec} K \to X$, the properness of X gives us an induced map $s_P : \mathcal{B} \to \Delta_P \hookrightarrow \mathbf{P}_{\mathcal{B}}^n$. Let x denote a section of $\mathcal{L}$ such that $s_P^*(x) \neq 0$. Since $\pi|_{\Delta_P}$ is generically of degree 1, Proposition 7.5.14 implies

$$\pi_*(\widehat{c}_1(\pi^*\overline{\mathcal{H}}|_{\Delta_P}) \cdot \widehat{c}_1(\mathcal{L}|_{\Delta_P})) = \widehat{c}_1(\overline{\mathcal{H}}) \cdot \pi_*(\widehat{c}_1(\mathcal{L}|_{\Delta_P}))$$
$$= \widehat{c}_1(\overline{\mathcal{H}}) \cdot \operatorname{div}(s_P^*(x))).$$

Thus

$$h_{(\mathcal{X},\mathcal{L})}^{\overline{\mathcal{B}}} = \widehat{\deg}(\widehat{c}_1(H) \cdot \operatorname{div}(s_P^*(x))). \tag{2}$$

Write $\operatorname{div}(s_P^*(x)) = \sum_{\gamma\subset B} a_\gamma \gamma$, where $a_\gamma \in \mathbf{Z}$ and the sum is over irreducible divisors $(\gamma, g_\gamma) \subset B$.[1] The finite contribution to the Arakelov height comes from the cycle $\sum_\gamma a_\gamma(\gamma \cdot \overline{\mathcal{H}})$; we conflate the line bundle $\overline{\mathcal{H}}$ with the Weil divisor corresponding to it and the intersection is as in

[1] Here, $g_\gamma = -\log \|t_\gamma\|^2$ for some section t_γ of $\mathcal{O}(\gamma)$.

Definition 7.5.10. Write $\gamma \cdot H = \sum_i Q_i^\gamma$ as a sum of points. By definition,

$$
\begin{aligned}
&\widehat{c}_1(\overline{\mathcal{H}}) \cdot \operatorname{div}(s_P^*(x)) \\
&\quad = \left(\sum_{\gamma,i} a_\gamma Q_i^\gamma, [-\log \|s_P^*(x)|_{\overline{\mathcal{H}}}\|_\infty^2] + \sum_\gamma c_1(\overline{\mathcal{H}}) \wedge g_\gamma \right).
\end{aligned}
$$

Equation (2) thus implies

$$
\begin{aligned}
&\widehat{\deg}(\widehat{c}_1(\pi^*\overline{\mathcal{H}}|_{\Delta_P}) \cdot \widehat{c}_1(\mathcal{L}|_{\Delta_P})) \\
&\quad = \sum_{\gamma,i} a_\gamma \log |k(Q_i^\gamma)| + \int_{B(\mathbf{C})} -\log \|s_P^*(x)|_{\overline{\mathcal{H}}}\|_\infty + \sum_\gamma \int_{B(\mathbf{C})} c_1(\overline{\mathcal{H}}) \wedge g_\gamma \\
&\quad = \sum_\gamma a_\gamma \widehat{\deg}(\widehat{c}_1(\overline{\mathcal{H}}|_\gamma)) + \int_{B(\mathbf{C})} -\log \|s_P^*(x)\|_\infty c_1(\overline{\mathcal{H}}).
\end{aligned}
$$

Now let $P = [p_0 : \cdots : p_n]$ and suppose (without loss of generality) that $p_0 \neq 0$. Then x_0 is a non-vanishing section of $\mathcal{O}_{\mathbf{P}^n_{\mathcal{B}}}(1)$ around P. By Example 7.4.7, we have that $\|s_P^*(x_0)\|_\infty = \frac{|p_0|}{\max\{|p_0|,\ldots,|p_n|\}}$. Thus $-\log \|s_P^*(x_0)\|_\infty = \log \max_i\{|p_i|\} - \log |p_0|$.

For the finite places, note that the sections $x_0, \ldots, x_n$ generate $\mathcal{O}_{\mathbf{P}^n_{\mathcal{B}}}(1)$ and thus $s_P^*(x_0), \ldots, s_P^*(x_n)$ globally generate $s_P^*(\mathcal{O}_{\mathbf{P}^n_{\mathcal{B}}}(1))$. Now, since $s_P \colon \mathcal{B} \hookrightarrow \mathbf{P}^n_{\mathcal{B}}$ is an embedding, we have

$$
\begin{aligned}
a_\gamma = \operatorname{ord}_\gamma(s_P^*(x_0)) &= \operatorname{len}_{\mathcal{O}_{\mathcal{B},\gamma}}(\mathcal{O}_{\mathcal{B},\gamma}/s_P^* x_0) \\
&= \operatorname{len}_{\mathcal{O}_{\mathcal{B},\gamma}}(s_P^*\mathcal{O}(1)_\gamma/s_P^* x_0) \\
&= \operatorname{len}_{\mathcal{O}_{\mathcal{B},\gamma}}\left(\frac{\mathcal{O}_{\mathcal{B}} p_0 + \cdots + \mathcal{O}_{\mathcal{B}} p_n}{\mathcal{O}_{\mathcal{B}} p_0} \right) \\
&= \max_i\{-\operatorname{ord}_\gamma(p_i)\} + \operatorname{ord}_\gamma(p_0). \qquad \square
\end{aligned}
$$

Remark 7.6.8 (Northcott property) Moriwaki heights need not satisfy the Northcott property in general. However, if $\overline{\mathcal{H}}_i \to \mathcal{B}$ is ample for $1 \leq i \leq d$, then $h^{\overline{B}}_{(X,\mathcal{L})}$ is Northcott [23, Proposition 3.3.7 (4)] (the positivity assumptions *nef* and *big* are both implied by *ample*).

Acknowledgements We thank the organizers of the Stacks Project Workshop and Max Lieblich for their support and guidance. We thank Will Sawin for explaining geometric heights to us. We would also like to thank the referees for their detailed and helpful comments on the early draft of this chapter. The second-named author also thanks Brandon Alberts for many helpful conversations. The first-named author received support from Kirsten Wickelgren's NSF CAREER grant (DMS-1552730).

References

[1] S. Ju. Arakelov. An intersection theory for divisors on an arithmetic surface. *Izv. Akad. Nauk SSSR Ser. Mat.*, 38:1179–1192, 1974.

[2] V. V. Batyrev and Yu. I. Manin. Sur le nombre des points rationnels de hauteur borné des variétés algébriques. *Math. Ann.*, 286(1-3):27–43, 1990. doi:10.1007/BF01453564.

[3] M. F. Becker and S. MacLane. The minimum number of generators for inseparable algebraic extensions. *Bull. Amer. Math. Soc.*, 46:182–186, 1940. doi:10.1090/S0002-9904-1940-07169-1.

[4] Siegfried Bosch, Werner Lütkebohmert, and Michel Raynaud. *Néron models*, volume 21 of *Ergebnisse der Mathematik und ihrer Grenzgebiete. 3. Folge.* Springer, 1990. doi:10.1007/978-3-642-51438-8.

[5] Antoine Chambert-Loir. Heights and measures on analytic spaces. A survey of recent results, and some remarks. In *Motivic integration and its interactions with model theory and non-Archimedean geometry, Volume II*, volume 384 of *London Math. Soc. Lecture Note Ser.*, pages 1–50. Cambridge University Press, 2011.

[6] Antoine Chambert-Loir. Chapter VII: Arakelov geometry, heights, equidistribution, and the Bogomolov conjecture. In *Arakelov geometry and Diophantine applications*, volume 2276 of *Lecture Notes in Mathematics*, pages 299–328. Springer, 2021. doi:10.1007/978-3-030-57559-5_8.

[7] Brian Conrad. Chow's K/k-image and K/k-trace, and the Lang–Néron theorem. *Enseign. Math. (2)*, 52(1-2):37–108, 2006.

[8] Jan-Hendrik Evertse and Kálmán Győry. *Unit equations in Diophantine number theory*, volume 146 of *Cambridge Studies in Advanced Mathematics*. Cambridge University Press, 2015. doi:10.1017/CBO9781316160749.

[9] G. Faltings. Endlichkeitssätze für abelsche Varietäten über Zahlkörpern. *Invent. Math.*, 73(3):349–366, 1983. doi:10.1007/BF01388432.

[10] Gerd Faltings. Calculus on arithmetic surfaces. *Ann. Math. (2)*, 119(2):387–424, 1984. doi:10.2307/2007043.

[11] Jens Franke, Yuri I. Manin, and Yuri Tschinkel. Rational points of bounded height on Fano varieties. *Invent. Math.*, 95(2):421–435, 1989. doi:10.1007/BF01393904.

[12] Michael D. Fried and Moshe Jarden. *Field arithmetic*, volume 11 of *Ergebnisse der Mathematik und ihrer Grenzgebiete. 3. Folge.* Springer, third edition, 2008. Revised by Jarden.

[13] William Fulton. *Intersection theory*, volume 2 of *Ergebnisse der Mathematik und ihrer Grenzgebiete. 3. Folge.* Springer, second edition, 1998. doi:10.1007/978-1-4612-1700-8.

[14] Henri Gillet and Christophe Soulé. Arithmetic intersection theory. *Inst. Hautes Études Sci. Publ. Math.*, 72:93–174, 1990.

[15] Phillip Griffiths and Joseph Harris. *Principles of algebraic geometry*. Pure and Applied Mathematics. Wiley-Interscience, 1978.

[16] A. Grothendieck. Éléments de géométrie algébrique. IV. Étude locale des schémas et des morphismes de schémas. II. *Inst. Hautes Études Sci. Publ. Math.*, (24):231, 1965.

[17] Marc Hindry and Joseph H. Silverman. *Diophantine geometry*, volume 201 of *Graduate Texts in Mathematics*. Springer, 2000. An introduction. doi:10.1007/978-1-4612-1210-2.

[18] Daniel Huybrechts. *Complex geometry*. Universitext. Springer, 2005.

[19] Serge Lang. *Diophantine geometry*, volume 11 of *Interscience Tracts in Pure and Applied Mathematics*. Interscience Publishers, 1962.

[20] Serge Lang. *Introduction to Arakelov theory*. Springer, 1988. doi:10.1007/978-1-4612-1031-3.

[21] Qing Liu. *Algebraic geometry and arithmetic curves*, volume 6 of *Oxford Graduate Texts in Mathematics*. Oxford University Press, Oxford, 2002. Translated from the French by Reinie Erné, Oxford Science Publications.

[22] J. S. Milne. Algebraic number theory. *www.jmilne.org/math*, 2009.

[23] Atsushi Moriwaki. Arithmetic height functions over finitely generated fields. *Invent. Math.*, 140(1):101–142, 2000. doi:10.1007/s002220050358.

[24] Jürgen Neukirch. *Algebraic number theory*, volume 322 of *Grundlehren der mathematischen Wissenschaften*. Springer, 1999. Translated from the 1992 German original and with a note by Norbert Schappacher, With a foreword by G. Harder. doi:10.1007/978-3-662-03983-0.

[25] Emmanuel Peyre. Hauteurs et mesures de Tamagawa sur les variétés de Fano. *Duke Math. J.*, 79(1):101–218, 1995. doi:10.1215/S0012-7094-95-07904-6.

[26] Jean-Pierre Serre. *Local fields*, volume 67 of *Graduate Texts in Mathematics*. Springer, 1979. Translated from the French by Marvin Jay Greenberg.

[27] Jean-Pierre Serre. *Lectures on the Mordell–Weil theorem*. Aspects of Mathematics. Friedr. Vieweg & Sohn, 1989. Translated from the French and edited by Martin Brown from notes by Michel Waldschmidt. doi:10.1007/978-3-663-14060-3.

[28] Joseph H. Silverman. *Advanced topics in the arithmetic of elliptic curves*, volume 151 of *Graduate Texts in Mathematics*. Springer, 1994. doi:10.1007/978-1-4612-0851-8.

[29] The Stacks Project Authors. *The Stacks project*. https://stacks.math.columbia.edu.

[30] Emmanuel Ullmo. Positivité et discrétion des points algébriques des courbes. *Ann. Math. (2)*, 147(1):167–179, 1998. doi:10.2307/120987.

[31] Paul Vojta. *Diophantine approximations and value distribution theory*, volume 1239 of *Lecture Notes in Mathematics*. Springer, 1987. doi:10.1007/BFb0072989.

[32] Paul Vojta. Siegel's theorem in the compact case. *Ann. Math. (2)*, 133(3):509–548, 1991. doi:10.2307/2944318.

[33] André Weil. L'arithmétique sur les courbes algébriques. *Acta Math.*, 52(1):281–315, 1929. doi:10.1007/BF02547409.

[34] Xinyi Yuan and Shou-Wu Zhang. Adelic line bundles over quasi-projective varieties, 2021. arXiv:2105.13587v3.

[35] Shou-Wu Zhang. Equidistribution of small points on abelian varieties. *Ann. Math. (2)*, 147(1):159–165, 1998. doi:10.2307/120986.

8

An explicit self-duality

Nikolas Kuhn
Max-Planck-Institut für Mathematik

Devlin Mallory
University of Utah

Vaidehee Thatte
King's College London

Kirsten Wickelgren
Duke University

Abstract We provide an exposition of the canonical self-duality associated to a presentation of a finite, flat, complete intersection over a Noetherian ring, following work of Scheja and Storch.

8.1 Introduction

Consider a finite flat ring map $f\colon A \to B$ and assume that A is Noetherian. Coherent duality for proper morphisms provides a functor $f^! : D(\operatorname{Spec} A) \to D(\operatorname{Spec} B)$ on derived categories. The assumptions on f imply that $f^! A$ is isomorphic to the sheaf on B associated to $\operatorname{Hom}_A(B, A)$. See for example [12, Tag 0AA2]. If we assume moreover that $f\colon \operatorname{Spec} B \to \operatorname{Spec} A$ is a local complete intersection morphism, then $f^! A$ is locally free [12, Tag 0B6V, Tag 0FNT]. Thus there exists an isomorphism

$$\operatorname{Hom}_A(B, A) \cong B \tag{1}$$

of B-modules under additional hypotheses; for example, we might assume that B is local.[1]

There are many choices for the isomorphism (1). (The set of these

[1] An alternative point of view on the equivalence $f^! A \simeq B$ is that a factorization $A \xrightarrow{p} A[x_1, \ldots, x_n] \xrightarrow{i} B$ of f into a regular immersion and structure map for $\mathbb{A}^n_A$ allows one to compute $f^! A$ as $i^! p^! A \simeq i^! (A[x_1, \ldots, x_n][n]) \simeq \det N_i^* [-n][n] \simeq B$, where N_i^* denotes the conormal bundle of the regular immersion $\operatorname{Spec} B \hookrightarrow \mathbb{A}^n_A$. See, for example, [4, Ideal Theorem p. 6, III, particularly Corollary 7.3].

isomorphisms forms a B^*-torsor.) An explicit presentation of B as

$$B = A[x_1, \ldots, x_n]/(f_1, \ldots, f_n) \tag{2}$$

singles out a particular choice, which satisfies certain nice properties such as compatibility with base change and the trace. In addition to the advantages of having a canonical choice (e.g., gluing such isomorphisms together), this choice is closely related to the degree map in $\mathbb{A}^1$-homotopy theory due to F. Morel. See Remark 8.1.1.

In this exposition, we follow the approach of [11] to construct this canonical isomorphism for B a finite flat A-algebra equipped with a presentation (2).

The approach is as follows. Consider the ideals

$$(f_1 \otimes 1 - 1 \otimes f_1, \ldots, f_n \otimes 1 - 1 \otimes f_n) \subset (x_1 \otimes 1 - 1 \otimes x_1, \ldots, x_n \otimes 1 - 1 \otimes x_n)$$

of $A[x_1, \ldots, x_n] \otimes A[x_1, \ldots, x_n]$. One writes

$$f_j \otimes 1 - 1 \otimes f_j = \sum a_{ij}(x_i \otimes 1 - 1 \otimes x_i).$$

and defines the element $\Delta \in B \otimes_A B$ as the image of $\det(a_{ij})$ under the morphism $A[x_1, \ldots, x_n] \otimes A[x_1, \ldots, x_n] \to B \otimes_A B$. This is shown to be independent of the choice of a_{ij}. There is a canonical A-module morphism

$$\chi\colon B \otimes_A B \to \mathrm{Hom}_A(\mathrm{Hom}_A(B, A), B).$$

Let I denote the kernel of multiplication $B \otimes_A B \to B$, or in other words the image of $(x_1 \otimes 1 - 1 \otimes x_1, \ldots, x_n \otimes 1 - 1 \otimes x_n)$. One can check that χ restricts to an isomorphism

$$\chi\colon \mathrm{Ann}_{B \otimes_A B}\, I \to \mathrm{Hom}_B(\mathrm{Hom}_A(B, A), B)$$

of B-modules and identify the annihilator as $\mathrm{Ann}_{B \otimes_A B}\, I \cong \Delta$. Finally, one can show that

$$\chi(\Delta) =: \Theta \in \mathrm{Hom}_B(\mathrm{Hom}_A(B, A), B)$$

provides the desired isomorphism of B-modules $\Theta\colon \mathrm{Hom}_A(B, A) \to B$ guaranteed by the general theory of coherent duality. This is Theorem 8.3.4 (or [11, Satz 3.3]) and the main result. For the compatibility of Θ with base change and the trace see [11, pp. 183–184 and Section 4] respectively.

Our arguments largely follow the outline of [11], although we make more use of Koszul homology in some proofs than the original proof did, and we provide a self-contained proof of Lemma 8.2.4; the goal in large part is to provide an English-language reference for this material. See also [7, Appendices H and I].

Remark 8.1.1 One motivation for providing an explicit description of this isomorphism is to describe the resulting A-valued bilinear form on B. This form is defined via

$$\langle b, c\rangle \mapsto \Theta^{-1}(b)(c) = \eta(bc) \in A,$$

where $\eta = \Theta^{-1}(1)$. The form $\langle -, -\rangle$ has been used to give a notion of degree [3] and [2, Some remaining questions (3)]. For example, it computes the local $\mathbb{A}^1$-Brouwer degree of Morel [5, 1], and is useful in quadratic enrichments of results in enumerative geometry [8, 6, 9, 10].

8.2 Commutative algebra preliminaries

Lemma 8.2.1 *[11, 1.2] Let A be a Noetherian ring and suppose that $f_1, \ldots, f_n$ and $g_1, \ldots, g_n$ are sequences satisfying the following hypotheses:*

(i) $\mathfrak{b} = (g_1, \ldots, g_n) \subset \mathfrak{a} = (f_1, \ldots f_n)$.
(ii) *If $\mathfrak{p}$ is a prime such that $\mathfrak{a} \subset \mathfrak{p}$, then the sequence $f_1, \ldots, f_n$ is a regular sequence in $A_{\mathfrak{p}}$, as is $g_1, \ldots, g_n$.*

Write $g_i = \sum_{i=1}^{n} a_{ij} f_j$, and let (a_{ij}) be the resulting matrix of coefficients. Set

$$\Delta := \det (a_{ij})$$

and define $\overline{\Delta}$ to be the image of Δ under the map $A \to A/\mathfrak{b}$. Then:

(a) *The element $\overline{\Delta}$ is independent of the choices of a_{ij}.*
(b) *We have an equality (of $A/\mathfrak{b}$-ideals):*

$$(\overline{\Delta}) = \mathrm{Fit}_{A/\mathfrak{b}}(\mathfrak{a}/\mathfrak{b}),$$

where Fit *denotes the 0th Fitting ideal.*
(c) *We have an equality of ideals:*

$$(\overline{\Delta}) = \mathrm{Ann}_{A/\mathfrak{b}}(\mathfrak{a}/\mathfrak{b}),$$

and

$$\mathfrak{a}/\mathfrak{b} = \mathrm{Ann}_{A/\mathfrak{b}}(\overline{\Delta}).$$

Remark 8.2.2 We comment on condition (ii). If $(A, \mathfrak{p})$ is a local ring and $\mathfrak{a} \subset \mathfrak{p}$, then condition (ii) is equivalent to asking that $f_1, \ldots, f_n$ and $g_1, \ldots, g_n$ are regular sequences. In general, condition (ii) asks only that they are regular sequences after localizing at primes containing $\mathfrak{a}$ (e.g., they may not be regular sequences on A).

Proof of Lemma 8.2.1 First, we may assume that A is a local ring and each of the f_i and g_i are in the maximal ideal $\mathfrak{m}$.

(a) Write $g_i = \sum_{i=1}^n b_{ij} f_j$. We want to show that $\det(a_{ij}) - \det(b_{ij})$ is in $\mathfrak{b}$. It suffices to consider the case where $a_{ij} = b_{ij}$ for all j and for $i = 1, \ldots, n-1$, as this allows us to change the presentation of one g_i at a time, and thus of all of them. Define

$$c_{ij} = \begin{cases} a_{ij} = b_{ij} & i = 1, \ldots, n-1, \\ a_{ij} - b_{ij} & i = n, \end{cases}$$

By cofactor expansion along the jth row, we have that

$$\det(a_{ij}) - \det(b_{ij}) = \det(c_{ij}).$$

But now

$$(c_{ij}) \begin{pmatrix} f_1 \\ f_2 \\ \vdots \\ f_{n-1} \\ f_n \end{pmatrix} = \begin{pmatrix} g_1 \\ g_2 \\ \vdots \\ g_{n-1} \\ 0 \end{pmatrix}$$

By Cramer's rule, for all $k = 1, \ldots, n$ we have that

$$\det(c_{ij}) \cdot f_k \in (g_1, \ldots, g_{n-1}),$$

which means

$$\det(c_{ij}) \cdot \mathfrak{a} \in (g_1, \ldots, g_{n-1}).$$

But $g_n \in \mathfrak{a}$ and hence

$$\det(c_{ij}) \cdot g_n \in (g_1, \ldots, g_{n-1}),$$

which means that $\det(c_{ij}) \in (g_1, \ldots, g_n) = \mathfrak{b}$ since $g_1, \ldots, g_n$ is a regular sequence.

(b) First observe that

$$\mathrm{Fit}_A(\mathfrak{a}/\mathfrak{b}) \bmod \mathfrak{b} = \mathrm{Fit}_{A/\mathfrak{b}}(\mathfrak{a}/\mathfrak{b}).$$

Therefore, to prove the claim, it suffices to prove that

$$\mathrm{Fit}_A(\mathfrak{a}/\mathfrak{b}) = \Delta + I,$$

where $I \subset \mathfrak{b}$.

To prove this claim, note that the Fitting ideal of the A-module $\mathfrak{a}/\mathfrak{b}$ is computed by a presentation:

$$A^{\oplus n} \oplus A^{\oplus \binom{n}{2}} \xrightarrow{T} A^{\oplus n} \to \mathfrak{a}/\mathfrak{b} \to 0,$$

where T is given by:

$$(a_{ij}) \times d_2^{\mathrm{Kosz}}.$$

In other words, the first n columns of the matrix T has are just given by a_{ij} and, the last $\binom{n}{2}$ columns are composed of the usual Koszul relations among the f_i. (Note that the sequence $f_1, \ldots, f_n$ is regular in our local ring, so the corresponding Koszul complex produces a resolution of $\mathfrak{a}$ [12, Tag 062F].)

Now, the Fitting ideal is given by the $n \times n$-minors of the matrix of T. The first minor is Δ. If Δ' is another $n \times n$-minor, then it is the determinant of a matrix T', which is composed of some r columns of (a_{ij}) and $n - r$ columns of d_1^{Kosz}; without loss of generality we may assume T' contains the first r columns of (a_{ij}) (if not, simply reorder the g_i, using that the ring A is local and thus regularity of the sequence of g_i is preserved). Applying T' to (f_k) we get

$$(T') \begin{pmatrix} f_1 \\ f_2 \\ \vdots \\ f_n \end{pmatrix} = \begin{pmatrix} g_1 \\ \vdots \\ g_r \\ 0 \\ \vdots \\ 0 \end{pmatrix}$$

We again conclude that $\Delta' f_i = \det(T') f_i \in \mathfrak{b}$ for each $i = 1, \ldots, n$. Thus,

$$\Delta' \cdot \mathfrak{a} \in (g_1, \ldots, g_{n-1}),$$

and in particular

$$\Delta' \cdot g_n \in (g_1, \ldots, g_{n-1}),$$

which by regularity of the g_i means that $\Delta' \in \mathfrak{b}$ and thus $\mathrm{Fit}_A(\mathfrak{a}/\mathfrak{b}) = \Delta + I$ with $I \subset \mathfrak{b}$.

(c) First, we claim that we have an isomorphism:

$$\mathrm{Ann}_{A/\mathfrak{b}}(\mathfrak{a}/\mathfrak{b}) \cong \mathrm{Tor}_n^A(A/\mathfrak{b}, A/\mathfrak{a}).$$

We will abbreviate Tor_j^A by Tor_j and $\otimes_A$ by $\otimes$ in what follows. To prove this, we deploy the Koszul complex. (As noted above, a regular sequence is Koszul-regular by [12, Tag 062F].) We thus have a quasi-isomorphism

$$K_\bullet(f_1, \ldots, f_n) \simeq A/\mathfrak{a}.$$

Therefore the Tor group above is computed as the kernel of $1 \otimes d_n^{\mathrm{Kosz}}$ in the complex $A/\mathfrak{b} \otimes K_\bullet(f_1, \ldots, f_n)$:

$$0 \to A/\mathfrak{b} \xrightarrow{(f_1,\ldots,f_n)} (A/\mathfrak{b})^{\oplus n}.$$

Indeed, the cohomology of this small complex is the desired annihilator and thus we obtain the desired isomorphism.

On the other hand, we claim that $\mathrm{Tor}_n(A/\mathfrak{a}, A/\mathfrak{b}) \cong \Delta \cdot A/\mathfrak{b}$. To see this, note that we have a short exact sequence of A-modules:

$$0 \to \mathfrak{a}/\mathfrak{b} \to A/\mathfrak{b} \to A/\mathfrak{a} \to 0.$$

We claim that the induced long exact sequence splits into short exact sequences

$$0 \to \mathrm{Tor}_j(A/\mathfrak{b}, \mathfrak{a}/\mathfrak{b}) \to \mathrm{Tor}_j(A/\mathfrak{b}, A/\mathfrak{b}) \to \mathrm{Tor}_j(A/\mathfrak{b}, A/\mathfrak{a}) \to 0$$

for $j \geq 1$. Indeed, via the Koszul complex for $A/\mathfrak{b}$, we see that, for $j \geq 1$:

$$\mathrm{Tor}_j(A/\mathfrak{b}, \mathfrak{a}/\mathfrak{b}) \cong (\mathfrak{a}/\mathfrak{b})^{\binom{n}{j}}, \qquad \mathrm{Tor}_j(A/\mathfrak{b}, A/\mathfrak{b}) \cong (A/\mathfrak{b})^{\binom{n}{j}}, \tag{1}$$

and the map $\mathrm{Tor}_j(A/\mathfrak{b}, \mathfrak{a}/\mathfrak{b}) \to \mathrm{Tor}_j(A/\mathfrak{b}, A/\mathfrak{b})$ is identified with the direct sum of copies of the injection $\mathfrak{a}/\mathfrak{b} \hookrightarrow A/\mathfrak{b}$. To conclude, the functoriality of the Koszul complex [12, Tag 0624] yields a morphism of complexes

$$A/\mathfrak{b} \otimes K_\bullet(g_1, \ldots, g_n) \to A/\mathfrak{b} \otimes K_\bullet(f_1, \ldots, f_n),$$

where the left end is as follows:

$$\begin{array}{ccc} A/\mathfrak{b} & \xrightarrow{0} & (A/\mathfrak{b})^{\oplus n} \\ \overline{\Delta}\downarrow & & \downarrow \\ A/\mathfrak{b} & \xrightarrow{(f_1,\ldots,f_n)} & (A/\mathfrak{b})^{\oplus n}. \end{array} \tag{2}$$

Since the map $\mathrm{Tor}_j(A/\mathfrak{b}, A/\mathfrak{b}) \to \mathrm{Tor}_j(A/\mathfrak{b}, A/\mathfrak{a})$ is a surjection, we conclude that

$$\mathrm{Tor}_n(A/\mathfrak{b}, A/\mathfrak{a}) \cong \mathfrak{I}(\overline{\Delta}) \cong \Delta \cdot A/\mathfrak{b}$$

as desired.

For the second claim, note that the ideal $\mathrm{Ann}_{A/\mathfrak{b}}(\overline{\Delta})$ is obtained as the kernel of the left vertical map in (2), and is thus isomorphic to $\mathrm{Tor}_n(A/\mathfrak{b}, \mathfrak{a}/\mathfrak{b})$, which we already know is isomorphic to $\mathfrak{a}/\mathfrak{b}$ by (1).

□

A module M over a ring R is said to be reflexive if the natural map $R \to \mathrm{Hom}_R(\mathrm{Hom}_R(M, R), R)$ is an isomorphism [12, Tag 0AUY]. There is a form of the following lemma in the Stacks project ([12, Tag 0AVA]), but it assumes that A is integral and that $A = B$. The following is [11, 1.3].

Lemma 8.2.3 *Let A be a Noetherian ring and B a finite flat A-algebra. A finite B-module M is reflexive if and only if the following conditions hold:*

(i) *If $\mathfrak{p} \subset A$ is a prime ideal with* $\operatorname{depth} A_\mathfrak{p} \leq 1$, *then $M_\mathfrak{p}$ is a reflexive $B_\mathfrak{p}$-module.*
(ii) *If $\mathfrak{p} \subset A$ is a prime ideal with* $\operatorname{depth} A_\mathfrak{p} \geq 2$, *then* $\operatorname{depth}_{A_\mathfrak{p}}(M_\mathfrak{p}) \geq 2$.

Proof The property of being reflexive is preserved under any localization of B [12, Tag 0EB9], and can be checked locally on B [12, Tag 0AV1]. Therefore, the reflexivity of M implies (i). Reflexivity also implies (ii): any regular sequence in $A_\mathfrak{p}$ is a regular sequence on $B_\mathfrak{p}$ by flatness. Let a_1, a_2 be a length-2 regular sequence on $A_\mathfrak{p}$. Let N be any $B_\mathfrak{p}$-module. Then a_1 is a non-zerodivisor on $\mathrm{Hom}_{B_\mathfrak{p}}(N, B_\mathfrak{p})$. The cokernel of multiplication by a_1 is a submodule of $\mathrm{Hom}_{B_\mathfrak{p}}(N, B_\mathfrak{p}/a_1 B_\mathfrak{p})$, on which a_2 is a non-zerodivisor. This shows the claim. (Note: This is almost [12, Tag 0AV5], except that we take Hom_B but want the A-depth.)

Conversely, suppose M is not reflexive. We assume for the sake of contradiction that properties (i) and (ii) hold. Since reflexivity can be checked locally, there is some minimal $\mathfrak{p} \subset A$ among all prime ideals of A for which $M_\mathfrak{p}$ is a not a reflexive $B_\mathfrak{p}$-module. Without loss of generality, we may assume that A is local with maximal ideal $\mathfrak{p}$. Since $M_\mathfrak{p}$ is not reflexive, we must have that $\operatorname{depth} A_\mathfrak{p} \geq 2$ and therefore $\operatorname{depth}_{A_\mathfrak{p}}(M_\mathfrak{p}) \geq 2$. We consider the exact sequence

$$0 \to \operatorname{Ker} \varphi \to M \to \mathrm{Hom}_B(\mathrm{Hom}_B(M, B), B) \to \operatorname{Coker} \varphi \to 0,$$

where φ is the canonical map to the double dual. By assumption, φ becomes an isomorphism after localization at any prime of A different from $\mathfrak{p}$. It follows that $\operatorname{Ker} \varphi$ and $\operatorname{Coker} \varphi$ have finite length. Since $\operatorname{depth}_A M \geq 1$, there exists some $x \in A$ which is a non-zerodivisor on M.

But then x is a non-zerodivisor on the finite-length module $\operatorname{Ker}\varphi$, which therefore must vanish. Since $\operatorname{Hom}_B(\operatorname{Hom}_B(M,B),B)$ is reflexive (as a B-module), it has A-depth ≥ 2 by the forward implication of the lemma. The exact sequence

$$0 \to M \to \operatorname{Hom}_B(\operatorname{Hom}_B(M,B),B) \to \operatorname{Coker}\varphi \to 0$$

then shows that $\operatorname{depth}_{A_\mathfrak{p}} \operatorname{Coker}\varphi \geq 1$, by the standard behavior of depth in short exact sequences [12, Tag 00LX]. Therefore the cokernel must vanish, which shows that M is reflexive. □

Lemma 8.2.4 *[11, 1.4] Let A be a Noetherian ring and let B be a finite flat A-algebra. Let M be a finite B-module, which is projective as an A-module. If* $\operatorname{Hom}_B(M,B)$ *is projective as a B-module, then M is projective as a B-module. In particular, if* $\operatorname{Hom}_B(M,B)$ *is free, then M is free.*

Proof It is enough to show that M is reflexive. We are therefore reduced to checking the conditions (i) and (ii) of Lemma 8.2.3. Clearly, (ii) holds, since M is projective over A. It remains to check (i). We may therefore assume that A is a Noetherian local ring with $\operatorname{depth} A \leq 1$, and we want to show that M is projective as a B-module. Since B is finite flat over A, we have $\operatorname{depth} B_\mathfrak{m} = \operatorname{depth} A$ for every maximal ideal $\mathfrak{m}$ of B [12, Tag 0337].

Throughout, we will write $N^* := \operatorname{Hom}_B(N,B)$ for a B-module N. Consider the map

$$\varphi\colon M \to M^{**}.$$

Let $C := \operatorname{Coker}\varphi$. Taking a presentation of M, we obtain an exact sequence

$$0 \to U \to F \to M \to 0$$

with F free. Consider the dual sequence

$$0 \to M^* \to F^* \to U^*,$$

and let $Q := \operatorname{Im}(F^* \to U^*)$. Since M^* is projective by assumption, Q has projective dimension 0 or 1 as a B-module.

We have the commutative diagram

$$\begin{array}{ccccccc} F & \longrightarrow & M & & & & \\ \downarrow{\scriptstyle\sim} & & \downarrow & & & & \\ F^{**} & \longrightarrow & M^{**} & \longrightarrow & \operatorname{Ext}^1_B(Q,B) & \longrightarrow & 0 \end{array}$$

with exact lower row. Since $F \to M$ is a surjection, we see that $C = \operatorname{Ext}^1_B(Q, B)$. Suppose that depth $A = 0$. Apply the Auslander–Buchsbaum formula [12, Tag 090V] to the $B_{\mathfrak{m}}$-module $Q_{\mathfrak{m}}$ for each maximal ideal $\mathfrak{m}$. We find that $Q_{\mathfrak{m}}$ has projective dimension zero, i.e., it is projective. Therefore $C_{\mathfrak{m}} = 0$ and $C = 0$.

Now suppose that depth $A = 1$. Then $\operatorname{depth}_{B_{\mathfrak{m}}} U^*_{\mathfrak{m}} \geq 1$ by [12, Tag 0AV5], whence

$$\operatorname{depth}_{B_{\mathfrak{m}}} Q_{\mathfrak{m}} \geq 1$$

by [12, Tag 00LX]. Again by the Auslander–Buchsbaum theorem, we find that $Q_{\mathfrak{m}}$ is projective and that $C = 0$.

We have shown that in any case $M \to M^{**}$ is surjective. Since M^{**} is projective, this implies $M \simeq M^{**} \oplus N$ for some B-module N. It follows that $N^* = 0$ and that N is again free as an A-module.

By assumption both M and M^{**} are free over the local ring A. A surjection of finite free A-modules is an isomorphism if they have the same rank. To show that two finite free modules have the same rank, we may localize at a minimal prime ideal $\mathfrak{q}$ of A, so that $B_{\mathfrak{q}}$ is also a zero-dimensional ring. Over the Artinian ring $B_{\mathfrak{q}}$, $\operatorname{Hom}_{B_{\mathfrak{q}}}(N_{\mathfrak{q}}, B_{\mathfrak{q}}) = 0$ implies $N_{\mathfrak{q}} = 0$. (To see this, note that we may assume that B is local, with maximal ideal $\mathfrak{m}$. Then $N_{\mathfrak{q}} \to \mathfrak{m}N_{\mathfrak{q}}$ is nonzero by Nakayama's lemma. Since $B_{\mathfrak{q}}$ has finite length, there is a nonzero element annihilated by $\mathfrak{m}$, whence a B-homomorphism $B/\mathfrak{m} \to B_{\mathfrak{q}}$.) Thus $M_{\mathfrak{q}}$ and $M^{**}_{\mathfrak{q}}$ have the same rank, and therefore $M \to M^{**}$ is an isomorphism. □

8.3 The explicit isomorphism

Recall that a ring map $A \to B$ is a *relative global complete intersection* if there exists a presentation $A[x_1, \ldots, x_n]/(f_1, \ldots, f_c) \cong B$, and every nonempty fiber of $\operatorname{Spec} B \to \operatorname{Spec} A$ has dimension $n - c$ [12, Tag 00SP]. Note that in this case the f_i form a regular sequence [12, Tag 00SV].

We note that a global complete intersection is flat [12, Tag 00SW], and thus syntomic. We will be interested in the situation where $A \to B$ is furthermore assumed to be a *finite* flat global complete intersection.

Construction 8.3.1 Suppose that $A \to B$ is a finite flat global complete intersection. Choose a presentation

$$A[x_1, \ldots, x_n] \xrightarrow{\pi} B \cong A[x_1, \ldots, x_n]/(f_1, \ldots, f_n).$$

Consider the commutative diagram

$$\begin{array}{ccc} A[x_1,\ldots,x_n] \otimes_A A[x_1,\ldots,x_n] & \xrightarrow{m_1} & A[x_1,\ldots,x_n] \\ {\scriptstyle \pi\otimes\pi}\downarrow & & \downarrow{\scriptstyle \pi} \\ B \otimes_A B & \xrightarrow{m} & B, \end{array} \tag{1}$$

with m_1, m the obvious multiplication maps. We note that the elements

$$\{f_j \otimes 1 - 1 \otimes f_j\}_{j=1,\ldots,n}$$

are all in $\ker(m_1)$, which is generated by $x_i \otimes 1 - 1 \otimes x_i$ for $i = 1, \ldots, n$, whence we have a relation

$$f_j \otimes 1 - 1 \otimes f_j = \sum_{i=1}^{n} a_{ij}(x_i \otimes 1 - 1 \otimes x_i).$$

Define $\Delta := (\pi \otimes \pi)(\det(a_{ij})) \in B \otimes_A B$. Moreover define $I := \ker m$.

Proposition 8.3.2 *The following properties of Δ hold:*

(a) *The element Δ is independent of the choice of a_{ij}.*
(b) *We have an equality of $B \otimes_A B$-ideals:*

$$(\Delta) = \mathrm{Fit}_{B\otimes_A B}\, I.$$

(c) *We have an equality of ideals*

$$(\Delta) = \mathrm{Ann}_{B\otimes_A B}\, I, \qquad \mathrm{Ann}_{B\otimes_A B}(\Delta) = I.$$

Proof Consider the ring map

$$\pi \otimes 1 \colon A[x_1,\ldots,x_n] \otimes_A A[x_1,\ldots,x_n] \to B \otimes_A A[x_1,\ldots,x_n].$$

Since

$$f_i \otimes 1 - 1 \otimes f_i = \sum_{i=1}^{n} a_{ij}(x_i \otimes 1 - 1 \otimes x_i)$$

in $A[x_1,\ldots,x_n] \otimes_A A[x_1,\ldots,x_n]$, we have that

$$-1 \otimes f_i = \sum_{i=1}^{n} a_{ij}(\pi(x_i) \otimes 1 - 1 \otimes x_i)$$

in $B \otimes_A A[x_1,\ldots,x_n]$.

Note that Δ is the image of $\det(a_{ij})$ under the obvious morphism $B \otimes_A A[x_1,\ldots,x_n] \to B \otimes_A B$, and that if $\mathfrak{a}$ is the ideal generated by the $\pi(x_i) \otimes 1 - 1 \otimes x_i$ and $\mathfrak{b}$ the ideal generated by the $(-1 \otimes f_i)$, then I is $\mathfrak{a}/\mathfrak{b}$. The desired properties will then follow immediately from applying

Lemma 8.2.1 to $\mathfrak{b} = (-1 \otimes f_i) \subset (\pi(x_i) \otimes 1 - 1 \otimes x_i) = \mathfrak{a}$, once we show that the conditions of the lemma are satisfied. It suffices to show that each is a regular sequence.

We claim that $\{-1 \otimes f_j\} \subset B \otimes_A A[x_1, \ldots, x_n]$ is a regular sequence. Indeed, since relative global complete intersections are flat [12, Tag 00SW] and regular sequences are preserved under flat morphisms, this follows by regularity of the f_i in $A[x_1, \ldots, x_n]$ and the flatness of $A \to B$. It is immediate also that the $(\pi(x_i) - x_i)$ form a regular sequence in $B[x_1, \ldots, x_n]$ as well (the $\pi(x_i)$ are just elements b_i of B, and $(x_i - b_i)$ is always a regular sequence in $B[x_1, \ldots, x_n]$).

Thus, the proposition follows by Lemma 8.2.1. □

Now, retain our setup from Construction 8.3.1. There is a canonical map of A-modules

$$\chi\colon B \otimes_A B \to \mathrm{Hom}_A(\mathrm{Hom}_A(B, A), B) \qquad \chi(b \otimes c) = (\phi \mapsto \phi(b)c).$$

The A-modules $B \otimes_A B$ and $\mathrm{Hom}_A(\mathrm{Hom}_A(B, A), B)$ each carry two natural B-module structures:

(i) B acts on $B \otimes_A B$ as multiplication on either the left or right factor (i.e., either $a(b \otimes c) = ab \otimes c$ or $a(b \otimes c) = b \otimes ac$).

(ii) B acts on $\mathrm{Hom}_A(\mathrm{Hom}_A(B, A), B)$ as either pre- or post-composing a homomorphism by multiplication (i.e., either $a\phi\colon \psi \mapsto \phi(a\psi)$ or $a\phi\colon \psi \mapsto a\phi(\psi)$).

Lemma 8.3.3 *The map χ induces a B-module isomorphism*

$$\mathrm{Ann}_{B \otimes_A B}\, I \cong \mathrm{Hom}_B(\mathrm{Hom}_A(B, A), B).$$

Proof We note first that this map is an isomorphism of A-modules, for which it suffices to check that it is bijective. Since B is a projective A-module we have that B is canonically isomorphic to $B^{\vee\vee}$ (where we denote by $^\vee$ the A-module dual), so that we have isomorphisms of A-modules

$$B \otimes_A B \cong (B^\vee)^\vee \otimes_A B \cong \mathrm{Hom}_A(B^\vee, B) = \mathrm{Hom}_A(\mathrm{Hom}_A(B, A), B);$$

one can check that χ is simply the composition of these canonical isomorphisms.

It can be immediately checked that the morphism χ is in fact a B-bimodule homomorphism for the B-module structures on $B \otimes_A B$ and $\mathrm{Hom}_A(\mathrm{Hom}_A(B, A), B)$ given by right multiplication and post-composition.

Now, we note the following:

(i) The largest submodule of $B \otimes_A B$ where the two B-module structures agree is $\mathrm{Ann}_{B \otimes_A B} I$: this follows since an element $r \in B \otimes_A B$ is annihilated by all $a \otimes 1 - 1 \otimes a$ exactly when $(a \otimes 1)r = (1 \otimes a)r$ for all a, which occurs exactly when the action of every a on r is the same under the two B-module structures.

(ii) The largest submodule of $\mathrm{Hom}_A(\mathrm{Hom}_A(B, A), B)$ where the two B-module structures agree is

$$\mathrm{Hom}_B(\mathrm{Hom}_A(B, A), B) \subset \mathrm{Hom}_A(\mathrm{Hom}_A(B, A), B);$$

this is clear since the condition that pre- and post-multiplying by elements of B gives the same result is exactly B-linearity.

Putting this together, we have that χ induces an isomorphism *of B-modules*

$$\chi\colon \mathrm{Ann}_{B \otimes_A B} I \to \mathrm{Hom}_B(\mathrm{Hom}_A(B, A), B),$$

which was our desired claim. □

Theorem 8.3.4 *The map $\chi(\Delta)\colon \mathrm{Hom}_A(B, A) \to B$ is an isomorphism of B-modules.*

Proof Applying Lemma 8.3.2(c) we have that $\mathrm{Ann}_{B \otimes_A B} I = \Delta(B \otimes_A B)$, and further that $\mathrm{Ann}_{B \otimes_A B} \Delta(B \otimes_A B) = I$. Thus, we obtain

$$\begin{aligned} \mathrm{Ann}_{B \otimes_A B} I &= \Delta(B \otimes_A B) \\ &\cong \Delta(B \otimes_A B)/\mathrm{Ann}_{B \otimes_A B} \Delta \\ &= \Delta(B \otimes_A B)/I \\ &\cong m(\Delta)B. \end{aligned}$$

Applying Lemma 8.3.3, we have then that $\mathrm{Hom}_B(\mathrm{Hom}_A(B, A), B)$ is a free B-module with basis $\chi(\Delta)$. Applying Lemma 8.2.4, this implies that $\mathrm{Hom}_A(B, A)$ is a free B-module of rank 1. We must then have that the B-module homomorphism $\chi(\Delta)\colon \mathrm{Hom}_A(B, A) \to B$ is an isomorphism, as desired. □

Acknowledgements Kirsten Wickelgren was partially supported by NSF CAREER DMS 2001890 and NSF DMS 2103838.

References

[1] Thomas Brazelton, Robert Burklund, Stephen McKean, Michael Montoro, and Morgan Opie. The trace of the local $\mathbb{A}^1$-degree. *Homology Homotopy Appl.*, 23(1):243–255, 2021. doi:10.4310/hha.2021.v23.n1.a1.

[2] David Eisenbud. An algebraic approach to the topological degree of a smooth map. *Bull. Amer. Math. Soc.*, 84(5):751–764, 1978. doi:10.1090/S0002-9904-1978-14509-1.

[3] David Eisenbud and Harold I. Levine. An algebraic formula for the degree of a C^∞ map germ. *Ann. Math. (2)*, 106(1):19–44, 1977. doi:10.2307/1971156.

[4] Robin Hartshorne. *Residues and duality*, volume 20 of *Lecture Notes in Mathematics*. Springer, 1966. Lecture notes of a seminar on the work of A. Grothendieck, given at Harvard 1963/64, with an appendix by P. Deligne.

[5] Jesse Leo Kass and Kirsten Wickelgren. The class of Eisenbud–Khimshiashvili–Levine is the local $\mathbb{A}^1$-Brouwer degree. *Duke Math. J.*, 168(3):429–469, 2019. doi:10.1215/00127094-2018-0046.

[6] Jesse Leo Kass and Kirsten Wickelgren. An arithmetic count of the lines on a smooth cubic surface. *Compos. Math.*, 157(4):677–709, 2021. doi:10.1112/s0010437x20007691.

[7] Ernst Kunz. *Introduction to plane algebraic curves*. Birkhäuser, 2005. Translated from the 1991 German edition by Richard G. Belshoff.

[8] Marc Levine. Aspects of enumerative geometry with quadratic forms. *Doc. Math.*, 25:2179–2239, 2020.

[9] Stephen McKean. An arithmetic enrichment of Bézout's Theorem. *Math. Ann.*, 379(1-2):633–660, 2021. doi:10.1007/s00208-020-02120-3.

[10] Sabrina Pauli. Quadratic types and the dynamic euler number of lines on a quintic threefold, 2020. arXiv:2006.12089v2.

[11] Günter Scheja and Uwe Storch. Über Spurfunktionen bei vollständigen Durchschnitten. *J. Reine Angew. Math.*, 278/279:174–190, 1975.

[12] The Stacks Project Authors. *The Stacks project*. https://stacks.math.columbia.edu.

9

Tannakian reconstruction of coalgebroids

Yifei Zhao
University of Münster

Abstract A theorem of Schäppi on the reconstruction of an affine category scheme (dually, a coalgebroid) over a general commutative ring from its category of finite-rank representations is explained.

9.1 Introduction

Let k be a field. It is well known from the works of Saavedra [3], Deligne and Milne [2, Chapter II]), and Deligne [1] that affine group schemes G over k can be reconstructed from their category of finite-rank representations $\mathrm{Rep}(G)^{\square}$ equipped with the forgetful functor T to k-vector spaces. This process is known as Tannakian reconstruction and the pair $(\mathrm{Rep}(G)^{\square}, T)$ is the prototype of a Tannakian category.

When k is any commutative ring, it is still possible to ask whether an affine group scheme over k can be reconstructed via the same formalism. The present chapter is an exposition of a theorem of Schäppi ([4]), which gives a formal criterion for an affine category scheme over k to be reconstructible. (The notion of an affine category scheme generalizes that of an affine groupoid, as well as a monoid scheme, and is dual to the notion of a coalgebroid.)

Theorem 9.1.1 *Let k be a commutative ring. Suppose* $\mathrm{Spec}(R_1)$ *is an affine category scheme acting on* $\mathrm{Spec}(R_0)$*, both defined over k. Let R_1-*$\mathrm{Mod}^{\square}$ *denote the category of R_1-comodules whose underlying R_0-module is finite and projective.*

Then $\mathrm{Spec}(R_1)$ *is reconstructible if and only if the canonical map*

$$\operatorname*{colim}_{f\,:\,V\to R_1} V \to R_1, \tag{1}$$

where the colimit is taken over all R_1-comodule maps $f\colon V \to R_1$ with $V \in R_1\text{-Mod}^{\square}$, is bijective.

In solving the Tannakian reconstruction problem for a field R_0 (or more generally a Dedekind domain, see [5, 8]), one obtains the result by expressing R_1 as an increasing union of its R_1-subcomodules $f\colon V \hookrightarrow R_1$, where each V belongs to $R_1\text{-Mod}^{\square}$. Theorem 9.1.1 tells us that the appropriate weakening of this condition is given by dropping the injectivity of f.

The original article of Schäppi proves a more general statement than that formulated above, where instead of working over k, one works with $\mathcal{V}$-enriched categories for an arbitrary cosmos $\mathcal{V}$ ([4, Theorem 7.5.1]). In the classical setting, however, considerably less categorical machinery is needed in order to prove Theorem 9.1.1, so a self-contained account of this special case is perhaps useful.

The proof presented here goes as follows. We start in Section 9.2 by defining the category of affine category schemes $\mathrm{Cat}(\mathrm{Sch}_k^{\mathrm{aff}})$ and the 2-category Tan_k of Tannakian triples. Then we prove an adjunction between them:

$$\mathrm{Tan}_k \underset{\mathrm{Rep}^{\square}}{\overset{\mathrm{H}^{\otimes}}{\rightleftarrows}} \mathrm{Cat}(\mathrm{Sch}_k^{\mathrm{aff}})^{\mathrm{op}}. \tag{2}$$

The counit of this adjunction goes from an affine category scheme $\mathrm{Spec}(R_1)$ to (the spectrum of) a special type of colimit, denoted by $\mathrm{coend}(V^{\vee} \otimes_k V)$, taken over $V \in R_1\text{-Mod}^{\square}$. The reconstructibility of $\mathrm{Spec}(R_1)$ can thus be reformulated as the bijectivity of the map

$$\mathrm{coend}(V^{\vee} \otimes_k V) \to R_1. \tag{3}$$

In Section 9.3, we prove that this coend is isomorphic to $\mathrm{colim}_{f\colon V \to R_1} V$ (over R_1), thus obtaining Theorem 9.1.1 (which appears later as Corollary 9.3.5). We explain how Theorem 9.1.1 recovers, in the case of Dedekind domains, the classical reconstruction theorem.[1]

In Section 9.4, we explain another problem which is properly contextualized by the Tannakian adjunction (2). This is the problem of embedding a finite-type affine group scheme $\mathrm{Spec}(R_1)$ over $\mathrm{Spec}(k)$ into $\underline{\mathrm{Aut}}_k(M)$ for a finite projective k-module M. We show that $\mathrm{Spec}(R_1)$ is embeddable if and only if (3) is surjective (Lemma 9.4.3). In particular, this

[1] The interpretation of Tannakian formalism as an adjunction goes back at least to Street ([7, Section 16]). The relevant coend is already present in [1] and features prominently in most subsequent works.

implies that a finite-type reconstructible affine group scheme is always embeddable.

9.2 Tannakian adjunction

Let k be a ring. All rings (resp. algebras) in this chapter are assumed commutative and unital. The goal of this section is to set up an adjunction between affine category schemes and Tannakian triples. We first define the appropriate categories and functors between them. The main result appears as Proposition 9.2.13.

9.2.1 Affine category schemes

Denote by $\mathrm{Sch}_k^{\mathrm{aff}}$ the category of affine schemes over k. Let $\mathrm{Cat}(\mathrm{Sch}_k^{\mathrm{aff}})$ be the category of category objects in $\mathrm{Sch}_k^{\mathrm{aff}}$, defined with respect to fiber product over k.

Concretely, an object $\mathcal{X}$ of $\mathrm{Cat}(\mathrm{Sch}_k^{\mathrm{aff}})$ consists of affine schemes X_0 and X_1, equipped with morphisms satisfying the axioms of a category:

$$e\colon X_0 \to X_1, \qquad X_1 \underset{p_i}{\overset{p_t}{\rightrightarrows}} X_0, \qquad c\colon X_1 \underset{p_i, X_0, p_t}{\times} X_1 \to X_1.$$

Informally, X_0 is the affine scheme of "objects" of $\mathcal{X}$, and X_1 is the affine scheme of "morphisms"; e is the map sending an object to the identity arrow on it, p_t (resp. p_i) is the map sending an arrow to its codomain (resp. domain), and c is the composition law. The subscript on p_t (resp. p_i) stands for "terminal" (resp. "initial").

A morphism $\mathcal{X} \to \mathcal{Y}$ consists of a pair of morphisms $f\colon X_0 \to Y_0$ and $f_1\colon X_1 \to Y_1$ commuting with the above structures.

Remark 9.2.1 The notion of an affine category scheme is dual to that of a coalgebroid, i.e., a k-algebra R_0, an $(R_0 \otimes_k R_0)$-algebra R_1 endowed with a comultiplication map $\Delta\colon R_1 \to R_1 \otimes_{R_0} R_1$, and a counit map $\epsilon\colon R_1 \to R_0$ satisfying the counital and coassociative conditions.

In the dictionary between affine category schemes and coalgebroids, we will always associate to p_t (resp. p_i) the left (resp. right) R_0-structure of R_1.

Remark 9.2.2 Affine groupoid schemes can be characterized as those affine category schemes $\mathcal{X}$ for which the map below is an isomorphism:

$$(c, \mathrm{pr}_2)\colon X_1 \underset{p_i, X_0, p_t}{\times} X_1 \xrightarrow{\sim} X_1 \underset{p_i, X_0, p_i}{\times} X_1. \tag{1}$$

Indeed, it is easy to construct the antipode morphism $X_1 \to X_1$ from the inverse of (1).

Affine monoid schemes over k are those affine category schemes $\mathcal{X}$ such that the structural map $X_0 \to \mathrm{Spec}(k)$ is an isomorphism. Affine group schemes over k are those affine category schemes which are at once affine groupoid and monoid schemes.

9.2.2 Tannakian triples

Let $\mathrm{Cat}_k^{\otimes}$ denote the 2-category of symmetric monoidal k-linear categories. Its 1-morphisms are symmetric monoidal k-linear functors and its 2-morphisms are natural transformations which respect the symmetric monoidal structure. Let $\mathrm{Cat}_k^{\otimes,\mathrm{sm}} \subset \mathrm{Cat}_k^{\otimes}$ denote its full 2-subcategory of essentially small symmetric monoidal k-linear categories.

Remark 9.2.3 What we call a "2-category" is what some authors refer to as a "bicategory" or "weak 2-category." Namely, we require the composition law of Hom-categories (also called *horizontal composition*) to be associative only up to homotopy. Similarly, we prefer the term "functor" over "pseudofunctor" or "2-functor."

For $X \in \mathrm{Sch}_k^{\mathrm{aff}}$, the category of quasi-coherent sheaves $\mathrm{QCoh}(X)$ is naturally an object of $\mathrm{Cat}_k^{\otimes}$. Its full subcategory consisting of dualizable objects, to be denoted $\mathrm{QCoh}(X)^{\square}$, inherits a symmetric monoidal k-linear structure and belongs to $\mathrm{Cat}_k^{\otimes,\mathrm{sm}}$. For $X = \mathrm{Spec}(R)$, the category $\mathrm{QCoh}(X)^{\square}$ is identified with that of finite projective R-modules, which we alternatively denote by $R\text{-}\mathrm{Mod}^{\square}$.

For the purpose of defining the Tannakian adjunction, it is convenient to introduce a notion of Tannakian triples which is more general than that of Tannakian categories introduced in [1]. We call a *Tannakian triple* a triple (X_0, C, T) with $X_0 \in \mathrm{Sch}_k^{\mathrm{aff}}$, $C \in \mathrm{Cat}_k^{\otimes,\mathrm{sm}}$, and $T\colon C \to \mathrm{QCoh}(X)^{\square}$ a morphism in $\mathrm{Cat}_k^{\otimes}$.

Tannakian triples are naturally organized into a 2-category. To wit, given two Tannakian triples (X_0, C, T) and $(Y_0, \mathcal{D}, U)$, objects of their Hom-category

$$\mathrm{Hom}((X_0, C, T), (Y_0, \mathcal{D}, U)) \tag{2}$$

consist of a morphism $f\colon Y_0 \to X_0$, a functor $F\colon C \to \mathcal{D}$ in $\mathrm{Cat}_k^{\otimes}$, and a 2-morphism h witnessing the commutativity of the following diagram in $\mathrm{Cat}_k^{\otimes}$:

$$\begin{array}{ccc} \mathcal{C} & \xrightarrow{T} & \mathrm{QCoh}(X_0)^{\square} \\ \downarrow{\scriptstyle F} & & \downarrow{\scriptstyle f^*} \\ \mathcal{D} & \xrightarrow{U} & \mathrm{QCoh}(Y_0)^{\square} \end{array} \tag{3}$$

Morphisms in the Hom-category (2) from (f, F, h) to (f', F', h') exist only when $f = f'$ and are specified by a 2-morphism $\eta\colon F \Rightarrow F'$ compatible with h, h'. We omit the description of the horizontal composition in Tan_k.

Remark 9.2.4 If U reflects isomorphisms then the category (2) is a groupoid. If U is faithful, then (2) is equivalent to a set (viewed as a discrete category).

The goal of this section is to construct an adjunction of 2-categories

$$\mathrm{Tan}_k \underset{\mathrm{Rep}^{\square}}{\overset{\mathrm{H}^{\otimes}}{\rightleftarrows}} \mathrm{Cat}(\mathrm{Sch}_k^{\mathrm{aff}})^{\mathrm{op}}. \tag{4}$$

Let us first give an informal summary of (4). The functor $\mathrm{Rep}^{\square}$ associates to an affine category scheme $\mathcal{X}$ the category $\mathrm{Rep}^{\square}(\mathcal{X})$ of representations whose underlying quasi-coherent sheaf on X_0 is dualizable, and $T\colon \mathrm{Rep}(\mathcal{X})^{\square} \to \mathrm{QCoh}(X_0)^{\square}$ is the forgetful functor.

The functor $\mathrm{H}^{\otimes}$ sends a Tannakian triple $(X_0, \mathcal{C}, T)$ to the affine category scheme whose "scheme of morphisms" X_1 represents the presheaf over $X_0 \times X_0$ sending an affine scheme S to $\mathrm{Hom}(T_{1,S}, T_{2,S})$, where $T_{i,S}$ (for $i = 1, 2$) denotes the 1-morphism in $\mathrm{Cat}_k^{\otimes}$ given by the composition

$$\mathcal{C} \xrightarrow{T} \mathrm{QCoh}(X_0)^{\square} \xrightarrow{\mathrm{pr}_i^*} \mathrm{QCoh}(S)^{\square}$$

($\mathrm{pr}_i\colon S \to X_0$ is the structural map), and the Hom is understood as that between two 1-morphisms in the 2-category $\mathrm{Cat}_k^{\otimes}$.

Remark 9.2.5 Note that the left-hand side of (4) is a 2-category while the right-hand side is a 1-category. The adjunction (4) includes, in particular, the assertion that the Hom-category in Tan_k from $(X_0, \mathcal{C}, \mathcal{T})$ to the Tannakian triple associated to an affine category scheme is equivalent to a set. This fact, however has already been observed in Remark 9.2.4.

9.2.3 The functor Rep$^{\square}$

In this subsection, we define the functor Rep$^{\square}$ appearing in the adjunction (4). Given an affine category scheme $\mathcal{X}$, a *representation* of $\mathcal{X}$ consists of an object $V \in \mathrm{QCoh}(X_0)$ equipped with a morphism $\alpha\colon p_i^*V \to p_t^*V$ which satisfies:

(i) the counit condition for its pullback along $e\colon X_0 \to X_1$, i.e., that the following diagram commutes:

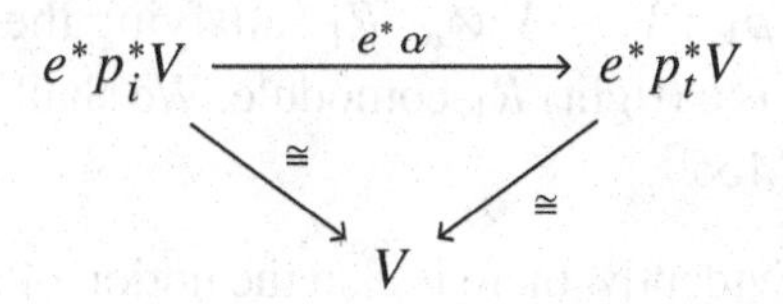

(ii) the cocycle condition relating its pullbacks along the three maps pr_1, pr_2, and c from $X_1 \underset{p_i, X_0, p_t}{\times} X_1$ to X_1, i.e., that the following diagram commutes:

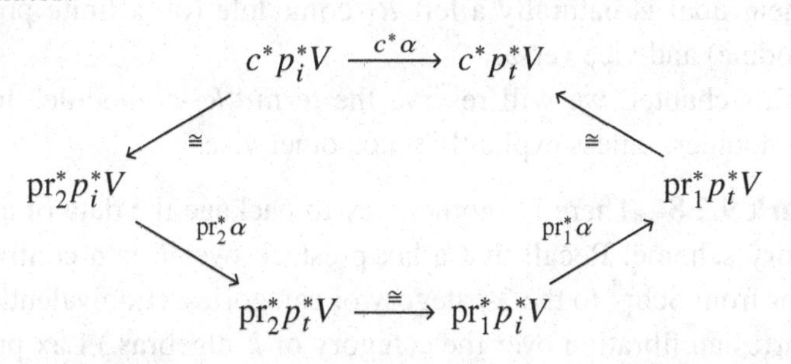

Representations of $\mathcal{X}$ naturally form a category, to be denoted by $\mathrm{Rep}(\mathcal{X})$. It is equipped with a functor $T\colon \mathrm{Rep}(\mathcal{X}) \to \mathrm{QCoh}(X_0)$ which remembers the underlying quasi-coherent sheaf. Let $\mathrm{Rep}(\mathcal{X})^{\square}$ be the full subcategory of $\mathrm{Rep}(\mathcal{X})$ consisting of objects whose image under T lies in $\mathrm{QCoh}(X_0)^{\square}$. Then the triple $(X_0, \mathrm{Rep}(\mathcal{X})^{\square}, T)$ is a Tannakian triple in the sense of Section 9.2.2.

It is clear that given any morphism $\mathcal{X} \to \mathcal{Y}$, there is an induced commutative diagram in $\mathrm{Cat}_k^{\otimes}$:

$$\begin{array}{ccc} \mathrm{Rep}(\mathcal{Y})^{\square} & \longrightarrow & \mathrm{QCoh}(Y_0)^{\square} \\ \big\downarrow F & & \big\downarrow f^* \\ \mathrm{Rep}(\mathcal{X})^{\square} & \longrightarrow & \mathrm{QCoh}(X_0)^{\square} \end{array} \tag{5}$$

The functor F is formed by pulling back the datum $p_i^*V \to p_t^*V$ for any $V \in \mathrm{Rep}(\mathcal{Y})$ along $f\colon X_0 \to Y_0$. Given morphisms $\mathcal{X} \to \mathcal{Y}$ and $\mathcal{Y} \to \mathcal{Z}$, there is a natural 2-isomorphism relating the composition of diagram (5) and the diagram associated to the composition $\mathcal{X} \to \mathcal{Z}$. Thus, the association of the Tannakian category $(X^0, \mathrm{Rep}(\mathcal{X})^\square, T)$ to $\mathcal{X}$ is a contravariant functor, which we will (with some abuse of notation) denote by the same letters:

$$\mathrm{Rep}^\square\colon \mathrm{Cat}(\mathrm{Sch}_k^{\mathrm{aff}})^{\mathrm{op}} \to \mathrm{Tan}_k.$$

Remark 9.2.6 In the dual algebroid description of Remark 9.2.1, an object of $\mathrm{Rep}(\mathcal{X})^\square$ can be viewed as a dualizable R_0-module V equipped with a morphism $\rho_V : V \to V \otimes_{R_0} R_1$ satisfying the counit and cocycle conditions, i.e., V is a (right) R_1-comodule. We shall alternatively denote $\mathrm{Rep}(\mathcal{X})^\square$ by $R_1\text{-}\mathrm{Mod}^\square$.

Remark 9.2.7 Evidently, there is also the notion of a "corepresentation" of $\mathcal{X}$, which consists of $V \in \mathrm{QCoh}(X_0)$ together with a morphism $p_t^*V \to p_i^*V$ satisfying analogous conditions. A corepresentation of $\mathcal{X}$ is equivalent to a left R_1-comodule. For any object in $R_1\text{-}\mathrm{Mod}^\square$, its R_0-linear dual is naturally a left R_1-comodule (on a finite projective R_0-module) and vice versa.

In this chapter, we will reserve the term "R_1-comodule" for *right* R_1-comodules, unless explicitly stated otherwise.

Remark 9.2.8 There is another way to package the data of an affine category scheme. Recall that a lax prestack over k is a contravariant functor from $\mathrm{Sch}_k^{\mathrm{aff}}$ to the 2-category of categories. (Equivalently, it is a co-Cartesian fibration over the category of k-algebras.) Lax prestacks naturally form a 2-category. An example of a lax prestack is the functor $\mathrm{QCoh}(-)$ (or its variant $\mathrm{QCoh}(-)^\square$). For any lax prestack Y, we denote by $\mathrm{QCoh}(Y)^\square$ the category $\mathrm{Hom}(Y, \mathrm{QCoh}(-)^\square)$.

Given an affine category scheme $\mathcal{X}$, one obtains a lax prestack (to be denoted by X_{-1}) whose value at $S \in \mathrm{Sch}_k^{\mathrm{aff}}$ is the category with objects $X_0(S)$ and morphisms $X_1(S)$. The category $\mathrm{Cat}(\mathrm{Sch}_k^{\mathrm{aff}})$ embeds fully faithfully into that of *morphisms* of lax prestacks, by sending $\mathcal{X}$ to the morphism $X_0 \to X_{-1}$. The functor $\mathrm{Rep}^\square$ is none other than the association from $\mathcal{X}$ to the triple consisting of X_0, $\mathrm{QCoh}(X_{-1})^\square$, and the pullback functor $T\colon\ \mathrm{QCoh}(X_{-1})^\square \to \mathrm{QCoh}(X_0)^\square$.

9.2.4 A representability lemma

In the definition of the functor $\mathrm{H}^\otimes$ in the adjunction (4), the key step is to construct the affine scheme X_1. We will deduce its existence from a general representability result.

Let $\mathcal{C} \in \mathrm{Cat}_k^{\otimes,\mathrm{sm}}$, $S_0 \in \mathrm{Sch}_k^{\mathrm{aff}}$, and T_1, T_2 be 1-morphisms $\mathcal{C} \to \mathrm{QCoh}(S_0)^{\square}$ in $\mathrm{Cat}_k^{\otimes,\mathrm{sm}}$. For an affine S_0-scheme S, we denote by $T_{i,S}$ (for $i = 1, 2$) the 1-morphism

$$T_{i,S} : \mathcal{C} \to \mathrm{QCoh}(S)^{\square},$$

given by composition with pullback along $S \to S_0$. We use the notation $\mathrm{Hom}(T_{1,S}, T_{2,S})$ to mean Hom formed in $\mathrm{Cat}_k^{\otimes,\mathrm{sm}}$. The goal of this subsection is to show that the presheaf that associates $\mathrm{Hom}(T_{1,S}, T_{2,S})$ to S is representable by an affine scheme over S_0.

Remark 9.2.9 The representing object, viewed as a commutative algebra in $\mathrm{QCoh}(S_0)$, arises from a general categorical construction known as a "coend." Instead of defining coends in general, we will confine ourselves to the particular situation at hand.

To any $c \in \mathcal{C}$, we may attach an object $T_1(c)^{\vee} \otimes_{S_0} T_2(c)$ in the category $\mathrm{QCoh}(S_0)$. For any morphism $d \to c$ in $\mathcal{C}$, there is a diagram

$$\begin{array}{ccc} T_1(c)^{\vee} \otimes_{S_0} T_2(d) & \longrightarrow & T_1(c)^{\vee} \otimes_{S_0} T_2(c) \\ \downarrow & & \\ T_1(d)^{\vee} \otimes_{S_0} T_2(d) & & \end{array} \tag{6}$$

Let $\mathrm{coend}(T_1(c)^{\vee} \otimes_{S_0} T_2(c))$ denote the universal object in $\mathrm{QCoh}(S_0)$ equipped with a map from $T_1(c)^{\vee} \otimes_{S_0} T_2(c)$ for each $c \in \mathcal{C}$, such that, for each morphism $d \to c$, the two maps it receives from $T_1(c)^{\vee} \otimes_{S_0} T_2(d)$ via (6) are equal. This description makes it evident that $\mathrm{coend}(T_1(c)^{\vee} \otimes_{S_0} T_2(c))$ is the cokernel of a morphism in $\mathrm{QCoh}(S_0)$ given by

$$\bigoplus_{(d \to c) \in \mathcal{C}} T_1(c)^{\vee} \otimes_{S_0} T_2(d) \to \bigoplus_{c \in \mathcal{C}} T_1(c)^{\vee} \otimes_{S_0} T_2(c),$$

and in particular ensures its existence.[2]

Furthermore, $\mathrm{coend}(T_1(c)^{\vee} \otimes_{S_0} T_2(c))$ has a commutative algebra structure, coming from the symmetric monoidal structures of T_1 and T_2. More precisely, given any $c, c' \in \mathcal{C}$, the object

$$(T_1(c)^{\vee} \otimes_{S_0} T_2(c)) \otimes_{S_0} (T_1(c')^{\vee} \otimes_{S_0} T_2(c')) \tag{7}$$

is naturally isomorphic to $T_1(c \otimes c')^{\vee} \otimes_{S_0} T_2(c \otimes c')$. Hence, (7) maps to $\mathrm{coend}(T_1(c)^{\vee} \otimes_{S_0} T_2(c))$ via the structural map associated to $c \otimes c' \in \mathcal{C}$. The universal property of the tensor product of $\mathrm{coend}(T_1(c)^{\vee} \otimes_{S_0} T_2(c))$

[2] The object $\mathrm{coend}(T_1(c)^{\vee} \otimes_{S_0} T_2(c))$ can also be regarded as a colimit of $T_1(c)^{\vee} \otimes_{S_0} T_2(d)$ over the twisted arrow category of $\mathcal{C}$.

with itself then determines its multiplication map. We omit the description of the counit.

Lemma 9.2.10 *The presheaf over S_0 sending an affine S_0-scheme S to* $\mathrm{Hom}(T_{2,S}, T_{1,S})$ *is representable by the affine S_0-scheme with structure sheaf* $\mathrm{coend}(T_1(c)^\vee \otimes_{S_0} T_2(c))$.

Proof For an affine S_0-scheme S, morphisms

$$\mathrm{coend}(T_1(c)^\vee \otimes_{S_0} T_2(c)) \to \mathcal{O}_S \tag{8}$$

as quasi-coherent $\mathcal{O}_{S_0}$-modules can be described using the universal property of the coend. Namely, such a morphism consists of an $\mathcal{O}_{S_0}$-linear morphism $T_2(c) \to T_1(c) \otimes_{S_0} \mathcal{O}_S$ for every $c \in C$ together with a compatibility condition for every $d \to c$ in C. Equivalently, these data are captured by a natural transformation

$$T_2 \otimes_{S_0} \mathcal{O}_S \to T_1 \otimes_{S_0} \mathcal{O}_S \tag{9}$$

between k-linear functors from C to $\mathrm{QCoh}(S)$. It suffices to observe that the morphism (8) preserves the commutative algebra structures if and only if the corresponding natural transformation (9) is compatible with the symmetric monoidal structures on the two functors. □

9.2.5 *The functor* $\mathrm{H}^\otimes$

The goal of this subsection is to define the other functor $\mathrm{H}^\otimes$ appearing in the adjunction (4):

$$\mathrm{H}^\otimes : \mathrm{Tan}_k \to \mathrm{Cat}(\mathrm{Sch}_k^{\mathrm{aff}})^{\mathrm{op}}.$$

Indeed, let (X_0, C, T) be a Tannakian triple. We shall functorially attach to it an affine category scheme $\mathcal{X}$ over k.

Let $S_0 := X_0 \times_k X_0$ and T_i be the 1-morphism $C \to \mathrm{QCoh}(X_0 \times_k X_0)^\square$ in $\mathrm{Cat}_k^{\otimes,\mathrm{sm}}$ given by composing T with pr_i^* (for $i = 1, 2$). By Lemma 9.2.10, the presheaf over $X_0 \times X_0$ sending S to $\mathrm{Hom}(T_{2,S}, T_{1,S})$ is representable by the affine scheme X_1 whose structure sheaf is

$$\mathrm{coend}(T(c)^\vee \otimes_k T(c)) \in \mathrm{QCoh}(X_0 \times X_0), \tag{10}$$

equipped with the commutative algebra structure specified above. We record the structural maps of X_1 below:

$$(p_t, p_i) : X_1 \to X_0 \times_k X_0. \tag{11}$$

Next, we will upgrade (11) to an affine category scheme. Namely, we need to describe the following data:

(i) the unit section $e\colon X_0 \to X_1$, and
(ii) the composition law $c\colon X_1 \underset{p_i, X_0, p_t}{\times} X_1 \to X_1$.

To define e, we note that an affine $(X_0 \times_k X_0)$-scheme S with $\mathrm{pr}_1 = \mathrm{pr}_2$ determines a canonical section of $\mathrm{Hom}(T_{2,S}, T_{1,S})$, i.e., the identity. To define c, consider an affine scheme S over $X_0 \times_k X_0 \times_k X_0$. Denote by

$$T_{i,S}\colon C \to \mathrm{QCoh}(S)^{\square} \quad (\text{for } i = 1, 2, 3)$$

the composition of $T_i\colon C \to \mathrm{QCoh}(X_0 \times_k X_0 \times_k X_0)^{\square}$ with the pullback functor to $\mathrm{QCoh}(S)^{\square}$. The composition of natural transformations defines a map of sets, functorial in S:

$$\mathrm{Hom}(T_{2,S}, T_{1,S}) \times \mathrm{Hom}(T_{3,S}, T_{2,S}) \to \mathrm{Hom}(T_{3,S}, T_{1,S}). \qquad (12)$$

The left-hand side is represented by $X_1 \underset{p_i, X_0, p_t}{\times} X_1$ and the right-hand side is represented by the base change of X_1 along the projection:

$$\mathrm{pr}_{13}\colon X_0 \times_k X_0 \times_k X_0 \to X_0 \times_k X_0.$$

Hence, we obtain from (12) the map c lying over $X_0 \times_k X_0 \times_k X_0$.

Remark 9.2.11 Let us give an explicit description of the coalgebroid structure on the coend (10). We write R_0, R_1 for the k-algebras corresponding to X_0, X_1 (so R_1 is synonymous with (10)). For any $c \in C$, there is a canonical map $T(c)^{\vee} \otimes_k T(c) \to R_1$ in $(R_0 \otimes_k R_0)$-Mod, which dualizes to a morphism in R_0-Mod:

$$\rho_c\colon T(c) \to T(c) \otimes_{R_0} R_1, \qquad (13)$$

where the R_0-structure on the right-hand side is the right R_0-action on R_1. (Similarly, one has $\rho_c^{\vee}\colon T(c)^{\vee} \to R_1 \otimes_{R_0} T(c)^{\vee}$.) For any morphism $c \in C$, there is a contraction map

$$T(c)^{\vee} \otimes_k T(c) \to R_0, \qquad (14)$$

which defines the counit $\epsilon\colon R_1 \to R_0$. The comultiplication map $\Delta\colon R_1 \to R_1 \otimes_{R_0} R_1$ is the unique morphism making the following diagram commute for all $c \in C$:

$$\begin{array}{ccc} T(c)^{\vee} \otimes_k T(c) & \xrightarrow{\text{can}} & R_1 \\ \downarrow{\scriptstyle \rho_c^{\vee} \otimes \rho_c} & & \downarrow{\scriptstyle \Delta} \\ R_1 \otimes_{R_0} T(c)^{\vee} \otimes_k T(c) \otimes_{R_0} R_1 & \xrightarrow{(14)} & R_1 \otimes_{R_0} R_1 \end{array} \qquad (15)$$

Lemma 9.2.12 *The morphism ρ_c equips $T(c)$ with the structure of an R_1-comodule.*

Proof This follows from the commutativity of (15). □

To address the functoriality of the construction of the affine category scheme attached to (X_0, C, T), let us consider another Tannakian triple $(Y_0, \mathcal{D}, U)$. We need to construct a functor from their Hom-category (cf. Section 9.2.2)

$$\mathrm{Hom}((X_0, C, T), (Y_0, \mathcal{D}, U)) \tag{16}$$

to the set $\mathrm{Hom}(\mathcal{Y}, \mathcal{X})$, where $\mathcal{X}$ and $\mathcal{Y}$ are the corresponding affine category schemes.

Note that an object (f, F, h) in the Hom-category (16) already supplies the morphism of schemes $f \colon Y_0 \to X_0$. In order to build the morphism $f_1 \colon Y_1 \to X_1$ commuting with the structural maps to $Y_0 \times_k Y_0$, respectively $X_0 \times_k X_0$, it suffices to describe a map of sets functorial in affine $(Y_0 \times_k Y_0)$-schemes S:

$$\mathrm{Hom}(U_{2,S}, U_{1,S}) \to \mathrm{Hom}(T_{2,S}, T_{1,S}). \tag{17}$$

The required maps (17) are supplied by the commutative diagram (3). More precisely, they are defined by precomposing a given natural transformation $U_{2,S} \to U_{1,S}$ with $F : C \to \mathcal{D}$ and identifying the functors via h (for $i = 1, 2$):

$$h \colon U_{i,S} \circ F \xrightarrow{\sim} T_{i,S}.$$

It is clear that the resulting map $f_1 \colon Y_1 \to X_1$ is compatible with the unit and composition morphisms. Next, a morphism $\eta \colon F \Rightarrow F'$ in (16) induces a commutative diagram of functors (for $i = 1, 2$):

$$\begin{array}{ccc} U_{i,S} \circ F & \xrightarrow{\eta} & U_{i,S} \circ F' \\ & \searrow^{h} & \downarrow h' \\ & & T_{i,S} \end{array}$$

Hence, the induced maps (17) coming from (f, F, h) and (f, F', h') are equal. This shows that we have a well-defined functor from (16) to $\mathrm{Hom}(\mathcal{Y}, \mathcal{X})$ and the composition laws are verified.

9.2.6 The adjunction

We are now ready to establish the adjunction (4). Let (Y_0, C, T) be a Tannakian triple and let $\mathcal{X}$ be an affine category scheme. For notational simplicity, we will record (Y_0, C, T) only by the functor T.

Proposition 9.2.13 *There is a canonical equivalence of categories*

$$\mathrm{Hom}_{\mathrm{Tan}_k}(T, \mathrm{Rep}(\mathcal{X})^{\square}) \xrightarrow{\sim} \mathrm{Hom}_{\mathrm{Cat}(\mathrm{Sch}_k^{\mathrm{aff}})}(\mathcal{X}, \mathrm{H}^{\otimes}(T)). \tag{18}$$

Note that both sides of (18) are equivalent to sets (see Remark 9.2.4).

Proof By definition, an object in the category on the left-hand side of (18) consists of a morphism $f \colon X_0 \to Y_0$, a functor $F \colon C \to \mathrm{Rep}(\mathcal{X})^{\square}$ in $\mathrm{Cat}_k^{\otimes}$, and a commutativity witness h of the diagram in $\mathrm{Cat}_k^{\otimes}$:

$$\begin{array}{ccc} C & \xrightarrow{T} & \mathrm{QCoh}(Y_0)^{\square} \\ \downarrow{\scriptstyle F} & & \downarrow{\scriptstyle f^*} \\ \mathrm{Rep}(\mathcal{X})^{\square} & \longrightarrow & \mathrm{QCoh}(X_0)^{\square} \end{array} \tag{19}$$

Let Y_1 denote the affine scheme associated to $\mathrm{H}^{\otimes}(T)$. Namely, it represents the functor which associates to every affine $Y_0 \times Y_0$-scheme S the set $\mathrm{Hom}(T_{2,S}, T_{1,S})$. Regarding X_1 as a $(Y_0 \times_k Y_0)$-scheme, there is a canonical element of $\mathrm{Hom}(T_{2,X_1}, T_{1,X_1})$ coming from the commutative diagram (19). This element defines a map $X_1 \to Y_1$ compatible with the category scheme structures.

Conversely, consider a morphism $\mathcal{X} \to \mathrm{H}^{\otimes}(T)$, viewed as morphisms $f \colon X_0 \to Y_0$ and $f_1 \colon X_1 \to Y_1$ making the corresponding diagrams commute. By Lemma 9.2.12, every object $T(c)$ (for $c \in C$) acquires the canonical structure of an object in $\mathrm{Rep}(\mathcal{Y})^{\square}$. Hence $f^*T(c)$ canonically upgrades to an object in $\mathrm{Rep}(\mathcal{X})^{\square}$. This gives rise to the commutative diagram (19). We omit the verification that these constructions are mutual inverses and are functorial in $\mathcal{X}$ and T. □

9.3 Reconstruction

We continue to let k denote a ring. Throughout this section, we fix an affine category scheme $\mathcal{X}$ over k.

The goal of this section is to investigate when $\mathcal{X}$ can be reconstructed from its associated Tannakian triple. We first prove a formal criterion for the reconstructibility of $\mathcal{X}$ (Corollary 9.3.5). Then we deduce some consequences from it when additional conditions are satisfied (flatness, finiteness, etc.)

9.3.1 The counit

The counit of the adjunction (4) is a morphism of affine category schemes:

$$\mathcal{X} \to \mathrm{H}^{\otimes}(\mathrm{Rep}(\mathcal{X})^{\square}). \tag{1}$$

We say that $\mathcal{X}$ is *reconstructible* if (1) is an isomorphism.

Remark 9.3.1 Likewise, we call a Tannakian triple (Y_0, C, T) *reconstructible* if the unit of the adjunction (4) in Section 9.2 is an equivalence when evaluated at (Y_0, C, T). It follows formally that (4) restricts to an equivalence between reconstructible affine category schemes and reconstructible Tannakian triples.

Let us recall some notation. Denote by R_0, R_1 the k-algebras involved in the coalgebroid associated to $\mathcal{X}$ (see Remark 9.2.1). The category of representations of $\mathcal{X}$ is identified with the category R_1 -Mod of (right) R_1-comodules. Its full subcategory $\mathrm{Rep}(\mathcal{X})^{\square}$ corresponds to R_1 -Mod$^{\square}$, and is characterized by the property of having finite projective underlying R_0-modules.

The next lemma is a tautological reformulation of the reconstructibility of the affine category scheme $\mathcal{X}$.

Lemma 9.3.2 *The following are equivalent:*

(i) *$\mathcal{X}$ is reconstructible;*
(ii) *The coend over $V \in R_1$ -Mod$^{\square}$ maps isomorphically to R_1,*

$$\mathrm{coend}(V^{\vee} \otimes_k V) \xrightarrow{\sim} R_1. \tag{2}$$

The map (2) is defined by dualizing the R_1-coaction on V.

Proof This follows from the description of the affine scheme Y_1 associated to $\mathrm{H}^{\otimes}(\mathrm{Rep}(\mathcal{X})^{\square})$ given in Section 9.2.5. □

9.3.2 Calculation of the coend

Consider the category of maps $f: V \to R_1$ of R_1-comodules with source $V \in R_1$ -Mod$^{\square}$. One can form the colimit of V in the category R_1 -Mod indexed by the above category of maps. It is equipped with a tautological map to R_1:

$$\operatorname*{colim}_{f\colon V\to R_1} V \to R_1. \tag{3}$$

In addition to the termwise R_0-module structure on $\operatorname{colim}_{f\colon V\to R_1} V$, it admits another R_0-structure such that (3) carries it to the left R_0-structure on R_1. To describe this second R_0-structure, note that any morphism $f\colon V \to R_1$ can be left-multiplied by an element $r \in R_0$ to form a morphism $r \cdot f\colon V \to R_1$. Thus any $r \in R_0$ induces an endomorphism of $\operatorname{colim}_{f\colon V\to R_1} V$, obtained as the identity map from the copy of V corresponding to the index f to that corresponding to $r \cdot f$. We refer to these two R_0-structures on $\operatorname{colim}_{f\colon V\to R_1} V$ as the *right*, respectively *left* R_0-structures. Observe that these R_0-structures coincide over k, effectively turning $\operatorname{colim}_{f\colon V\to R_1} V$ into an $(R_0 \otimes_k R_0)$-module.

Our next goal is to construct a canonical map:

$$\operatorname*{colim}_{f\colon V\to R_1} V \to \operatorname{coend}(V^\vee \otimes_k V) \tag{4}$$

which is compatible with both the left and right R_0-structures as well as the (right) R_1-comodule structures. Furthermore, it will render the following diagram commutative:

$$\begin{array}{ccc} \operatorname*{colim}_{f\colon V\to R_1} V & \xrightarrow{(4)} & \operatorname{coend}(V^\vee \otimes_k V) \\ & \searrow^{(3)} & \downarrow{\scriptstyle (2)} \\ & & R_1 \end{array} \tag{5}$$

We begin with a preparatory observation.

Lemma 9.3.3 *For any $V \in R_1\text{-Mod}^\square$, composition with the counit $\epsilon\colon R_1 \to R_0$ defines a bijection of R_0-modules*

$$\operatorname{Hom}_{R_1\text{-Mod}}(V, R_1) \xrightarrow{\sim} V^\vee. \tag{6}$$

Here, the R_0-structure on the left-hand side comes from the left R_0-action on R_1.

Proof The inverse map sends an element $\varphi\colon V \to R_0$ to the composition of the coaction map $V \to V \otimes_{R_0} R_1$ with φ. □

The required morphism (4) is defined as follows. For every $f\colon V \to R_1$ with $V \in R_1\text{-Mod}^\square$, corresponding to $f^\vee$ under (6), we consider the map $V \to V^\vee \otimes_k V$ which sends v to $f^\vee \otimes v$. It is clear that this definition is compatible with any change of R_1-comodules $W \to V$, thus inducing the map (4).

We now come to the main observation of this subsection.

Lemma 9.3.4 *The map* (4) *is an isomorphism.*

Proof It suffices to show that (4) is an isomorphism of $(R_0 \otimes_k R_0)$-modules, with both colimits computed as such.[3]

Let M be an $(R_0 \otimes_k R_0)$-module, viewed as an (R_0, R_0)-bimodule. The Hom-set from $\mathrm{coend}(V^\vee \otimes_k V)$ to M can be identified with that of natural transformations $T \to T \otimes_{R_0} M$ as R_0-linear functors from $R_1\text{-Mod}^\square$ to $R_0\text{-Mod}$. More concretely, it is a collection of R_0-linear maps (in reference to the right R_0-action on M)

$$V \to V \otimes_{R_0} M \text{ for all } V \in R_1\text{-Mod}^\square, \tag{7}$$

satisfying a compatibility condition for $W \to V$. The Hom-set from $\mathrm{colim}_{f\,:\,V \to R_1} V$ to M can be identified with that of a collection of R_0-linear maps (in reference to the right R_0-action on M):

$$V \to M \text{ for all } f^\vee \in V^\vee, \tag{8}$$

which depend R_0-linearly on $f^\vee$ (in reference to the left R_0-action on M) and satisfy a compatibility condition for $W \to V$. The passage from (7) to (8) induced from (4) is the composition with $f^\vee \otimes \mathrm{id}\colon V \otimes_{R_0} M \to M$. This procedure defines a bijection since $V \otimes_{R_0} M$ can be identified with $\mathrm{Hom}_{R_0}(V^\vee, M)$. □

Corollary 9.3.5 *The following are equivalent:*

(i) $\mathcal{X}$ *is reconstructible;*
(ii) *the morphism* (3) *is bijective.*

Proof This follows from Lemma 9.3.2 and Lemma 9.3.4. □

Remark 9.3.6 Instead of (4), we may consider a map:

$$\underset{f\,:\,V^\vee \to R_1}{\mathrm{colim}} V^\vee \to \mathrm{coend}(V^\vee \otimes_k V), \tag{9}$$

where the colimit is of *left* R_1-comodules $V^\vee$ with finite projective underlying R_0-modules, and the index category is over left R_1-comodule maps of them into R_1. The analogues of Lemma 9.3.4 and Corollary 9.3.5 assert that (9) is bijective and that $\mathcal{X}$ is reconstructible if and only if the canonical map from $\mathrm{colim}_{f\,:\,V^\vee \to R_1} V^\vee$ to R_1 is bijective.

[3] Of course, it is enough to prove the analogous statement for k-modules, but we find the statement for $(R_0 \otimes_k R_0)$-modules more satisfying. For instance, if one wants to generalize the result beyond the affine case, it is only sensible to prove the isomorphism over $X_0 \times_k X_0$.

9.3.3 Consequences of flatness

In this subsection, we use Corollary 9.3.5 to prove some sufficient conditions for $\mathcal{X}$ to be reconstructible, assuming certain flatness conditions on the structural maps of X_1. Let us recall that, by our convention, the morphism $p_t\colon X_1 \to X_0$ (resp. $p_i\colon X_1 \to X_0$) corresponds to the left (resp. right) R_0-action on R_1.

Remark 9.3.7 Let $f\colon W \to V$ be a morphism of R_1-comodules. Then its cokernel as a map of R_0-modules $\operatorname{coker}(f)$ inherits an R_1-comodule structure.

Suppose $p_t\colon X_1 \to X_0$ is flat. Then $\ker(f)$ inherits an R_1-comodule structure. In this case, R_1 -Mod forms an abelian category such that the forgetful functor R_1 -Mod $\to R_0$ -Mod is exact.

Lemma 9.3.8 *Suppose that*

(i) $p_t\colon X_1 \to X_0$ *is flat;*
(ii) *for every* R_1*-comodule* W *and* $w \in W$, *there exists some* $W_0 \in R_1$ -Mod$^{\square}$ *equipped with a* R_1*-comodule map* $W_0 \to W$ *whose image contains* w.

Then $\mathcal{X}$ *is reconstructible.*

Proof According to Corollary 9.3.5, we need to verify that (3) is bijective. It is clearly surjective under the hypothesis, so only injectivity remains to be demonstrated. Since the index category $f\colon V \to R_1$ contains finite coproducts, it suffices to show that for an individual morphism $f\colon V \to R_1$ with $V \in R_1$ -Mod$^{\square}$, an element $v \in V$ satisfying $f(v) = 0$ is necessarily sent to zero in $\operatorname{colim}_{f\colon V\to R_1} V$. Let $W := \operatorname{Ker}(f) \subset V$. Since p_t is flat, W inherits an R_1-comodule structure (see Remark 9.3.7). Let W_0 be an object in R_1 -Mod$^{\square}$ equipped with a map $W_0 \to W$ whose image contains $v \in W$. The commutative diagram

$$\begin{array}{ccc} W_0 & \longrightarrow & 0 \\ \downarrow & & \downarrow \\ V & \xrightarrow{f} & R_1 \end{array}$$

shows that any element in W_0 has zero image in $\operatorname{colim}_{f\colon V\to R_1} V$. In particular, v has zero image in $\operatorname{colim}_{f\colon V\to R_1} V$. □

Lemma 9.3.9 *Suppose that every finite* R_0*-submodule of* R_1 *is contained in an* R_1*-subcomodule* $W \subset R_1$ *which belongs to* R_1 -Mod$^{\square}$. *Then* $\mathcal{X}$ *is reconstructible.*

Proof Proceeding as the proof of the above lemma, we need to verify that (3) is injective, which again reduces to showing that, for any morphism $f: V \to R_1$ with $V \in R_1\text{-Mod}^{\square}$, an element $v \in V$ satisfying $f(v) = 0$ vanishes in $\operatorname{colim}_{f: V \to R_1} V$. Since $f(V) \subset R_1$ is a finite R_0-submodule, we may find a subcomodule $W \subset R_1$ containing $f(V)$, with $W \in R_1\text{-Mod}^{\square}$. The fact that f factors through a map $\tilde{f}: V \to W$ lying over R_1 such that $\tilde{f}(v) = 0$ implies that v vanishes in $\operatorname{colim}_{f: V \to R_1} V$. □

Remark 9.3.10 The hypothesis of Lemma 9.3.9 is equivalent to the combination of the following conditions:

(i) The category of injective morphisms $f: V \hookrightarrow R^1$ of R_1-comodules with $V \in R_1\text{-Mod}^{\square}$ is filtered.
(ii) The colimit over the above category $\operatorname{colim}_{f: V \hookrightarrow R_1} V$ maps isomorphically to R_1.

In particular, the hypothesis implies that R_1 is a filtered colimit of flat R_0-modules, so the map $p_i: X_1 \to X_0$ (corresponding to the right R_0-structure of R_1) must be flat.

9.3.4 Consequences of finiteness

In this subsection, we explain how a finiteness condition allows us to reformulate the reconstructibility of the affine category scheme $\mathcal{X}$. We begin with a classical observation of Serre ([5, Proposition 2]; see also [6, 07TU]).

Lemma 9.3.11 *Suppose that R_0 is Noetherian and $p_t: X_1 \to X_0$ is flat. Let W be an R_1-comodule. Then every finite R_0-submodule of W is contained in an R_1-subcomodule $W_0 \subset R_1$ which is finite over R_0.*

Proof Let $M \subset W$ be a finite R_0-submodule. Write $\rho_W: W \to W \otimes_{R_0} R_1$ for the coaction map. Since $\rho_W(M)$ is a finite R_0-submodule of $W \otimes_{R_0} R_1$, it is contained in $N \otimes_{R_0} R_1$ for some finite R_0-submodule $N \subset W$. Write W_0 for the fiber product of R_1-comodules (see Remark 9.3.7 for its formation):

$$\begin{array}{ccc} W_0 & \longrightarrow & N \otimes_{R_0} R_1 \\ \downarrow & & \downarrow \\ W & \xrightarrow{\rho_W} & W \otimes_{R_0} R_1 \end{array} \tag{10}$$

Since (10) is also a fiber product of R_0-modules, we have $M \subset W_0$. On the other hand, it holds that $W_0 \subset N$, since the image of W_0 under

$\mathrm{id} \otimes \epsilon \colon W \otimes_R R_1 \to W$ is contained in $(\mathrm{id} \otimes \epsilon)(N \otimes_{R_0} R_1) \subset N$. Because R_0 is Noetherian, this implies that W_0 is a finite R_0-module. □

Suppose R_0 is Noetherian. Lemma 9.3.11 allows us to reformulate the criterion of reconstructibility in Lemma 9.3.8. Namely, in this situation, we may replace condition (ii) by the following:

(ii′) for every R_1-comodule W finite over R_0, there exists some $W_0 \in R_1\text{-Mod}^{\square}$ together with a surjection of R_1-comodules $W_0 \twoheadrightarrow W$.

Remark 9.3.12 When $\mathcal{X}$ is an affine groupoid scheme (see Remark 9.2.2), condition (ii′) is equivalent to what some authors call the "resolution property" of the classifying stack of $\mathcal{X}$.

We finish this section with a reconstruction theorem for affine category schemes over a Dedekind domain.

Corollary 9.3.13 *Suppose that R_0 is a Dedekind domain*[4] *and both $p_t \colon X_1 \to X_0$ and $p_i \colon X_1 \to X_0$ are flat. Then $\mathcal{X}$ is reconstructible.*

Proof According to Lemma 9.3.9, it is enough to show that every finite R_0-submodule M of R_1 is contained in an R_1-subcomodule which belongs to $R_1\text{-Mod}^{\square}$. Since R_0 is Noetherian and p_t is flat, Lemma 9.3.11 shows that M is contained in an R_1-subcomodule $W \subset R_1$ which is finite over R_0. Since p_i is flat, R_1 is torsion-free as a right R_0-module. Thus W is also torsion-free. Finally, every finite torsion-free R_0-module is projective because R_0 is a Dedekind domain ([6, 0AUW]). □

9.4 The embedding problem

We continue to let k denote a ring. In this section, we clarify the relationship between the reconstructibility of an affine monoid (resp. group) scheme of finite type over $\mathrm{Spec}(k)$ and the problem of closely embedding it into the endomorphism monoid (resp. automorphism group) scheme of a finite projective k-module. The results of this section depend minimally on those of Section 9.3.

9.4.1 Affine monoid schemes

Let $\mathcal{X}$ be an affine monoid scheme over k. In other words, $\mathcal{X}$ is an affine category scheme for which the structural map $X_0 \to \mathrm{Spec}(k)$ is an

[4] In [8, (5.13)], Wedhorn claims the analogous result when R_0 is a Prüfer ring, but the present author could not follow the proof (specifically (5.6) of *op. cit.*).

isomorphism (see Remark 9.2.2). It is customary to view $X_1 \to \mathrm{Spec}(k)$ as the affine monoid scheme, with the tacit understanding that the additional data in $\mathcal{X}$ are included.

For any finite projective k-module M, the presheaf $\underline{\mathrm{End}}_k(M)$ on $\mathrm{Spec}(k)$ sending an affine k-scheme S to the monoid of endomorphisms of $M \otimes_k \mathcal{O}_S$ in $\mathrm{QCoh}(S)$ (or, equivalently, k-linear maps $M^\vee \otimes_k M \to \mathcal{O}_S$) is representable by an affine k-scheme with structure sheaf $\mathrm{Sym}_k(M^\vee \otimes_k M)$. Furthermore, the structure of an X_1-representation on M is equivalent to a map of affine monoid k-schemes:

$$X_1 \to \underline{\mathrm{End}}_k(M). \tag{1}$$

As usual, we write R_1 for the k-algebra $\mathcal{O}_{X_1}$. The map on k-algebras corresponding to (1) is the map

$$\mathrm{Sym}_k(M^\vee \otimes_k M) \to R_1 \tag{2}$$

induced from the (dual of the) coaction map $M^\vee \otimes_k M \to R_1$ (viewed as a map of k-modules).

We now relate the embedding problem to the behavior of the counit map of the Tannakian adjunction evaluated at X_1. Note that the latter is described by the morphism (2) of k-algebras

$$\mathrm{coend}(V^\vee \otimes_k V) \to R_1, \tag{3}$$

where the coend is taken over $V \in \mathrm{Rep}(\mathcal{X})^\square$.

Lemma 9.4.1 *Suppose that $X_1 \to \mathrm{Spec}(k)$ is of finite type. The following are equivalent:*

(i) *the map* (3) *is surjective;*
(ii) *there exists some finite projective k-module M and a closed immersion of affine monoid schemes $X_1 \to \underline{\mathrm{End}}_k(M)$.*

Proof Suppose (ii) is verified. Then the map (2), induced by some $M \in \mathrm{Rep}(\mathcal{X})^\square$, is surjective. On the other hand, there is a canonical map of k-modules from $M^\vee \otimes_k M$ to $\mathrm{coend}(V^\vee \otimes_k V)$. The induced map on k-algebras makes the following diagram commute:

$$\begin{array}{ccc} \mathrm{Sym}_k(M^\vee \otimes_k M) & \longrightarrow & \mathrm{coend}(V^\vee \otimes_k V) \\ & \searrow^{(2)} & \downarrow{\scriptstyle (3)} \\ & & R_1 \end{array}$$

This implies that (3) is surjective as well.

Conversely, assume that (3) is surjective. By definition, the coend can be viewed as the cokernel of a map of k-modules (see Section 9.2.4):

$$\bigoplus_{(W \to V) \in \mathrm{Rep}(\mathcal{X})^{\square}} V^{\vee} \otimes_k W \to \bigoplus_{V \in \mathrm{Rep}(\mathcal{X})^{\square}} V^{\vee} \otimes_k V.$$

Hence, the map given by assembling all coaction maps

$$\bigoplus_{V \in \mathrm{Rep}(\mathcal{X})^{\square}} V^{\vee} \otimes_k V \to R_1$$

is also surjective. Since $\mathrm{Rep}(\mathcal{X})^{\square}$ admits finite direct sums and R_1 is a finitely generated k-algebra, there exists an individual $M \in \mathrm{Rep}(\mathcal{X})^{\square}$ such that the image of the coaction map $M^{\vee} \otimes_k M \to R_1$ contains a set of k-algebra generators of R_1. In other words, the map (2) induced from this M is surjective. □

9.4.2 Affine group schemes

We continue with the notation of Section 9.4.1. The goal of this subsection is to prove an analogue of Lemma 9.4.1 for the case when $X_1 \to \mathrm{Spec}(k)$ is an affine group scheme.

Suppose M is a finite projective k-module. The affine group scheme $\underline{\mathrm{End}}_k(M)$ contains a subpresheaf $\underline{\mathrm{Aut}}_k(M)$ whose S-points are automorphisms of $M \otimes_k \mathcal{O}_S$ in $\mathrm{QCoh}(S)$. The inclusion $\underline{\mathrm{Aut}}_k(M) \subset \underline{\mathrm{End}}_k(M)$ is an open immersion ([6, 00O0]). The monoid operation on $\underline{\mathrm{End}}_k(M)$ turns $\underline{\mathrm{Aut}}_k(M)$ into a group scheme over $\mathrm{Spec}(k)$.

Lemma 9.4.2 *There exists a closed immersion $\underline{\mathrm{Aut}}_k(M) \to \underline{\mathrm{End}}_k(M')$ of monoid schemes for some finite projective k-module M'.*

In particular, $\underline{\mathrm{Aut}}_k(M)$ is affine.

Proof Without loss of generality, we assume that $\mathrm{Spec}(k)$ is connected. Then M is a finite locally free k-module of constant rank r ([6, 00NX]). Let $\det(M)$ denote $\wedge^r(M)$, which is finite locally free of rank 1. Write $M' := M \oplus \det(M)$. An S-point α of $\underline{\mathrm{Aut}}_k(M)$ gives rise to an S-point of $\underline{\mathrm{End}}_k(M')$ represented by the following matrix:

$$\begin{pmatrix} \alpha & 0 \\ 0 & \det(\alpha)^{-1} \end{pmatrix}.$$

This defines an isomorphism between $\underline{\mathrm{Aut}}_k(M)$ and a closed monoid subscheme of $\underline{\mathrm{End}}_k(M')$. □

Lemma 9.4.3 *Suppose that $X_1 \to \mathrm{Spec}(k)$ is an affine group scheme of finite type. The following are equivalent:*

(i) *the map* (3) *is surjective;*
(ii) *there exists some finite projective k-module M and a closed immersion of affine group schemes $X_1 \to \underline{\mathrm{Aut}}_k(M)$.*

Proof We will apply Lemma 9.4.1. Suppose (ii) is satisfied. By Lemma 9.4.2, there exists a finite projective k-module M' and a closed immersion of affine monoid schemes $X_1 \to \underline{\mathrm{End}}_k(M')$, so Lemma 9.4.1 gives statement (i).

Conversely, if (i) is satisfied, then there exists a finite projective k-module M and a closed immersion of affine monoid schemes $X_1 \to \underline{\mathrm{End}}_k(M)$. Since X_1 is an affine group scheme, its image is contained in the open subscheme $\underline{\mathrm{Aut}}_k(M)$. □

In particular, when k is a Dedekind domain, every affine group scheme which is flat and of finite type over $\mathrm{Spec}(k)$ admits a closed immersion into some $\underline{\mathrm{Aut}}_k(M)$ (see Corollary 9.3.13).

Acknowledgements The author thanks his team leader Aise Johan de Jong and other participants at the Stacks Project Workshop in 2020 – Simon Felten, Amelie Flatt, Quentin Guignard, and Shubhodip Mondal – for the wonderful opportunity of learning about Tannakian categories together. He thanks A. J. de Jong as well as the anonymous referee for helpful comments on previous drafts of this chapter.

References

[1] P. Deligne. Catégories tannakiennes. In *The Grothendieck Festschrift, Vol. II*, volume 87 of *Progr. Math.*, pages 111–195. Birkhäuser, 1990.
[2] Pierre Deligne, James S. Milne, Arthur Ogus, and Kuang-yen Shih. *Hodge cycles, motives, and Shimura varieties*, volume 900 of *Lecture Notes in Mathematics*. Springer, 1982.
[3] Neantro Saavedra Rivano. *Catégories Tannakiennes*, volume 265 of *Lecture Notes in Mathematics*. Springer, 1972.
[4] Daniel Schäppi. The formal theory of Tannaka duality. *Astérisque*, (357):viii+140, 2013.
[5] Jean-Pierre Serre. Groupes de Grothendieck des schémas en groupes réductifs déployés. *Inst. Hautes Études Sci. Publ. Math.*, (34):37–52, 1968.
[6] The Stacks Project Authors. *The Stacks project*. `https://stacks.math.columbia.edu`.

[7] Ross Street. *Quantum groups*, volume 19 of *Australian Mathematical Society Lecture Series*. Cambridge University Press, Cambridge, 2007. doi:10.1017/CBO9780511618505.

[8] Torsten Wedhorn. On Tannakian duality over valuation rings. *J. Algebra*, 282(2):575–609, 2004. doi:10.1016/j.jalgebra.2004.07.024.